Flood Risk

ice
Institution of Civil Engineers
publishing

Flood Risk

Planning, Design and Management of Flood Defence Infrastructure

Edited by

Paul B. Sayers
Sayers and Partners, Oxfordshire, UK

Published by ICE Publishing, 40 Marsh Wall, London E14 9TP.

Full details of ICE Publishing sales representatives and distributors can be found at: www.icevirtuallibrary.com/info/printbooksales

Also available from ICE Publishing

Land Drainage and Flood Defence Responsibilities, Fourth edition. Institution of Civil Engineers. ISBN 978-0-7277-3389-4
Future Flooding and Coastal Erosion Risks.
C. Thorne. ISBN 978-0-7277-3449-5
Flood Risk Management.
G. Flemming. ISBN 978-0-7277-3112-8
Coasts, Marine Structures and Breakwaters: Adapting to Change.
W. Allsop. ISBN 978-0-7277-4129-5
Innovative Coastal Zone Management: Sustainable Engineering for a Dynamic Coast.
A Schofield. ISBN 978-0-7277-5749-4

www.icevirtuallibrary.com

A catalogue record for this book is available from the British Library

ISBN: 978-0-7277-4156-1

ICE Publishing is a division of Thomas Telford Ltd, a wholly-owned subsidiary of the Institution of Civil Engineers (ICE).

Associate Commissioning Editor: Victoria Thompson
Production Editor: Imran Mirza
Market Development Executive: Catherine de Gatacre

Typeset by Academic + Technical, Bristol
Printed and bound by CPI Group (UK) Ltd, Croydon CR0 4YY

Contents

Foreword

I am delighted to provide the forward to this publication that provides an excellent reference covering many of the aspects associated with the planning, design and management of flood risk infrastructure. The book builds upon a significant body of research and development that has allowed us to take an increasingly holistic approach to analysing the performance of infrastructure systems and taking appropriate action to reduce flood risk.

As we continue to transition towards flood risk management, projects such as the UK Foresight Future Flooding remind us that flood defences will continue to play a major role. This book usefully combines a focus on engineered assets but within the context of a broader, portfolio-based response to flood risk. Appropriately planning for future change – in climate and land use – and delivering engineered solutions that work more closely with the natural functioning of our rivers and coasts will be important challenges. The chapters within this book provide a window on how this might be done and challenge engineers, planners and designers to think on a 'whole system' and 'whole life' basis.

River catchments are dynamic systems that evolve over time but can also be subject to sudden changes due to flood events. Similarly, the body of knowledge relating to flood defence planning, design and management also evolves over time and this book has captured the current position of the flood risk asset management community's knowledge. The book covers a wide range of topics and these are explored as both general principles and methodologies which are then illustrated via a number of more detailed case studies.

Chapter 1 on twenty-first century flood risk management highlights the issues currently facing those tasked with designing and maintaining our flood risk infrastructure in light of climate change. The work presented in this chapter provides an essential summary for those new to this area as well as those wishing to update their knowledge.

Floods are a relatively infrequent event, but it's essential that we study flood events to learn the lessons. Chapter 2 does this by reviewing a number of recent flood events from around the world and presenting the causes, impacts and lessons learned. Lessons learned is a common theme that runs through the book and is an essential way of capturing and sharing the author's analysis of the flood related issues within their areas of expertise.

A key aspect of maintaining flood risk infrastructure is inspecting the assets that comprise your asset system. Chapter 3 does this in detail and it is interesting to note the mixture of visual inspection and condition grading, which is widely used in the UK, and remote sensing techniques. The subject of remote sensing for flood risk asset management was considered in detail by the Flood Risk Management Research Consortium 'Infrastructure' work package (led by Paul Sayers), and the work presented in Chapter 3 provides a good starting point for those wishing to develop their knowledge on this topic. The question of how to use this information is explored in detail in the chapters that consider the topics of developing a hierarchical approach to asset management decision making as well as how to represent the performance of your flood risk assets in system based models. Many of these topics have been explored in details in projects such as Performance-based Asset Management Systems (PAMS) as well as its predecessors and successor projects. As someone tasked with leading an applied research theme which has considered many of the topics addressed in Chapters 4, 5, 6 and 9, it is particularly pleasing to see work on topics such as failure modes, fault trees and reliability and their utility in asset performance assessment being drawn together and presented in such a clear and concise manner. It is good to see that the past R&D, which has underpinned much of the work in these chapters, has matured to the stage that it can be communicated to such a wide audience via this book.

Flood risk systems analysis will increasingly need to consider flooding from all sources and, for this to happen effectively, practitioners will need to have a common approach to asset management planning and design. The chapters on Urban Drainage (Chapter 7) and Dams and Reservoirs (Chapter 8) will further help to increase the wider flood risk management community's knowledge of these particular types of flood risk infrastructure.

The time horizons for assessing various flood risk management interventions are often very long and there is often a degree of uncertainty associated with many of the variables that shape the asset management intervention appraisal process. How then should practitioners allow for this uncertainty when preparing scenarios and carrying out modelling to try and determine the most effective interventions? This question has been the subject of a significant body of research work over many years and the work presented in Chapter 9 provides the reader with an understanding of how best to explicitly represent uncertainty in an analysis and how this can lead to adaptive and resilient solutions. Many of these techniques are explored in more detail by way of the specific example of development of a long term strategy for managing flood risk on the Thames Estuary. The chapter on this (Chapter 12) highlights the type of data, information and analysis needed to develop an adaptable and resilient approach to flood risk management given the complexity of the existing infrastructure in the Thames Estuary.

Whilst the book does have a strong engineering feel to it, there is also a recognition that the planning, design and management of flood risk infrastructure effects both people and the environment. The opportunities to optimise environmental and flood risk benefits has been highlighted in the book. Understanding how to do this effectively is likely to become ever more important as the Water Framework, and more recently the need to take an Ecosystem Services approach, play an increasing role in flood risk asset management work on the ground.

To conclude, I feel that the work presented in this book provides an essential one-stop reference manual for the design management and operation of flood and coastal risk management infrastructure which complements the existing body of knowledge on this topic.

Dr Geoff Baxter

Research Scientist and manager of the Sustainable Asset Management Theme of the Defra/ EA Joint Flood and Coastal Erosion Risk Management R&D Programme Environment Agency

List of contributors

Editor

P.B. Sayers *Sayers and Partners, Oxfordshire, UK*

Paul B. Sayers is a Partner at an independent consultancy specialising in the management of the water environment and its associated risks. Previously a Director at HR Wallingford Ltd, Paul has over 20 years' UK and international experience in all aspects of flood risk management.

The performance of engineered flood defence infrastructure has always been a focus of Paul''s work. He was instrumental in the introduction of reliability based appraisal methods into the UK, as well as the development of the RASP methods and the associated toolsets that are now routinely used to support national and regional flood risk assessment and planning. He led the quantified analysis of climate impacts within the UK Foresight Future Flooding study, and has contributed to various asset-specific engineering guides, covering topics as diverse as dams, seawalls and channels.

The development of robust long-term policies and strategies for managing all sources of flood risk, and developments in analysis and management of urban flood systems that allow urban infrastructure investments to be prioritised based on risk, have also been key areas of interest.

Paul is an advisor to the Sustainable Asset Management theme of the Joint Defra/Environment Agency Flood and Coastal Erosion Risk Management research programme, and led the infrastructure management research within the UK Flood Risk Management Research Consortium. He is an associate editor for the CIWEM *Journal of Flood Risk Management* and acts as a reviewer for many other publications.

Contributors

S. Arthur *School of the Built Environment, Heriot-Watt University, Edinburgh, UK*
A. Davis *US Army Engineer R&D Center, Vicksburg, USA*
J. Davis *US Army Engineer R&D Center, Vicksburg, USA*
S. Fitzgerald *Harris County Flood Control District, Harris County, USA*
J. Flikweert *Royal Haskoning, Peterborough, UK*
J. Francis *Environment Agency, UK*
G.E. Galloway *University of Maryland, Maryland, USA*
J.W. Hall *University of Oxford, Oxford, UK*
M. Hardwick *Royal Haskoning, Peterborough, UK*
A. Hughes *Atkins, Epsom, UK*
A. Kortenhaus *Technische Universität Braunschweig, Braunschweig, Germany*
G. Long *The University of Nottingham, Nottingham, UK*
W. McBain *Arup, Leeds, UK*
C. Oliver *Royal Haskoning, Peterborough, UK*

D.N. Porter *Rivers Agency, Belfast, UK*
M. I. Smith *The University of Nottingham, Nottingham, UK*
D. Stewart *Torbay Council, Torbay, UK*
O. Tarrant *Environment Agency, UK*
C.R. Thorne *University of Nottingham, Nottingham, UK*
N. Wallerstein *School of the Built Environment, Heriot-Watt University, Edinburgh, UK*
M.J. Wallis *HR Wallingford, Wallingford, UK*
P. Winfield *Royal Haskoning, Peterborough, UK*

Abbreviations

AES	afflux estimation system
BRM	blockage risk model
BRP	blockage risk procedure
CDOG	Culvert Design and Operation Guide
CFRD	concrete faced rockfill dam
CIRIA	Construction Industry Research and Information Association
CIWEM	Chartered Institution of Water and Environmental Management
Defra	Department for Environment, Food and Rural Affairs
DEM	digital elevation model
EA	Environment Agency
FEH	*Flood Estimation Handbook*
FHWA	(USDOT) Federal Highway Administration
FRMRC	Flood Risk Management Research Consortium (www.floodrisk.org.uk)
GBFS	ground blast furnace slag
GIS	geographical information system
HEC RAS	(USACE) Hydraulic Engineering Centre – River Analysis System
MDM	(NISRA) Multiple Deprivation Measure
NIRA	Northern Ireland Rivers Agency
NISRA	Northern Ireland Statistics and Research Agency
OPW	Office of Public Works
PFA	pulverised fuel ash
RCC	roller compacted concrete
SAWA	Strategic Alliance for Water Management Actions
SEPA	Scottish Environment Protection Agency
SPP	Scottish Planning Policy
TSSG	*Trash and Security Screen Guide*
UKWIR	UK Water Industry Research
USACE	United States Army Corps of Engineers
USDOT	United States Department of Transport (FHWA)

Flood Risk
ISBN: 978-0-7277-4156-1

http://dx.doi.org/10.1680/fr.41561.001

Introduction

Much of the developed world has already invested heavily in flood defence assets. The UK is no exception, with an estimated £20 billion of sunk investment in river defences alone (Environment Agency, 2009). The wide variety of asset types (i.e. the embankments, dams, walls, barriers and gates, sewers and pumps, as well as the coasts and river channels themselves) and, uniquely to flood risk management, the interaction between each asset and its physical surroundings (including other assets) makes understanding the performance of existing flood defence infrastructure a considerable challenge. Uncertainties in engineering parameters and profound unknowns in future conditions further complicate the task of design and management.

No longer is infrastructure seen as a silver bullet for dealing with flood issues, a seemingly central belief within the flood-control paradigms of the 1950s to 1980s. Today, it is recognised as something that must be integrated within a broader portfolio of non-structural and policy responses to help deliver multiple outcomes (for ecosystems, the economy and society) while promoting resilience to natural disasters (an approach understood by many early civilisations).

As has always been the case, good engineering is based on evidence. As we transition from flood defence to the evolving approach of flood risk management, new demands are being placed upon those who design, manage and plan flood defence infrastructure (Box 1). In particular there is a need to understand, as completely as reasonably practicable, the risks and opportunities of action and inaction, and the associated costs, and to recognise and manage uncertainty more explicitly.

The chapters in this book explore some of the challenges and present some of the tools and techniques available to assist the engineers, planners and asset managers in making informed risk-based design and planning decisions. This introduction provides an overview of some of these, and provides a signpost to the more detailed discussions contained within this book.

Delivering better infrastructure: rising to the challenge

Throughout the 1990s and 2000s a series of devastating flood defence failures and near misses occurred. With every failure and flood event, new lessons were learnt, collectively providing some powerful insights into the shortcomings of existing approaches and presenting valuable lessons for the future (see Chapters 2 and 13). In response, around the world innovative tools and techniques are being developed to better support asset managers in overcoming the challenges they face (Havinga and Kok, 2005; USACE,

Box 1 A comparison of characteristics – a traditional engineering standards approach and a risk-based approach
Based on personal communication between Paul Sayers and David Bowles

Characteristics of an engineering approach	Characteristics of a risk-based approach
■ Safety judged by **compliance** with engineering standards and good practice ■ Focus on **design loading** events ■ **Engineers** are prime safety 'decision-makers' through **conservative assumptions** (loads and responses) ■ Consequences of failure not considered in detail ■ Achieving compliance sometimes assumed as removing '**all risk**' ■ Generally not **well integrated with other activities**	■ Safe enough judged with reference to the **understanding of risk** and **tolerable risk** guidelines ■ Considers **all significant loading** events and **failure modes** ■ Engineers provide **unbiased evidence** on risk to decision-makers ■ Consequences of failure explicitly considered ■ Seeks to **identify, understand and manage residual risk** ■ Is **well integrated with other management processes**

2008; Environment Agency, 2010). This is in recognition that structural responses (including the ongoing management of existing assets) will continue to play a major role in flood risk management into the future (e.g. 'A safe public and reduced economic losses by means of reliable levees – part of an integrated solution to flooding', taken from the USACE, National Committee on Levee Safety, October 2008 (USACE, 2008), and the findings from the Foresight Future Flooding Studies (Evans *et al.*, 2004a,b)).

A number of challenges continue to emerge from these activities, including the drive for the outcomes described below.

Better evidence on individual assets – to support design and assessment

In England and Wales, the Environment Agency has stated that it will have succeeded in its asset management role when it knows exactly:

> what assets we have; where they are; what standard of protection they provide; how they were constructed; their current engineering integrity; and, how they work together to provide a flood defence system.
>
> (Tim Kersley, Head of Asset Management, Environment Agency, 2008)

Similarly, seemingly basic requirements can be seen to exist around the world and across sectoral disciplines (within rail, road, etc.), and are a central thrust of the USACE National Levee Safety Program (USACE, 2008).

Understanding the performance of an individual asset under a range of loads is, therefore, a vital first step towards understanding how best to manage that asset. The geometry and structural components of the asset, together with the loading it experiences (e.g. waves, water levels) and the associated probability of failure are all important. Inspection methods (see Chapter 3) linking intrusive and non-intrusive and reliability analysis (see, for example, Chapters 4–6) provide vital aids to the asset manager in developing this understanding. As risk assessment and reliability analysis, explicitly recognising uncertainty, increasingly find their way into engineering design – although not yet common practice, international standards offer the designer the opportunity to apply reliability methods to assess performance – the more general question of how safe is safe enough remains a subject of debate (see Hall, 2011).

Better planning and design decision-making

All infrastructure asset managers seek to make *good* decisions; decisions that minimise whole-life costs and maximise environmental gain while ensuring that communities are appropriately protected from flooding both now and in the future. This involves the explicit recognition of all sources of flooding, and managing flood risk in a comprehensive manner. This includes improving our understanding of the following.

- *The role that an individual asset plays within a larger asset system*, highlighting those assets, within a system of assets, that contribute most to residual risk (Sayers *et al.*, 2010). This includes major structures such as dams (Chapter 8) through to individual trash screens (Chapter 9).
- *The impact of uncertainty on the decisions made*, making robust choices and building in adaptive capacity to deal appropriately with severe uncertainties in future conditions (see, for example, Chapter 11) as well as explicitly including uncertainties within the geometry and structural parameters within the reliability analysis (see, for example, Chapters 4–6), enabling critical uncertainties (e.g. within toe level, crest level and asset condition, as well as the modelling methods) to be highlighted.
- *Identifying optimal investment strategies*. Where to act? When to action, now or later? And, how to act – collect more data, undertake more analysis or intervene? Formal optimisation methods are increasingly being used to support major investment, and are likely to be become commonplace as their applicability and practicality for use in more routine decisions increases (see Chapter 11).

This whole-system, whole-life, approach to design, management and planning is in contrast to the more constrained engineering standards approach used in the past (see Box 2).

Better multifunctional performance

Increasingly there is a well-recognised need to work with natural processes in all aspects of flood management. This is particularly true in the planning and design of flood defences. Defra policy requires 'understanding of natural processes to ensure that the impacts of different options are properly appraised and opportunities to work with nature to reduce risk are identified'. Equally, the European Water Framework Directive requires 'practicable steps that can be taken to mitigate adverse impact from physical modifications to our watercourses, coasts and estuaries'. In August 2011, Caroline Spelman (the Secretary of State) launched Biodiversity 2020 – A Strategy for England's Wildlife and

Box 2 Moving away from an engineering-standards to a risk-based approach

Within a traditional engineering/safety standards based approach, the design process and decision-making procedure are relatively simple and follow along the lines of (Hall and Penning-Rowsell, 2010)

- establishing the appropriate design standard (e.g. the '100-year return period' river level) either based on the land use of the area protected, or reasons of uniformity or tradition
- estimating the design load, such as the water level or wave height, within the specified return period
- designing the structures to withstand that load (i.e. crest level, structural strength)
- incorporating safety factors, such as freeboard allowances, to account for local uncertainties, using local guides
- incorporating warning systems – often based on forecasting source terms, such as in-river water levels or at-sea waves/surge, with levels that would trigger action for the warned area (and not flood or risks forecasts).

Such an approach has a number of shortcomings, with the potential to lead to

- unfairness – protecting some and not others
- *inefficiency of investment* – by providing design standards above the minimum for economic efficiency in some areas and below minimum in others, as the additional benefits accrued through the provision of a design standard above the minimum at one location are usually less than if the additional money had been spent elsewhere (this typically occurs because the costs tend to increase much more quickly than the associated benefits).

A flood risk management process proceeds as a continuous and iterative process (Sayers *et al.*, 2002) and along the lines of the following.

- Identifying all possible sources and pathways of flooding and a range of potential strategies (including a portfolio of structural and non-structural responses) and possible future scenarios (reflecting plausible change in climate, demographic, funding, etc.).
- Evaluating the performance of each strategic alternative against multicriteria representing societal preferences for economic efficiency, ecosystem benefits and social equality across the potential futures.
- Considering investing proportionately greater resources to protect the vulnerable and deliver ecosystem benefits.
- Identifying preferred short-term actions and a coherent, but flexible, longer term strategy, and then continue to monitor and adapt the strategy as the reality of the future becomes known.

A risk-based approach to flood management can, therefore, be seen to have a number of advantages by promoting robust, resilient and flexible strategies, but raises questions of: (i) how to make design choices that appropriately minimise the chance of failure, (ii) what level of residual risk is acceptable at any given location, and (ii) how much additional investment should be provided above a minimum level for maximum efficiency.

Ecosystem Services, which includes a priority water management action (3.7) to 'promote approaches to flood and erosion management which conserve the natural environment and improve biodiversity'. These requirements challenge traditional flood defence approaches. For example, the role of and approach to sediment and debris within channels are being challenged, with the need to better understand the geomorphological response of the river channel and the role of vegetation and debris in promoting biodiversity, and not simply to seek to remove material, being increasingly recognised as a valuable part of a flood defence system (see Chapter 14).

Better integration into broader planning objectives

Flood risk management is only one component of a broader planning framework. The creation of surface water planning and the increasing role of the local planning authorities at the forefront of delivering flood risk management, within the context of a strategic regional and national overview provided by the Environment Agency recognises this need (see, for example, Chapter 1). The increasing recognition that flood water must be managed in an integrated manner has challenged traditional practice based on the separate consideration of pluvial (urban drainage) and fluvial and coastal flooding sources (see Chapters 1 and 7).

Conclusion

Delivering resilient infrastructure is much more than simply reducing the chance of damage through the provision of 'strong' structures, and adaptive management is much more than simply 'wait and see'. Both are purposeful approaches to design that are inherently risk based and, importantly, seek to actively manage uncertainty. Accepting that the future as unknown, although widely recognised as important, is yet to become a routine consideration within flood risk management. Accepting this premise has a number of profound implications; both in terms of the way strategies are developed and the way infrastructure is designed. Equally, adopting an integrated view of the often diverse sources of flooding, the responses available to manage the associated flood risk, and embedding these within a broader planning framework provides a significant, yet achievable, challenge.

The chapters in this book provide engineering students, early career engineers and scientists, as well as more senior engineers involved in the design, management and planning of the flood defence infrastructure with a range of useful insights on the demands and emerging science of a risk-based approach. A single book cannot cover all aspects in detail, however, and we have taken care to provide signposts to supporting information, tools and methods to support the reader in learning more about specific subjects.

REFERENCES

Environment Agency (2009) *Long Term Investment Strategy (LTIS)*. Environment Agency, London, UK.

Environment Agency (2010) *PAMS (Performance-based Asset Management System) – Phase 2, Outcome Summary Report*. Environment Agency, Bristol. Flood and Coastal Defence R&D Programme Science Report No. SC040018/SR1.

Evans E, Ashley R, Hall J *et al.* (2004a) *Foresight Future Flooding, Scientific Summary: Volume 1: Future Risks and their Drivers*. Office of Science and Technology, London, UK.

Evans E, Ashley R, Hall J *et al.* (2004b) *Foresight Future Flooding, Scientific Summary: Volume 2: Managing Future Risks*. Office of Science and Technology, London, UK.

Hall J (2011) How safe is safe enough? *Journal of Flood Risk Management* **4(4)**: 211–212.

Hall J and Penning-Rowsell E (2010) Setting the scence for flood risk management. In *Flood Risk Science and Management* (Faulkner PA (ed.)). Wiley-Blackwell, London, pp. 3–16.

Havinga FJ and Kok M (2005) *Flood Risks and Safety in the Netherlands (Floris). Floris Study – Full Report*. Ministry of Public Works, Delft, The Netherlands. Report No. DWW-2006-014.

Sayers PB, Hall JW and Meadowcroft IC (2002) Towards risk-based flood hazard management in the UK. *ICE Journal of Civil Engineering* **150(5)**: 36–42.

Sayers PB, Wallis M, Simm J, Baxter G and Andryszewski T (2010) Towards the next generation of risk-based asset management tools. In *Flood Risk Science and Management* (Pender G (ed.)). Blackwell, London, UK.

USACE (United States Army Corps of Engineers) (2008) National Committee on Levee Safety. National Levee Safety Program. Available at: http://www.leveesafety.org (accessed 2 February 2012).

van Gelder P, Buijs F, ter Horst W *et al.* (2008) Reliability analysis of flood defence structures and systems in Europe. FloodRISK08. Proceedings of the European Conference on Flood Risk Management, Research into Practice, Oxford, UK.

Flood Risk
ISBN: 978-0-7277-4156-1

http://dx.doi.org/10.1680/fr.41561.007

Chapter 1
Twenty-first century flood risk management

Will McBain
Arup, Leeds, UK

Climate change is likely to result in increasing demand for infrastructure to reduce flood risk. Given sufficient resources, the vast majority of flooding problems can be controlled. However, there is a limit to the rate at which resources, and in particular fossil fuels, can be used sustainably. Flood defence infrastructure can also have wider negative consequences for society and the environment. European policy and legislation is, therefore, increasingly focused on the promotion of river basin scale planning, appropriate environmental and sustainability assessment, and optimised use of the wide array of techniques available to manage flood risk. Flood defence infrastructure is just one important means of managing flood risk by reducing the likelihood of floods occurring. Measures that reduce the severity of the consequences of flooding include spatial planning, insurance, awareness raising, flood forecasting and warning, civil contingency planning, the establishment of community flood plans, and the implementation of physical interventions to improve the resilience of the built environment in flood zones. Avoiding the creation of new flood risks and planning for the consequences, as well as the causes, of flooding will allow the long-term benefits of flood risk management expenditure to be optimised.

1.1. Introduction

Managing water resources to control flooding, improve soil fertility, irrigate crops and provide water supplies for domestic and industrial use is an ancient endeavour. Much of the basic technology used by civil engineers today dates back to pre-Roman times. For early civilisations on the flood plains of the rivers Nile, Yangtze and Euphrates, flooding will have been a constant threat, but the risks were outweighed by the opportunities that these great rivers provided to society for agricultural production, water supply, fishing, navigation and trade. Towns and cities frequently evolved around strategic crossing points at a time when rivers represented major natural barriers. The strategic benefit of living adjacent to rivers and coastlines has continued to influence settlement patterns into the modern era. The growth in international maritime trade in the nineteenth and twentieth centuries was particularly critical to the development of major estuarine and coastal conurbations. Currently, approximately 20% of the world's population occupies areas that are at potential risk of flooding, including those located in cities such as New York, New Orleans, London, Amsterdam, Paris, Dhaka, Bangkok, Shanghai and Jakarta.

The infrastructure required to allow this pattern of urban settlement to be sustained is very substantial. For example, properties within the flood plain of the tidal Thames Estuary, which includes significant parts of the City of London itself, are protected from all but the most extreme floods, by a system of barriers, floodwalls and levees valued at in excess of £3 billion.

Global rises in sea level, coupled with an intensified hydrological cycle, due to global warming, are likely to result in ever greater demand for infrastructure to protect communities, industry, agriculture and commerce from the threat of flooding. Civil engineers will take the lead role in planning, designing and constructing this infrastructure. The range of structural measures available to control flooding, as outlined in this book, is now formidable. Tidal surge barriers, major river and coastal levee-raising projects, flood-attenuation reservoirs and integrated urban storm water drainage schemes are currently under construction, or in the planning phases, at thousands of locations across the globe. Given sufficient resources, the vast majority of flooding problems can be controlled.

However, at no other time in history has mankind been so acutely aware of the fact that there is a limit to the rate at which resources, and in particular fossil fuels, can be used sustainably. There is also a limit to the rate at which natural cycles, such as the carbon cycle, can absorb the by-products of human activity. This realisation is driving policy and legislation in the direction of resource efficiency and carbon reduction. There is a prevailing need for leaner, efficient solutions that work with natural processes, rather than against them. Managed adaptation to the climate change threat will involve use of an array of options to manage flood risk and increase resilience, many of which are aimed specifically at avoiding the need for large-scale flood defence infrastructure.

Of particular importance, going forward, is the need to avoid a policy of allowing new development to occur in protected flood plains in the belief that infrastructure can eliminate the flood risk altogether. The problem of relying on conventional infrastructure-driven solutions to reducing flood risk, without complementary planning policies to prevent an intensification of development in flood plain areas, is nowhere better summed up than by the American geographer Gilbert F White, who wrote the following over 50 years ago

> ... the rate at which flood losses are being eliminated by construction of engineering or land-treatment works is of about the same magnitude as the rate at which new property is being subjected to damage ... The construction of new flood-protection works frequently has been the signal for accelerated movement into the flood plain. (White, 1960)

Unless it is promoted within the context of a mature approach to risk management, flood control infrastructure can create what is known as a 'moral hazard', isolating individuals and organisations from a true appreciation of the risks associated with occupying flood plain areas. In reality, there is nearly always a residual risk that low-lying flood plain areas will flood at some point – whether from surface water, groundwater, or from overtopping or breaching of flood defences by rivers or the sea. Planning for the consequences of natural disasters is, therefore, as important as addressing the causes. This has been illustrated again and again in recent years, most specifically in the aftermath to Hurricane Katrina in New Orleans in 2005, Cyclone Nagris in the Irrawaddy Delta in 2008, and the catastrophic floods in Pakistan in 2010.

This chapter aims to provide the wider context in which civil engineers engaged in designing and managing flood defence infrastructure in Europe, and specifically the UK, operate. It covers

- the key policies and legislation relevant to flood risk management
- flood risk concepts and commonly used terminology
- design standards, asset performance thresholds and climate change allowances
- the implications of uncertainty in estimating flood frequency
- the flood-management hierarchy (avoid, substitute, control, mitigate)
- the different types of measure that can be taken to manage all kinds of flood risks
- the management of residual risks.

An appreciation of this context is critical to the development of sustainable, long-term solutions to flooding problems.

1.2. Policy and legislative background

The failure of localised attempts to address flooding problems has underlined the need to take a strategic, catchment-based approach with high levels of transboundary cooperation (UNECE, 2009). There are equally compelling arguments to deal with other catchment functions on a river-basin scale, including water supply and hydroecology. Integrated water resources management (IWRM) is defined as a 'process which promotes the coordinated development and management of water, land and related resources, in order to maximise the resultant economic and social welfare in an equitable manner without compromising the sustainability of vital ecosystems' (Global Water Partnership, 2000). Satisfying this multifunctional approach to river-basin planning, the term 'integrated flood management' (IFM) has lately risen up the international agenda (WMO/APFM/GWP, 2004).

The need for a strategic approach is at the heart of the European Union Floods Directive (EEC, 2007). The Directive forms the legal component of the European Flood Action Programme, and is closely linked to the objectives of the Water Framework Directive (EEC, 2000). The Floods Directive requires European Member States to identify river basins and associated coastal areas that are at risk of flooding, before drawing up flood risk maps and management plans. This legislation also contributes to the EU's requirement to adapt to climate change. The Floods Directive outlines three stages by which flood risk will be reduced.

- Undertaking a preliminary flood risk assessment of river basins and associated coastal zones by 2011.
- Developing flood hazard maps and flood risk maps by 2013 for areas where there is a danger of flood damage occurring.
- Producing flood risk management plans for these zones by 2015.

The flood risk management plans will include measures both to reduce the probability of flooding and its potential consequences. Preventive measures will include avoiding the construction of human settlement in flood-prone areas. Protection measures will involve restoring flood plains and wetlands. Governments will also have to prepare plans, for example, providing instructions to the public on what to do in the event of flooding. Furthermore, risk assessments, maps and plans must be made available to the public, and the process will be repeated on a 6-year cycle to ensure that long-term developments are taken into account.

In response to the EU Flood Directive, Scotland introduced the Flood Risk Management (Scotland) Act 2009 (Scottish Government, 2009), which sets in place a statutory framework for delivering a sustainable and risk-based approach to managing flooding. This includes taking flood risk management plans into account when spatial development plans are prepared. A consolidated Scottish Planning Policy (SPP) on nationally important land-use planning matters came into effect in February 2010 (Scottish Government, 2010). This shorter, clearer and more focused statement of new national planning policy supersedes a large number of documents, including 'SPP 7 – Planning and Flooding' (Scottish Government, 2004). SPP divides flood risk into three categories, and outlines an appropriate planning response. It also requires development plans to protect the coastal environment and identify coastal areas likely to be suitable for development.

In England, the Flood and Water Management Act 2010 received Royal Assent on 8 April 2010 (UK Government, 2010). The Act creates a more comprehensive and risk-based regime for managing flooding and coastal erosion compared with the previous legislation, which mainly addresses the provision of hard-engineered protection measures. It also introduces an improved risk-based approach to reservoir safety, and encourages the uptake of sustainable drainage systems by removing the automatic right to connect surface water drainage systems to sewers. The Act complements planning policy in England as currently embodied in the National Planning Policy Framework (CLG, 2012a) and its accompanying Technical Guidance (CLG, 2012b). The policies in this framework advocate a sequential approach, whereby the local planning authorities responsible for allocating sites for new development should seek to locate vulnerable land uses in locations where the flood hazard is lowest. The policy recognises that wider sustainability criteria may occasionally override flood-risk considerations, and includes an Exception Test, which must be satisfied, in such circumstances. This is designed to ensure that the likelihood of flooding is controlled to acceptable levels and that residual risks are managed.

The Flood and Water Management Act 2010 also applies in Wales. Planning Policy Wales incorporates Technical Advice Note 15: Development and Flood Risk (Welsh Government, 2004), which is similar to SPP and the English National Planning Policy Framework, but with a greater emphasis given to assessing the consequences of flooding.

The Northern Ireland Assembly prepared secondary legislation for flood risk management. The approach to flood risk management until 2021 is outlined in two recent government documents *Living with Rivers and the Sea* (Northern Ireland Rivers Agency, 2008), and the Water Environment (Floods Directive) Regulations Northern Ireland 2009, which came into operation in December 2009 (UK Government, 2009). Under the Regulations, the Department of the Environment is required to publish appropriate flood risk management plans by 2015 (a response to the EU Flood Directive). The Department is responsible for spatial planning, and its policies on flood risk are set out in Planning Policy Statement 15 (PPS 15): Planning and Flood Risk (Northern Ireland Planning Service, 2006).

There is a substantial body of environmental legislation in place within EU member states designed to provide statutory protection to key natural habitats and endangered species, and to ensure that the wider environmental impacts of programmes and projects are properly assessed and mitigated. Key European legislation includes the Environmental Impact Assessment Directive (EEC, 1985), the Habitats Directive (EEC, 1992) and the Strategic Environmental Assessment (SEA) Directive (EEC, 2001), all of which have been transposed into UK legislation. Of perhaps greatest significance to those planning and designing flood defence infrastructure is the Water Framework Directive (EU, 2007a, 2011). The Water Framework Directive requires that Environmental Objectives be set for all surface and ground waters in the EU in order to enable them to achieve 'Good Ecological Status' (or Good Ecological Potential for those defined as Heavily Modified and Artificial Water Bodies) by a defined date. The key aims of the Water Framework Directive are to

- achieve at least Good Ecological Status/Potential for all water bodies by 2015; where this is not possible and subject to the criteria set out in the Directive, aim to achieve Good Ecological Status/Potential by 2021 or 2027
- prevent deterioration in the status of aquatic ecosystems, protect them, and improve the ecological condition of waters
- conserve habitats and species that depend directly on water
- meet the requirements of Water Framework Directive Protected Areas
- promote sustainable use of water as a natural resource
- progressively reduce or phase out the release of individual pollutants or groups of pollutants that present a significant threat to the aquatic environment
- progressively reduce the pollution of groundwater and prevent or limit the entry of pollutants
- contribute to mitigating the effects of floods and droughts.

The Water Framework Directive was transposed into UK law in 2003. This legislation should steer those responsible for designing and managing flood defence infrastructure in the direction of options that actively enhance the ecological status/potential of water bodies. Any activity that has the potential to have an impact on the ecological quality of a water body needs to be assessed to establish whether it could cause deterioration in the current Ecological Status (or Potential) classification. All activities that could impact on watercourses also need to be considered in terms of whether they will compromise the ability of the water body to reach Good Ecological Status or Good Ecological Potential by the date specified in the River Basin Management Plan.

There will, exceptionally, be circumstances where wider benefits outweigh the River Basin Management Plan objectives. If the initial assessment procedure predicts that an activity will cause deterioration in water body status or prevent a water body from achieving its ecological objectives, then a subsequent assessment is required against the conditions listed in Article 4.7 of the Water Framework Directive. Article 4.7 states that the following criteria must be met for the project to proceed.

- All practicable steps are taken to mitigate the adverse impact on the status of the water body.
- The reasons for the modifications or alterations are specifically set out and explained in the River Basin Management Plan and the objectives are reviewed every 6 years.
- The reasons for the modifications or alterations are of overriding public interest and/or the benefits to the environment and to society of achieving the River Basin Management Plan objectives are outweighed by the benefits to human health, the maintenance of human safety or to sustainable development.
- The beneficial objectives served by those modifications or alterations of the water body cannot, for reasons of technical feasibility or disproportionate cost, be achieved by other means, which are a significantly better environmental option (EU, 2007b).

It is therefore evident that, when tackling flooding problems, scheme developers should always consider options that will actively enhance the ecological status of water bodies first. River restoration, flood plain rehabilitation and wetland/saltmarsh re-establishment projects, if shown to be effective at reducing flood risk, can also have significant biodiversity, amenity and carbon sequestration benefits (Environment Agency, 2010). These elements are increasingly being incorporated into flood defence infrastructure projects, aided by techniques to place an economic valuation on environmental effects, such as those developed for the Environment Agency by Eftec (2010).

One final area of policy and legislation that is of critical importance to those designing and managing flood defence infrastructure relates to resource efficiency and carbon reduction – both of which are fundamental to sustainable development. Flood defence infrastructure can have significant benefit in terms of reducing the waste, resource use and greenhouse gas emissions associated with repairing damage caused by flooding. However, constructing, operating and maintaining flood defences also has the potential to generate waste and increase carbon emissions, both through the direct use of fossil fuels and indirectly via the embodied carbon within construction materials.

In the twenty-first century it is imperative that we use resources efficiently and seek to use renewable resources wherever possible. Resource efficiency involves 'using natural resources in the most effective way, as many times as possible, while minimising the impact of their use on the environment' (Waste and Resources Action Programme (WRAP), 2011). In the UK, a third of all waste is currently generated by the construction industry. In response to this problem, Site Waste Management Plans became mandatory in 2008 (UK Government, 2008a). The industry templates developed for these incorporate the facility to capture efforts by designers to design out waste. However, it is at the very early stage of options appraisal that the greatest benefits can be achieved, particularly if the solution avoids construction work altogether. Scheme appraisers should aim to achieve these early wins through the environmental appraisal process.

Resource efficiency is closely allied to initiatives to reduce carbon emissions. The UK Climate Change Act 2008 (UK Government, 2008b) is the world's first legally binding environmental sustainability framework. Passed in November 2008, the Act aims to lower greenhouse gas emissions by 80% by 2050 compared to the 1990 baseline. The Carbon Reduction Commitment is a mandatory emissions-reduction standard in the UK. It aims for a carbon reduction of 1.2 million tonnes by 2020, and an 80% cut in carbon emissions by 2050. With the (notable) exception of transportation fuels, the Carbon Reduction Commitment covers all forms of energy, including electricity, gas, fuel and oil. Again, those designing and managing flood defence infrastructure will need to be fully mindful of the carbon footprint of their projects, not least because of the resource and cost savings that 'low carbon infrastructure' can provide. Tools, such as the Environment Agency's 'Carbon Calculator', can be used to compare the relative merits of shortlisted options (Environment Agency, 2007).

1.3. Flood risk concepts and terminology

Having set the policy and legislative context, this section introduces some of the key concepts underpinning the assessment and management of flood risks. This theoretical understanding is a prerequisite for engineers involved in designing and managing flood defence infrastructure.

Risk, within the context of risk analysis, is generally understood to be a function of probability and consequences.

- *Probability* is the measure of our strength of belief that a given event will occur (e.g. a flood depth exceeding 1 m at a given location). It is considered over a specific timeframe (e.g. one tidal cycle, one month, one year, a lifetime). The concept of probability can be extended further to consider the chance of receptors being exposed to flooding and, therefore, of experiencing adverse consequences.
- *Consequences* express the degree of harm suffered by a receptor, or group of receptors, as a result of a given flood event. Consequences can be subdivided into two key components – exposure and vulnerability.

It is important to note that flood risk is not stationary in time. Climate change, land-use change, the deterioration of flood defence systems, and the degree of exposure and vulnerability of receptors can all influence flood risk over time.

The components of flood risk can be analysed using the source–pathway–receptor model, which has its origins in the analysis of contaminated land, but has been adapted for flood risk management purposes (Sayers *et al.*, 2002, 2003). 'Sources' constitute flood hazards (anything with the potential to cause harm through flooding). 'Pathways' represent the mechanisms by which the flooding hazard would cause harm.

Box 1.1 Flood sources, pathways and receptors (information updated from CIRIA Report C624 (Lancaster *et al.*, 2004)

Sources	Pathways	Receptors
Rainfall River flows Artificial drainage systems Extreme sea levels Wind-generated waves Tidal storm surges Tsunamis Lakes/reservoirs Canals Groundwater Mines/quarries	Inundation of flood plains Overland flow Overtopping or failure of river defences Breaching of natural or man-made coastal defences Failure of flood defence components such as barriers and gates Reservoir failure Inadequate drainage	People Domestic and commercial property Emergency services installations Infrastructure Agriculture Ecosystems

'Receptors' comprise the people, property, infrastructure and ecosystems potentially affected should a flood occur. The consequences of flooding for receptors are highly dependent on their degree of vulnerability. For people, vulnerability can be dependent on where they live, their age, income, education and disability, and on broader social and environmental factors such as level of preparedness and quality of emergency service response (Blaikie *et al.*, 1994).

Assessing and managing flood risk can involve consideration of all of the components listed in Box 1.1. Increasingly, modelling tools used in the assessment of flood risk incorporate all three of these components, linking hydrological models of flood sources to hydrodynamic models of flood pathways and GIS databases of economic and land-use data that characterise the relative vulnerability of the receptors. Integrated catchment models (ICMs) of this kind potentially allow a full array of flood management possibilities to be simulated (Blongewicz, 2000). The likelihood of a flood occurring is generally expressed in terms of annual probability – i.e. the probability that a flood of a given magnitude will occur, or be exceeded, in any one year. This can be expressed as a percentage (e.g. a 1% annual probability) or chance (e.g. 1 in 100). This is illustrated for a range of annual probabilities in Table 1.1. This definition has largely replaced the traditional, and potentially misleading, term 'return period'.

The risk r of an event of a given annual exceedance probability (AEP) (both expressed as a percentage) occurring in a period of M years is given by

$$r = [1 - (1 - \text{AEP}/100)^M] \times 100$$

A low probability flood may have a significant likelihood of occurring over an extended period. For example, a 1% (1 in 100) annual probability event has a 53% probability

Table 1.1 Flood frequency terminology

Annual probability	Annual chance
2%	1 in 50
1.3%	1 in 75
1%	1 in 100
0.5%	1 in 200
0.1%	1 in 1000

of occurring, or being exceeded, at least once within a 75-year period. It is therefore important, if adopting this as the design standard for a system of flood defences, to consider what the consequences would be of a design exceedance event. This is particularly important given the numerous potential sources of hydrological uncertainty discussed below.

An indication of the standards of protection that may be appropriate for different types of asset is provided in Table 1.2. For some assets, design standards may need to be considered with regard to the overall level of performance expected of the assets during a range of design events. For example, a railway station has numerous components, all of which have a varying degree of criticality for its continued operation. Figure 1.1 provides a rationale for the consideration of design levels versus performance levels that can be applied to the various components of an asset or system.

It can be seen from Figure 1.1 that, in principle, it is possible to set both design standards and performance levels for assets with varying degrees of criticality and according to the severity of the consequences of their failure. The application of this rationale is particularly useful in the consideration of the flood resilience of critical infrastructure assets, such as transport hubs, which have multiple components.

The design of flood risk management measures to improve the resilience of existing assets should take climate change into account over the anticipated lifetime of the asset. Several EU Member States now provide guidance on appropriate allowances for key parameters such as net sea level, rainfall and river flow. In the UK, the Department for Environment,

Table 1.2 Indicative standards of protection for various land use types (data taken from Report C688 (McBain *et al.*, 2010)

Land-use type	Indicative standard of protection (annual probability)
Residential/commercial property	<1% (>1 in 100)
Key emergency service installations	<0.1% (>1 in 1000)
Large raised reservoirs (above an urban area)	<0.01% (>1 in 10 000)

Figure 1.1 Design standards versus performance levels
Adapted by BG Clarke from SEAOC (1995); reproduced courtesy of BG Clarke and SEAOC

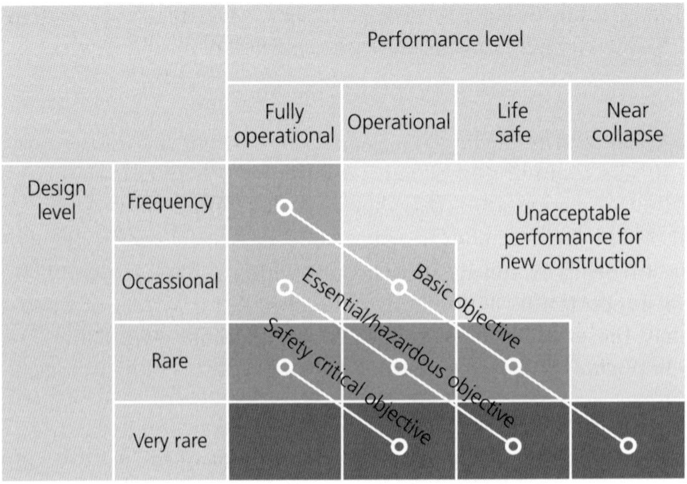

Food and Rural Affairs (Defra) issues updates based on the findings from the UK Climate Impacts Programme. Defra identifies two generic approaches to designing flood defence infrastructure to take account of climate change; the *precautionary approach*, which involves incorporating mitigation measures for potential climate change impacts now; and the *managed-adaptive approach*, which involves making provision for mitigation measures to be undertaken at a future date when there will also be greater certainty regarding the likely impacts of climate change on parameters such as river flow and rainfall. Options that have the greatest adaptive capacity should take precedence over those that are dependent on a particular future climatic scenario.

The identification of annual probabilities for design events of a given magnitude remains a very useful tool for use in the planning and design of flood risk management systems. There are, however, several sources of potential error in flood frequency estimation. Floods are often caused by a complex combination of circumstances, which makes statistical analysis very challenging. It is critical for engineers to understand, acknowledge and manage uncertainties, which can be derived from both sampling errors and modelling errors.

Sampling errors are those associated with using finite historical records of the variable of interest. Accurate records of river flow, for example, that allow frequency analysis of flood peaks to be undertaken, are seldom available for periods in excess of 40 years, even in countries with well-developed hydrometric infrastructure. Furthermore, the majority of gauging stations are designed primarily for the management of the water resources available for water supply, and may not be wholly reliable for the accurate estimation of extreme out-of-bank flood flows.

Where a good record of river flows is available, the assumption that the flood peak data within it represent a random sample from a larger population of flood peaks that have a specific, probability distribution function, can be associated with modelling errors of an even greater magnitude. Selection of an appropriate probability distribution is therefore critical, and requires a high degree of expert judgement, particularly where data series are short.

Two further sources of modelling error are those associated with the assumption that the probability distribution underlying the flood frequency estimates is static in time (the assumption of stationarity), and that each flood peak value within the sample is independent of the values in previous years (i.e. shows a lack of persistence). In reality, instances of both non-stationarity and serial persistence are likely to distort the flood frequency relationship. The principle of stationarity in particular is becoming increasingly untenable as evidence gathers regarding how changes in both land use and climate are affecting flood frequency.

The implications of these sources of uncertainty are fivefold. First, it is vital that sufficient time and well-qualified resource is allocated to develop as good an estimate of flood frequency as possible given the available evidence. Second, any estimate of flood frequency should be reinforced by historical review of wider data sources, including local history books and local civic authority records. Third, allowances for potential sources of uncertainty should be made within scheme designs through use of an appropriate freeboard allowance in the crest height of flood defences. Fourth, appropriate allowances should be made for any likely trends in future flood frequency, due to anticipated land use or climate changes, either by using additional, precautionary freeboard allowances or via a planned set of future managed adaptation measures. Design standards and climate change allowances are discussed further below. Fifth, and finally, the consequences of design exceedance events should be carefully considered by designers in order to minimise the impacts.

1.4. Flood risk management techniques and methods

The hierarchy given in Table 1.3 provides an indication of the effectiveness of different flood risk management measures. This hierarchy places spatial planning firmly at the top of the list of measures to reduce flood risk. Spatial plans that have due regard for flood risk represent the most effective means of minimising the exposure of human settlement to the flood hazard. Many flood-control projects require planning permission and are subject to the policies set out in these plans. The principles need to apply to all areas, including those already protected by flood defences, in order to prevent an intensification of development within such areas.

The concept of sources, pathways and receptors has been introduced in Box 1.1. The associated options that can be taken to avoid, substitute, control or mitigate flood risk are summarised in Table 1.4. These measures will vary in their effectiveness depending on the specific nature of the flooding problem. The challenge for those tasked with solving flooding problems is to identify the combination of measures that achieves defined objectives at the optimum whole-life cost, given the wider physical, social and

Table 1.3 The flood risk management hierarchy

Flood risk management measure	Description	Examples
Avoid	Locate land uses vulnerable to the impact of flooding within areas least likely to flood	Critical infrastructure such as hospitals and emergency services installations should be located in areas where the risk of flooding is negligible
Substitute	Substitute land uses incompatible with flooding with less vulnerable, or water-compatible, land uses	In the regeneration of a riverside urban area, existing housing within flood plain areas could be partially or wholly substituted with playing fields or amenity open space
Control	Reduce the likelihood of a site flooding by installing new infrastructure or restoring natural features. Implement a sustainable drainage strategy to control run-off from the site	Construction of floodwalls, barriers or levees to contain river or tidal waters. Attenuation of flood peaks using flood storage reservoirs. Restoration of coastal mangrove swamps, saltmarshes and natural delta systems to absorb wave energy
Mitigate	Reassess residual flood risks (those risks that remain after the control measures have been implemented) and implement mitigation measures to minimise the impact of these	Educate affected parties about the nature of the residual risk. Make sure the relevant insurances are in place. Provide a flood warning service. Ensure emergency evacuation plans have been developed

environmental constraints. This is what makes flood risk management such a fascinating profession.

1.5. Managing residual risks

Flood control schemes alter the nature of the residual risk and sometimes create new risks. When designing flood control infrastructure to protect flood plain areas, engineers should understand and communicate the nature of these residual risks. The aim should be to work closely with spatial and civil contingency planners, as well as the asset owners and communities themselves, to raise awareness, increase levels of preparedness, and enhance people's ability to cope and rapidly return to normal after a flood occurrence. There is a growing number of kite-marked products available that individuals and businesses can use to help exclude shallow floodwaters at property level (flood resistance products). Guidance is now

Table 1.4 Flood risk management measures (non-structural measures are shown in italics) (based on McBain *et al.*, 2010)

Source control *– measures which reduce the likelihood of high flows/water levels occurring*	***Pathway modifications*** *– measures which modify or block the pathways taken by floodwater to a site*	***Receptor resilience*** *– measures which reduce the vulnerability of receptors to the impacts of a flood*
■ *Spatial planning and land use policies* ■ Sustainable drainage: – detention basins – filter drains/strips – flow control systems – infiltration basins/ trenches – permeable paving – retention ponds – soakaways – swales – wetlands – green roofs/walls ■ Oversized pipes/ attenuation tanks within the drainage network ■ Rainwater harvesting ■ Attenuation reservoirs ■ River regulation – river restoration and flood plain rehabilitation	■ Ground raising ■ Construction of floodwalls and embankments ■ Construction of diversion channels or tunnels ■ Tidal surge barriers, gates and sluices ■ Removal or modification of existing structures ■ Demountable flood defences ■ Temporary flood defences ■ Designing drainage networks for exceedance (e.g. overland flow routing) ■ Managed realignment to 'make space for water' ■ Property or asset level flood resistance measures	■ *Flood risk identification and mapping* ■ *Planning policies and development control* ■ *Civil contingency planning* ■ *Awareness raising and enhanced preparedness* ■ *Flood forecasting and warning* ■ *Improved emergency response procedures* ■ *Desk top incident management exercises* ■ *Business continuity management* ■ *Risk transfer (e.g. flood insurance)* ■ *Feedback from lessons identified* ■ *Property or asset level flood resilience measures*

also widely available on the use of flood-resilient building materials, as well as on how to most effectively deal with the aftermath of flooding.

Breaking down the moral hazard frequently associated with flood defence infrastructure is the key to minimising residual risks. All communities and businesses that choose to occupy flood plains need to accept that occasional flooding will be inevitable, even if flood defences are in place. Community flood plans are an ideal means of acknowledging this risk and maximising business's and people's abilities to help themselves and each other. Public participation in the flood defence infrastructure planning process is therefore important, and can add considerable value in terms of minimising the risk of loss of life and optimising the return on investments made.

1.6. Conclusions

European policy and legislation is increasingly focused on the promotion of planning on a river-basin scale, appropriate environmental and sustainability assessment, and optimised use of the wide array of techniques available to manage flood risk. Flood defence infrastructure is one important means of managing flood risk, by reducing the likelihood of floods occurring in such areas. Such infrastructure alters the nature of the residual risk and sometimes creates new risks. Given the inherent uncertainty in flood frequency estimation, and the fact that floods which exceed the design standard of flood defences will inevitably occur, such infrastructure should always be used in conjunction with other measures. The most effective means of reducing flood risk is to avoid locating vulnerable land uses in flood hazard areas. Other measures include awareness raising, flood forecasting and warning, civil contingency planning, the establishment of community flood plans, and the implementation of physical interventions to improve the resilience of the built environment in flood zones. Avoiding the creation of new flood risks and planning for the consequences, as well as the causes, of flooding will allow the long-term benefits of flood risk management expenditure to be optimised.

REFERENCES

Blaikie P, Cannon T, Davis I *et al.* (1994) *At Risk: Natural Hazards, People's Vulnerability and Disasters.* Routledge, New York.

Blongewicz M (2000) The development of the integrated catchment modelling system using ArcView GIS. Paper presented at the European Conference on Advances in Flood Research, Potsdam.

CLG (Department for Communities and Local Government) (2012a) *National Planning Policy Framework: Planning, building and the environment.* Available at: http://www.communities.gov.uk/publications/planningandbuilding/nppf.

CLG (2012b) *Technical Guidance to the National Planning Policy Framework: Planning, building and the environment.* Available at: http://www.communities.gov.uk/publications/planningandbuilding/nppf.

EEC (European Economic Community) (1985) Council Directive of 27 June 1985 on the assessment of the effects of certain public and private projects on the environment. 85/337/EEC. Environmental Assessment. *Official Journal of the European Union* **L 175**: 40–48.

EEC (1992) Council Directive 92/43/EEC of 21 May 1992 on the conservation of natural habitats and of wild fauna and flora. *Official Journal of the European Union* **L 206**: 7–50.

EEC (2000) Directive 2000/60/EC of the European Parliament and of the Council of 23 October 2000 establishing a framework for Community action in the field of water policy. *Official Journal of the European Union* **L 327**: 1–73.

EEC (2001) Directive 2001/42/EC of the European Parliament and of the Council of 27 June 2001 on the assessment of the effects of certain plans and programmes on the environment. *Official Journal of the European Union* **L 197**: 30–37.

EEC (2007) Directive 2007/60/EC of the European Parliament and of the Council of 23 October 2007 on the assessment and management of flood risks. *Official Journal of the European Union* **L 288**: 27–34.

Eftec (2010) *Flood and Coastal Erosion Risk Management: Economic Valuation of Environmental Effects.* Available at: http://publications.environment-agency.gov.uk/PDF/GEHO0310BSFH-E-E.pdf (accessed 2 February 2012).

Environment Agency (2007) *Carbon Calculator for Construction Activities.* Available at: http://www.environment-agency.gov.uk/business/sectors/37543.aspx (accessed 2 February 2012).

Environment Agency (2010) *Working with Natural Processes to Manage Flood and Coastal Erosion Risk*. Available at: http://publications.environment-agency.gov.uk/PDF/GEHO0310BSFI-E-E.pdf (accessed 2 February 2012).

EU (European Union) (2007a) Directive 2007/60/EC of the European Parliament and of the Council of 23 October 2007 on the assessment and management of flood risks. *Official Journal of the European Communities* **L 288**: 27–34.

EU (2007b) *Common Implementation Strategy For The Water Framework Directive. Exemptions to the Environmental Objectives under the Water Framework Directive, allowed for new modifications or new sustainable human development activities (WFD Article 4.7 – Policy Paper)*. Available at: http://isearch.avg.com/search?cid=%7B485d35ed-8266-4ae5-bf84-a2bdd42c8f32%7D&mid=ec56762d5575b1869a826ac290968cbe-b602d594afd2b0b327e07a06f36ca6a7e42546d0&ds=AVG&v=10.0.0.7&lang=en&pr=pr&d=2012-01-19%2010%3A37%3A12&sap=tr&q=Exemptions%20To%20The%20Environmental%20Objectives%20Under%20The%20Water%20Framework%20Directive%2C%20Allowed%20for%20new%20modifications%20or%20new%20sustainable%20human%20development%20activitie (accessed 2 February 2012).

EU (2011) *A European Flood Action Programme*. Updated 25 March 2011. Available at: http://ec.europa.eu/environment/water/flood_risk/key_docs.htm (accessed 2 February 2012).

Global Water Partnership (2000) *What is IWRM?* Available at: http://www.gwp.org/en/The-Challenge/What-is-IWRM (accessed 2 February 2012).

Lancaster JW, Preene M and Marshall CT (2004) *Development and Flood Risk: Guidance for the Construction Industry*. CIRIA Publication C624. CIRIA, London.

McBain W, Wilkes DJ and Retter M (2010) *Flood Resilience and Resistance for Critical Infrastructure*. CIRIA Publication C688. Available at: http://www.ciria.org/service/Home/AM/ContentManagerNet/ContentDisplay.aspx?Section=Home&ContentID=15881 (accessed 2 February 2012).

Northern Ireland Planning Service (2006) Planning Policy Statement 15 (PPS15): Planning and Flood Risk. Available at: http://www.planningni.gov.uk/index/policy/policy_publications/planning_statements/pps15-flood-risk.pdf (accessed 2 February 2012).

Northern Ireland Rivers Agency (2008) *Living with Rivers and the Sea – Government Response to the Flood Management Policy Review*. Available at: http://www.dardni.gov.uk/riversagency/living-with-rivers-and-the-sea.09.097_rivers_agency_-_living_with_rivers_and_the_sea_report_-_published_august_2008.pdf.pdf (accessed 2 February 2012).

Sayers P, Hall J and Meadowcroft I (2002) Towards risk-based flood hazard management in the UK. *ICE Journal of Civil Engineering* **150(5)**: 36–42.

Sayers PB, Gouldby BP, Simm JD, Meadowcroft I and Hall J (2003) *Risk, Performance and Uncertainty in Flood and Coastal Defence – A Review*. Defra/Environment Agency. R&D Technical Report FD2302/TR1.

Scottish Government (2004) *Scottish Planning Policy (SPP) 7 Planning and Flooding*. Available at: http://www.scotland.gov.uk/Publications/2004/02/18880/32954 (accessed 2 February 2012).

Scottish Government (2009) Flood Risk Management (Scotland) Act 2009. Available at: http://www.legislation.gov.uk/asp/2009/6/pdfs/asp_20090006_en.pdf (accessed 2 February 2012).

Scottish Government (2010) *A Statement of the Scottish Government's Policy on Nationally Important Land Use Planning Matters*. Available at: http://www.scotland.gov.uk/Publications/2010/02/03132605/0 (accessed 2 February 2012).

SEAOC (1995) *Vision 2000 – A Framework for Performance Based Earthquake Engineering*, Vol. 1. Structural Engineers Association of California, Vision 2000 Committee. Sacramento, CA, USA.

UK Government (2008a) *Site Waste Management Plans Regulations 2008*. Available at: http://www.legislation.gov.uk/uksi/2008/314/contents/made (accessed 2 February 2012).

UK Government (2008b) Climate Change Act 2008. Available at: http://www.legislation.gov.uk/ukpga/2008/27/contents (accessed 2 February 2012).

UK Government (2009) *The Water Environment (Floods Directive) Regulations (Northern Ireland)*. Available at: http://www.legislation.gov.uk/nisr/2009/376/contents/made (accessed 2 February 2012).

UK Government (2010) Floods and Water Management Act 2010. Available at: http://www.legislation.gov.uk/ukpga/2010/29/contents (accessed 2 February 2012).

UNECE (United Nations Economic Commission For Europe) (2009) *Transboundary Flood Risk Management: Experiences from the UNECE Region*. United Nations, New York and Geneva. Available at: http://www.unece.org/fileadmin/DAM/publications/oes/Transboundary_Flood_Risk_Management_Final.pdf (accessed 2 February 2012).

Welsh Government (2004) *Technical Advice Note (TAN) 15: Development and Flood Risk*. Available at: http://wales.gov.uk/topics/planning/policy/tans/tan15/?lang=en (accessed 2 February 2012).

White GF (1960) Strategic aspects of urban flood plain occupance. *Journal of the Hydraulics Division, Proceedings of the American Society of Civil Engineers* **86(HY2)**: 89–102.

WMO/APFM/GWP (World Meteorological Organization/Associated Programme on Flood Management/Global Water Partnership) (2004) *Integrated Flood Management*. WMO Concept Paper No. 1047; APFM Technical Document No. 1, 2nd edn. Available at: http://apfm.info/pdf/concept_paper_e.pdf (accessed 2 February 2012).

WRAP (Waste and Resources Action Programme) (2011) *Resource Efficiency*. Available at: http://www.wrap.org.uk/wrap_corporate/about_wrap/resource_efficiency.html# (accessed 2 February 2012).

Flood Risk
ISBN: 978-0-7277-4156-1

http://dx.doi.org/10.1680/fr.41561.023

Chapter 2
Past failures and design lessons

David N. Porter
Rivers Agency, Belfast, UK

This chapter considers a number of different flood events that have caused widespread impacts, in order to draw some lessons from them for design engineers. In each example the causes and the impacts of the event are outlined. The summer floods of 2007 in the UK are considered, with a detailed examination of the risks associated with impoundments from the Ulley Reservoir. This event shows the risks of surface water flooding, the vulnerability of infrastructure, and likelihood of flood risk increasing with climate change. The significant flood event that occurred in August 2008 in Northern Ireland is used to show the limitations to the design capacity of a culvert system, and the risk of overtopping of flood defence structures. Finally, the catastrophic flooding caused by Hurricane Katrina shows that design standards can be misleading, that maintenance must be joined up if flood defence systems are to work, and that predictable events, such as hurricanes, demonstrate the need for early warning systems and forecasting as an active part of flood risk management.

2.1. Introduction

Flood defence structures are designed to operate in a defined loading environment, which is enshrined within the various design codes and standards. These standards have been developed based on the performance and characteristics of construction materials, academic research, our approach to risk and its management, and the experience of past performance. It is this final aspect that this chapter will focus on, as we need to have a good understanding of the actual performance of defence structures when loaded by the extremes of nature. By examining failure we can gain this understanding, and this will enable us to design a more robust and better performing defence structure.

2.2. The UK 2007 summer floods

2.2.1 Causes

During June and July 2007, the UK experienced extreme weather, with exceptionally high intensity rain events occurring in many different areas. The impacts were widespread, although the flooding in Northern Ireland and Scotland was on a lesser scale than that experienced in England and Wales. In many cases, these extreme rainfall events fell on already saturated ground, as they followed ‘the wettest ever May to July period since national records began in 1766’ (Cabinet Office, 2008).

There were two main weather fronts that caused these flood events. First, on Monday 25 June 'a slow-moving area of low pressure brought a prolonged period of heavy rain to northern and central England' (Met Office, 2007). This impacted on Hull, Sheffield and Doncaster; caused flooding in the counties of Derbyshire, Lincolnshire and Worcestershire; and was the trigger that caused the spillway failure at the Ulley Reservoir, which will be examined in more detail later in this chapter.

The second rain event took place on the 20 July was caused by a 'slow-moving depression centred over south-east England, drawing warm moist air from the continent across the UK' (Met Office, 2007). Impacts were greatest in the counties of Warwickshire, Worcestershire, Gloucestershire, Herefordshire, Shropshire and Oxfordshire. Critical infrastructure that was impacted on included: the Mythe water treatment works, which result in 350 000 people being without mains water; the road network, in particular the M5 motorway where 10 000 people were stranded; and the railway network, with 500 people stranded within Gloucester railway station.

This prolonged period of unsettled weather is untypical for the area, and was caused by the displacement of the polar jet stream to the south of the UK by an area of low pressure. The jet stream is a narrow ribbon of fast-moving air travelling at about 300 mph above 30 000 feet. It is normally north of the UK during the summer, which keeps the atmospheric pressure high and results in settled, predictable weather patterns. This stream of air naturally occurs where cold polar air from the north meets the warm subtropical air from the south. The jet stream is also used by airline pilots, as they can position their aircraft in the fast-flowing air, which will speed up progress while using less fuel. 'For much of summer 2007, the jet stream was strong and further south than normal, resulting in heavy rain producing weather systems crossing southern and central areas of the UK' (Cabinet Office, 2008).

2.2.2 Impacts

Given the widespread impacts of the 2007 floods, the UK Environment Agency carried out a detailed assessment of the economic costs in order to quantify the losses. Their final report estimated the impact in financial 'out-of-pocket' terms to be '£3.2 billion in 2007 prices' (Environment Agency, 2010), of which 75% was borne by householders. About 63% of the losses were covered by some form of insurance or attracted compensation, with 56% covered by private insurance. In the wider context, 'the floods that devastated England ranked as the most expensive in the world in 2007' (Cabinet Office, 2008).

In some respects, the economic figures, while providing a measure of the flood event, do not adequately communicate the human impact. Again quoting from Environment Agency information, 13 people died, 350 000 were without fresh mains water for over 2 weeks, the electricity supply to 150 000 properties was affected for an average of 15 hours, and 55 000 properties were flooded. The long-term impact on UK society is difficult to measure, but it is clear that this event will have wide-reaching impacts for years to come.

2.2.3 Lessons learnt

It is in the context of the extreme weather event, the impact on the economy and the human tally that we as engineers need to consider the lessons we can learn if we are to play our part in reconstructing thc built fabric of society, while designing for future infrastructure needs in a manner that proactively manages flood risk. The in-depth review of the summer floods of 2007 carried out by Sir Michael Pitt on behalf of government (Cabinet Office, 2008) provided 92 recommendations, which cover all aspects of the flood, including preparedness, emergency response and recovery. Here we will simply consider those areas that may be relevant to design engineers.

2.2.3.1 Climate change

From the outset, it is worth stating that a single weather event, even one that caused a flood as widespread as the one that occurred in the UK during the summer of 2007, is not in itself, an indication of climate change. Climate is the long-term pattern of weather, and a single wet summer, or even a number of such events occurring in close proximity, do not prove that the climate has changed. With time and the benefit of more data, it may be proven that the 2007 flooding was due to climate change, but at this stage it is too early to conclude with any degree of certainty. However, the predictions that are generated by climate science are indicating that these types of event are what we could reasonably expect to encounter if the theories are confirmed and if the consequences are as predicted. Given this position, it is entirely reasonable to conclude that the current level of flood defence structures, the capacity within our drainage culverts and the ability of the watercourses to discharge will all diminish with time. As engineers, we have to come to terms with the direct impacts of climate change, and build in measures to our structures that will deal with increasing flows. Work such as the Stern Review (HM Treasury, 2006), the Foresight (2004) report and *Future Flooding* and *Facing up to Rising Sea Levels: Retreat? Defend? Attack?* by Building Futures and the Institution of Civil Engineers (2010) all indicates that practical action is required if we as a society are to adopt to a different living environment.

2.2.3.2 Surface water

Historically, the focus of flood prevention measures has been on river or coastal flooding. It is not that surface water has not existed or that it was not a problem, but it tended to be a localised problem, and so generally did not cause widespread, headline-grabbing, floods. The summer floods of 2007 in the UK are a very clear indication that surface water is potentially a larger problem than previously experienced, and needs some thought if we are to manage the risk from this source. Surface water or pluvial flooding ‘occurs as a result of rainfall which overwhelms natural or engineered drainage systems resulting in water flowing overland and ponding in depressions in the ground’ (Rivers Agency, 2011).

Two lessons needed to be learned from the 2007 floods. First, there needs to be an overview of flood risk from all sources. Responsibility for most other sources of flooding is well defined and emergency procedures are in place to offer assistance, but the Pitt

Review (Cabinet Office, 2008) highlighted this gap in the flood risk management approach.

At the time of these flood events, surface water was, in real terms, not the responsibility of a single organisation. Individual councils, water companies and other drainage authorities all carried a portion of the risk, but there was no overview to ensure that all the systems worked in unison with each other, nor was there an overall plan to deal with known inadequacies within the network. Within England, and as a direct result of the summer 2007 floods, this role has now been allocated to the UK Environment Agency.

The existence of such an overview will improve the understanding and management of surface water flooding, but other engineering improvements are also needed if we are to minimise the risk of a recurrence of the flooding witnessed in, for example, Hull. Either the engineered drainage systems will need to be significantly upgraded, or our approach to surface water management will need to be radically rethought. The cost of a complete upgrade would be unaffordable in many areas, and we therefore need to consider our fundamental approach to urban drainage. Many systems cannot cope; they are too costly to upgrade; and there is a continued change in the run-off characteristic of land due to development. All this is placing greater pressure on drainage, and all before the uncertainty that exists due to the likely impacts of climate change is factored into the equation. The old method of collecting surface water and piping it in an enclosed system to a river or outlet needs to be rethought. As we go forward, greater emphasis needs to be placed on sustainable drainage systems (SuDS), but the scale of installation and the performance of these systems needs to be carefully considered. Small-scale SuDS that only perform adequately in low-intensity rain events will be totally overwhelmed during an extreme event, and so cannot really be considered to be a flood risk management tool if, when inundated, they still cause property to be impacted on. SuDS systems need to be designed with exceedance in mind. Roads, footways and other channels need to be capable of conveying flood water away from those at risk.

2.2.3.3 Flood plains

The natural flood plain of a river performs a vital role by providing attenuation of water during a flood. This is maintained until the flow characteristics of the watercourse are affected, say by the construction of a bridge or other restriction, or when something is built in the flood plain. The protection of flood plains is therefore the only long-term, sustainable option for the prevention of damage to property from flooding, and at all costs they should be preserved. In the real world, however, this is not always possible, but when development in these areas is unavoidable it should be designed with resistance to flooding built in. It is of paramount importance that the wider impacts are also considered, as the infilling of a flood plain to enable floor levels to be above the design flood level does not reduce the amount of water generated by a particular flood, it will simply displace it. If this displacement places others at risk the solution is not adequate. Again looking forward to the likely impacts of climate change, clearly flood plains will become of even greater value, and the frequency of the impacts of flooding on properties within flood plains will increase.

2.2.3.4 Critical infrastructure

Reservoirs, motorways, the rail network, electricity supply substations, the freshwater supply were all impacted during the 2007 flood event. It is clear, therefore, that future development of critical infrastructure must place greater emphasis on the impact of flooding. Poor design, inadequate contingency plans and ignorance of the risks must be a thing of the past if we are to assist society to maintain vital services during extreme conditions. With the advances in flood modelling and mapping, the necessary information should be available, and engineers must play their part in assessing these data and taking into consideration its consequences during the planning and design stages. We also need to consider the measures that can be taken to protect our existing infrastructure, either on a permanent or on a reactive basis. Localised flood banks or walls may be all that is required to ensure the continuity of supply of essential services, the failure of supply of which is often much more damaging and costly than the cost of the defence measures.

2.3. Lessons from the Ulley Reservoir

2.3.1 Causes

The Ulley Reservoir is situated near Sheffield in South Yorkshire, England. It covers some 35 acres and has a capacity of approximately 580 million litres of water, and construction was completed in 1874. The reservoir was originally designed to supply the town of Rotherham with fresh water, but in 1986 it was no longer needed and ownership transferred from Yorkshire Water to Rotherham Metropolitan Borough Council, who created the Ulley Country Park, which remains in place to this day.

During the summer floods of 2007 the Ulley Reservoir was a central feature, as the extreme weather, and in particular the heavy rainfall, caused the reservoir to reach dangerously high water levels. The design of reservoirs includes features that will accommodate this situation, with spillways being utilised to remove excess water in order to ensure the safety of the dam structure. In the case of Ulley, the spillway was activated, but during operation some of the masonry lining became dislodged, exposing the clay dam to erosion by the passage of water, which if left unrestricted would threaten the stability of the entire dam.

2.3.2 Impacts

As the water level continued to rise and the spillway discharged water, the puddle clay (Ulleyweb, 2007) core of the dam was exposed and started to become compromised. This was first noticed at about 8 pm on 25 June 2007, and Rotherham Metropolitan Borough Council engaged a multiagency emergency team, including an Institution of Civil Engineers (ICE) All Reservoirs Panel engineer to undertake an inspection (*New Civil Engineer*, 2007a). This inspection confirmed the fragile nature of the structure and the risk posed, and in response the emergency services undertook an evacuation of people within the dam-break inundation zone. These included 'about 100 homes in the villages of Catcliffe, Whiston and Canklow' (BBC, 2007). In addition, the M1 was closed between junctions 32 and 36, as was a section of Sheffield Parkway and 'the A618 from Whiston Cross Roads to Aughton' (Ulleyweb, 2007). There were also fears for the safety of high-voltage overhead power lines due to the proximity of a number of pylons to the dam, 'a major electricity sub-station, and the gas network connection to Sheffield' (ICE, 2009).

Emergency services personnel and engineers worked around the clock to draw down the water level in the reservoir by pumping, and to divert the flow from the damaged spillway. This enabled a temporary repair to be carried out to the '20 m long by 6 m deep hole' (*New Civil Engineer*, 2007a) in the embankment using geotextile and stone fill. After the immediate risk had been mitigated, attention was turned to the longer term solution. Complete demolition of the dam was even considered, due to liability concerns (*New Civil Engineer*, 2007b), but this option was discounted as it would mean losing a popular amenity facility. Therefore, permanent repairs were undertaken, including 'the construction of a new spillway and draw down system', 'the rehabilitation of the top of the dam core with a special watertight bentonite cement underground wall', and improvements to the wave protection of the dam wall, all at a cost of almost £5 million (Rotherham Metropolitan Borough Council, 2010).

2.3.3 Lessons learned

2.3.3.1 Design flaws

The key problem at Ulley was the spillway design, which routed excess water from the impoundment across the vulnerable front face of the downstream embankment toe. This exposed the dam structure to an unnecessary risk when the spillway was operational. This was not the first time that this failure mechanism had been the cause of an embankment failure, as only 2 years previous, on 19 June 2005, the Boltby Dam in North Yorkshire was compromised (Figure 2.1). In this case the reservoir was smaller: 'embankment was 100 m long, 19 m high and impounded approximately 13 000 m^2 of water' (Walker, 2008). Again heavy rain led to the operation of the spillway, its failure, and the subsequent removal of embankment cover material and the exposure of the puddle clay core. The Boltby Dam was not repaired or upgraded, but works were undertaken to discontinue its use by the formation of a massive V-notch weir in the structure, which reduced the impounded depth, and thus its volume, to a level that poses nominal risk.

In the examples above, the route of the spillways caused a problem, but the size of the spillways was also an issue. It is important to check the capacity of these features, as 'the design floods of a number of reservoirs were re-evaluated and the revised outflows were often larger than those used in the original spillway designs' (Chanson and Felder, 2010). The hydraulics of stepped spillways is complex, as the reaction of the water flow changes as the flow rate increases. 'At low flows the spillway would operate as a series of plunge pools', and this is referred to as 'nappe flow'. As the velocity increases, the water 'can "skim" down the spillway, effectively bouncing from step to step. Such a flow regime can lead to high hydrodynamic pressures and pressure fluctuations on parts of the spillway' (Mason and Hinks, 2008). This flow condition is known as 'skimming'.

The construction material also played a part in the failure of the spillway, and the resultant risk to the embankment. The material of choice at the time of the construction

Figure 2.1 Embankment erosion due to spillway damage, Boltby Reservoir (Walker, 2008)

of the Ulley Reservoir, and indeed many other such structures of that time, was masonry. The maintenance of the masonry is critical, in particular its embedment and pointing, if the ingress of water is to be minimised and the potential for block displacement reduced. The situation with pointing is further complicated as 'there is benefit to be gained by providing pressure relief to the vertical faces of steps, either through drainage holes or by not sealing all masonry joints' (Chanson and Felder, 2010).

2.3.3.2 Need for regulation

Beyond all doubt, the risks associated with reservoirs are clearly demonstrated by the failure at Ulley. The regulation of reservoirs to ensure public safety must, therefore, be robust, well managed and subject to ongoing review. The classification of Ulley before the summer 2007 floods was category C, on the four-point scale A–D, where A is the highest consequence category and D the lowest (Defra, 2009). Thus, it was considered to present 'negligible risk to human life but some property damage' (Environment Agency, 2007). With hindsight, this classification is obviously incorrect, given the potential risk to a significant number of properties (great enough for them to be evacuated in the early hours of the morning) and important, even critical, infrastructure (including the M1 motorway, which was shut for about 40 hours, costing an estimated £2.3 million) (Environment Agency, 2007).

Regulated reservoir structures can at least be considered to be in a reasonable state of repair, which removes some of the uncertainty connected with the risk posed by these water bodies. In addition, the powers associated with regulation enable government to step in and carry out inspection, remedial works and improvements in the absence of the 'undertaker' or owner fulfilling his duties under the legislation. Amendments to the current UK legislation is likely to ensure that the risk is better understood, as it will require the development of flood plans that include dam break inundation maps as part of the contingency plans.

Regulation also ensures that each structure is subject to independent inspection, which in the UK is completed by an ICE Panel Engineer. Given that the inspections are carried out by a third party, and that this person must be professionally qualified and approved by the ICE, the level of certainty surrounding the quality of the inspection is high. One potential drawback of the current system is that the Panel Engineer report is completed on a 10-yearly basis. although it does review the annual inspection reports produced by the supervising engineer. Ten years is a long time interval given the ageing nature of many of our reservoirs.

2.3.3.3 Old infrastructure

Ulley shows how a seemingly benign, decommissioned water storage facility that had been in place for over 130 years can be affected by an extreme natural event and result in widespread disruption, posing a risk to many lives and property and threatening other key pieces of infrastructure. As engineers, if we are to minimise the likelihood of events like the one experienced at the Ulley Reservoir, we need to ensure that the wider public knows and understands the value we offer them as we construct, maintain and operate critical infrastructure. Often the work of engineers is unseen, and therefore the true value is not considered, but events like this demonstrate the necessity of our skills.

2.4. August 2008 floods in Northern Ireland

2.4.1 Causes

Significant flooding occurred across Northern Ireland during August 2008 as a result of persistent heavy rain falling on already saturated ground. The most significant rainfall occurred on Saturday 16 August, but the preceding week was very wet. Indeed, 'some areas in Northern Ireland recorded their monthly rainfall averages – around 70 to 80 mm or three inches – between 11–15 August', and on 16 August 'more than two inches (55–65 mm) of rain fell widely across many central and eastern parts of Northern Ireland' (Met Office, 2011). An unregistered rain gauge in 'Ballyclare recorded 94 mm (3.75 inches) of rain during the day' (Northern Ireland Assembly, 2008), which is estimated to be a 150 year rainfall event. Figure 2.2 shows the 48-hour rainfall accumulations from the morning of 15 August 2008.

2.4.2 Impacts

The torrential rain falling on saturated catchments with already high river levels quickly caused widespread flooding, which impacted on some 1600 properties; 'approximately 50% were affected by river related flooding and 50% from other sources such as out of sewer and surface water runoff' (Rivers Agency, 2009a). In addition, numerous roads were closed due to flood waters, most notably the recently constructed Broadway underpass.

Figure 2.2 48-hour rainfall accumulations for Northern Ireland which caused the August 2008 floods
© Crown Copyright 2008

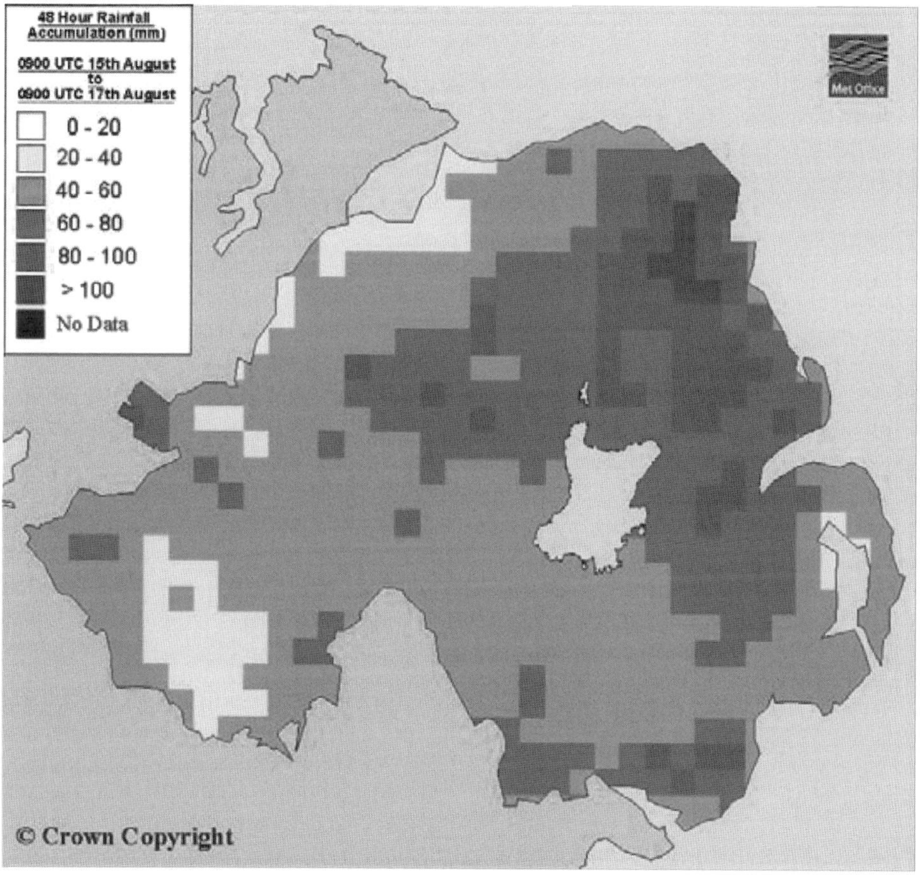

2.4.3 Lessons learned

For the purposes of this text, two areas of impact arising from the August 2008 floods will be used to demonstrate some lessons for design engineers. First, the Broadway underpass and, second, a small development called Meadowside, which was impacted by the Six Mile Water in Antrim.

2.4.3.1 Broadway underpass

The Broadway underpass was part of the £104 million scheme to upgrade the M1/ Westlink route, and included the removal of traffic-controlled junctions and replacing them with free-flow, grade-separated underpass arrangements. The section of road through the underpass was 'opened to traffic on Friday 4 July [2008], some 13 months ahead of schedule' (Northern Ireland Executive, 2008). Six weeks later the underpass

Figure 2.3 The Broadway underpass flooded on 16 August 2008
Photograph courtesy of Paul Smith © Paul Smith

was submerged to 0.5 m above the soffit of the roof slab by a volume of water estimated to be 50 000 m^3. The photograph in Figure 2.3 shows the underpass a few days after the flood event.

Critical infrastructure

This event again highlights the vulnerability of our infrastructure to flood impacts. In this example, the underpass started to fill with water at 16:30; 15 minutes later it was impassable, as it had filled to a depth of 2 m, and by 17:15 the water level had totally engulfed the entire structure (Amey, 2008). While it took just 45 minutes to fill the underpass with 50 000 m^3 of water, it required 60 hours of round the clock activity to restore normal operation, and during this time a principal route into the capital city was out of service and topping the international news headlines.

Design capacity

Given the enormity of the impact, an in-depth investigation into the causes was carried out. The report concluded that the rainfall event in the catchment was significantly less than the design capacity of the realigned culverts that were constructed as part of this project. It was also concluded that, although not the sole cause, 'a blockage of the trash screen has probably been a contributory cause of the flooding' (Amey, 2008). The ultimate capacity of the culvert system is, therefore, not determined by the theoretical conveyance capacity of the opening, but by some other limiting factor, such as the likelihood of blockage of a trash grille. Culvert systems are therefore like the proverbial chain, the strength of which is determined by its weakest link, so a design standard that may seem to offer a high degree of protection may in fact be misleading, as the standard may never be achieved. For more information on trash screens and culverts, see Chapter 9.

The value of routine maintenance

In order to mitigate the potential reduction in capacity due to grille blockage, many drainage authorities have instigated a system of routine grille maintenance. This is to ensure that debris is removed and that water is flowing freely. This function, while very rudimentary or even mundane, can often make the difference during a flood event. It is also important that designers consider the safe access of operatives to grilles during adverse conditions. A grille does not remain clean for long during high flow conditions, and critical systems may need constant attention at these times. Warning systems using water-level triggered alarms can offer an early alert of a developing problem. It is, however, necessary to consider the time for response. If this is too long, the flood event will have developed to a stage where any intervention may not be effective, and the damage will already have been done. To counteract this problem, lower alert levels can be set, but these are likely to result in a large number of false call-outs when the water rises to the given level but does not continue to rise and cause flooding. Another approach is to install a mechanised screen-cleaning device, which can be automatically or remotely triggered. These remove the time aspect of the response, but are more costly, require more technology and can sometimes be unreliable.

Design for exceedance

Where exceedance is foreseeable, this should be part of the design. Grille blockage is a very common cause of flooding, and so the structure should be designed with increased capacity and exceedance in mind. This may mean that a facility needs to be developed to take water from the watercourse at a higher level in the catchment when blockage occurs, provided that the water can be attenuated or can be routed using a flow path that does not cause flood problems elsewhere.

2.4.3.2 Six Mile Water – Meadowside

Meadowside is a small residential area approximately 1 km east of the town of Antrim. It is bounded to the north by the Six Mile Water, which has a channel width of 20–30 m. This river flows into Lough Neagh, which is approximately 2.6 km downstream of Meadowside. The houses are protected by a reinforced-concrete flood wall and clay-core flood banks, which were constructed in the 1990s following earlier flooding on 22 October 1987. During the August 2008 event, the flood defence structure was overtopped, but it did not suffer from structural failure (collapse, undermining or overturning). See Figure 2.4 for an aerial photo of the Meadows after the water level had peaked.

Residual risk

All flood defence structures are designed to withstand a particular size of flood event, and the level of protection offered varies by country, depending on each country's approach

Figure 2.4 Meadowside on 17 August 2008

to flood risk, the affordability of the project, the value of the properties and activities being protected, and consideration of environmental constraints. Irrespective of the design criteria, there always remains the possibility that a larger flood event will occur that overwhelms the defences and causes flooding. This is referred to as the 'residual risk'. This risk is present at all flood defence structures. There is no such thing as absolute protection, as failure is always a possibility.

The challenge for the engineer is to manage the risk, and this is done in a number of ways. First, the likelihood of material failure is addressed in the design codes by the application of factors of safety. Second, structural failure has its own set of design tests; for example, a flood wall is checked for overturning during the design process. Finally, the uncertainties associated with hydraulic modelling and the fact that we are dealing with natural events which may not always follow defined patterns of behaviour are mitigated by the use of freeboard. This involves increasing the level of protection by the addition of height to a flood wall or by raising the flood levels of proposed properties. The freeboard is typically 600 mm for river flood protection structures, but it may be increased where wave action is likely.

System

At Meadowside, the protection was provided by a '43 m section of reinforced concrete wall and a 200 m section of clay cored embankment' (Rivers Agency, 2009b). The general arrangement, the location of adjacent properties and the flood inundation area

Figure 2.5 Meadowside flood defence system. Dark grey line – the clay-core flood bank; light grey area – the area inundated by the 2008 flood

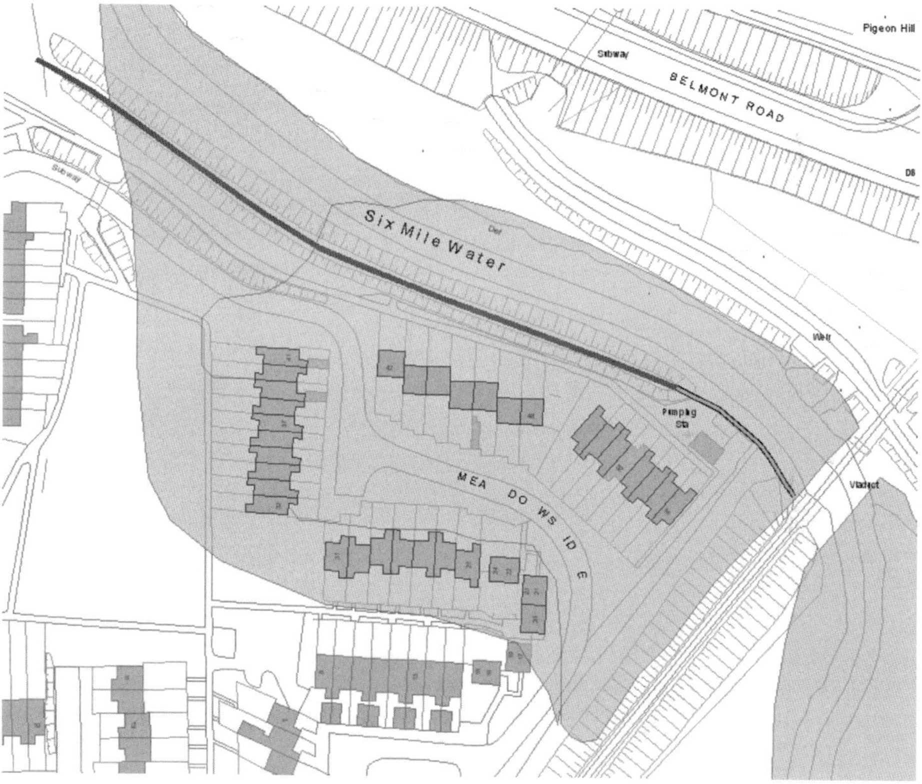

are shown in Figure 2.5. The report into the August 2008 flood event identified that the level of the reinforced-concrete section was similar to the 'as-built' drawings, or certainly within an acceptable tolerance. The clay-cored section of the flood bank had settled with time, and in places was lower than the adjacent concrete section, as it was not of a uniform height. Flood protection can be constructed using different techniques and materials, without affecting the level of protection offered; however, elements do not act alone, but as a system. This means that the level of protection offered by the defence system is that of the lowest section, as water will start to overtop at that point.

Back drainage

A flood defence is designed to keep water out, but it is, therefore, also very effective at keeping water in! Back drainage needs to be designed to manage this risk. It is often sufficient to build in one-way flow valves within the surface water or foul-drainage systems. This stops high water from the river back-flowing up the drainage systems and causing a flood problem. When the system is closed it stops this flow, but it also means that any surface water needs to be either stored behind the wall until the river has fallen to a level, when the flaps open and allow discharge, or in larger back drainage catchments a pump is provided to move the water over the wall or to another part of the drainage network. The system at Meadowside did not have a pump, but relied on attenuation behind the wall until the water level in the river had fallen. Calculations completed after the flood event show that the amount of rain falling within the catchment behind the flood wall was sufficient in itself to have caused some flooding. Therefore, both the overtopping of the defences and the back drainage contributed to the extent of the flood. These same calculations show that the flood water remained locked behind the defences for about 33 hours before the flow conditions in the main channel were such that the back drainage worked again using gravity (Rivers Agency, 2009b).

Communicating risk

The use of terminology when communicating the level of flood risk offered by any defence structure often causes problems, particularly when dealing with a non-technical audience. This problem is compounded as the industry is very familiar with a variety of risk language and often uses some of the terms interchangeably. The issue with trying to simplify the complex is that important details can often be lost in the process, and as a result the message can be changed dramatically.

The use of terms such as 'a 100 year event' can often lead the public to believe that when they have been subject to a flood they are then free from risk within their life time. This is a misunderstanding of probability and, therefore, of the risk. A numerical example may help to demonstrate the risk faced. The likelihood of exceedance or 'encounter probability' P_e is calculated using the equation (Kamphuis, 2000):

$$P_e = 1 - \left[1 - \left(\frac{1}{T}\right)\right]^n$$

where T is the flood return period (e.g. a 1 in 100 year event), and n is the length of period of interest in years. This generates the following results. A 100 year event is one that has a 1% chance, or probability, of occurring in any year. The results in Table 2.1 indicate that

Table 2.1 Encounter probability expressed as a percentage

Length of period of interest, *n*: years	Flood return period, *T*				
	1 in 10	1 in 50	**1 in 100**	1 in 200	1 in 1000
1	10.0	2.0	**1.0**	0.5	0.1
10	65.1	18.3	**9.6**	4.9	1.0
20	87.8	33.2	**18.2**	9.5	2.0
25	92.8	39.7	**22.2**	11.8	2.5
50	99.5	63.6	**39.5**	22.2	4.9
70	99.9	75.7	**50.5**	29.6	6.8
100	100.0	86.7	**63.4**	39.4	9.5

the probability of flooding in 100 years if you live in a 1 in 100 year flood area is 63.4%. This is a difficult concept, and quite frankly the numbers mean very little to the public. However, with further examination there are two results that may help to explain the risk posed and an individual's exposure.

Let us consider the same level of exposure, i.e. that of a 100 year flood, but this time place it in the context of an average life of 'three score years and ten' or 70 years. The likelihood of experiencing a flood during that period is 50.5%, which could be expressed as a 50–50 chance. So by simply placing the outputs in an easily understood context the task of communicating risk should be simplified.

To add to this understanding, and to look at another well-understood context for the same risk exposure, consider the current life of a UK mortgage of 25 years. This shows that the risk of being flooded is 22.2%, or about a one in five chance.

These two approaches do not require the engineering to be diluted, but do go some way to explaining a complex concept using a method that places the results in a real-life context. This would allow the well-established terminology used among mathematicians, scientists and engineers to be used as the basis for communicating risk to the public.

2.5. Hurricane Katrina

2.5.1 Causes

Hurricane Katrina formed on the 23 August 2005 'as a tropical storm off the coast of the Bahamas' (US Government, 2006a). 'Hurricane' is the regional term for a tropical cyclone that forms in the 'North Atlantic Ocean, the Northeast Pacific Ocean east of the dateline or the South Pacific Oceans east of 160E' (Atlantic Oceanographic and Meteorological Laboratory, 2011). A tropical cyclone is centred on a low pressure and is surrounded by numerous thunder storms, which produce the intense rainfall and strong winds, fuelled by the heat and energy released from moist rising air.

From its formation, Katrina grew over the next 7 days, peaking as a category 5 storm, with its effect being felt in the Gulf coast, including Mississippi, Louisiana and Alabama. On the morning of 29 August it made landfall at the south-eastern Gulf of Mexico as a category 3 storm, and set in motion a series of events that exposed vast numbers of Americans to extraordinary suffering (US House of Representatives, 2006). The hurricane generated winds in excess of 130 mph, with a storm surge of up to 27 feet impacting on 'nearly 93 000 square miles' (US Government, 2006b), an area in excess of the entire land mass of Great Britain.

2.5.2 Impacts

Hurricane Katrina resulted in widespread devastation 'which lead to infrastructure damage, flooding, civil disorder, fires, toxic chemical dispersion, disease risk and thousands of people isolated for days without water, food or medical care' (Litman, 2006). The human impacts were enormous: 'more than 1800 people died in the storm, mostly – nearly 1600 in Louisiana' (Kayen *et al.*, 2006). The most high-profile impacts were within the city of New Orleans, where the levees protecting the city were breached, resulting in flooding to 'approximately 75 per cent of the metropolitan area' (Knabb *et al.*, 2005). This crippled the city, hampering the evacuation and relief effort, and caused colossal economic loss, estimated to be £81 billion (CNN, 2009).

2.5.3 Lessons learned

There exists a huge amount of written material which examines all aspects of this natural disaster and draws attention to the many lessons that should be learnt. There are, however, a number of lessons that are highly relevant to flood defence engineers.

2.5.3.1 Design standards

It was a well-known documented fact that the 'levees protecting New Orleans were not built to survive the most severe hurricanes' (US House of Representatives, 2006). Indeed, there existed an operational plan to deal with their failure and the necessary unwatering of the city. The design standard used roughly equates to a fast-moving or 'moderate' category 3 hurricane (US House of Representatives, 2006). This has a 0.5% chance of occurrence in any given year, which can also be expressed as a 200 year event. With the benefit of hindsight, it is always possible to criticise the level of protection offered by any system of flood defences, but it must be understood that the risk, while mitigated by defences, is not removed. There always remains a residual risk. However, in this case the situation is slightly more complex, as the court heard, during the case taken against the US Army Corp of Engineers, that the available data used to calculate the 'Standard Project Hurricane' was from a very short time series, 'only from 1900 to 1956' (Trial transcript, 2009). From a limited dataset there is always a risk of underestimating an extreme event, and this compounds the problem of the residual risk of failure.

2.5.3.2 Maintenance

On 18 November 2009, a court found that the Army Corps of Engineers' failure to properly maintain a shipping channel that linked New Orleans, Louisiana, to the Gulf of Mexico contributed to some of the catastrophic flooding during Hurricane Katrina (US District Court, 2009). In addition, the report by the Select Bipartisan Committee to investigate the preparation for and response to Hurricane Katrina entitled *A Failure of Initiative* drew attention to the fact that 'responsibility for operating and maintaining the levees were split among many different organisations', going on to say that 'the number of organisations involved, and disagreements among them, makes accountability diffuse and creates potential gaps and weaknesses in parts of the flood protection system' (US House of Representatives, 2006).

The design standard was low, considering the geographical location of the area, the construction was flawed, with underlying poor ground conditions, and the maintenance responsibility was disjointed. The court found that these were contributory factors in the flooding of New Orleans.

2.5.3.3 Critical infrastructure

Given the widespread impacts of this hurricane and the resultant flood event, the impact on critical infrastructure was immense. The road infrastructure was totally compromised, and this had an immediate and direct impact on the evacuation of the city and the subsequent relief effort. The effects of these impacts on critical infrastructure were, in this case, much wider than just those on the local population, as this area of the Gulf coast is a key provider of crude oil and natural gas. Indeed, according to government data the area accounts for 'a quarter of total US oil output', and '92 percent of crude and 83 percent of natural gas production were shut down' (TerraDaily, 2009) as a result of this natural emergency. As a result, oil futures soared, and this influenced the worldwide oil price and required the US government to dip into its emergency oil reserves. This shows that the impacts of natural disasters of this scale can be felt on the world stage, even if the direct impacts are restricted to a local area. In addition, during this event 2.6 million customers across Louisiana, Mississippi, Alabama, Florida and Georgia suffered power shortages (US Department of Energy, 2005).

2.5.3.4 Flood warning and forecasting

The path of a hurricane is predictable, and therefore the impacts are foreseeable. This is similar for many such natural events, particularly flood events in large, slow responding catchments. Much of the criticism of government immediately after the impacts of Hurricane Katrina was of the emergency response, in particular the preparedness phase. If a flood event, irrespective of its source, is predictable, then those in charge, and in particular engineers, should ensure that systems are in place to foretell the event at an early a stage as possible. Like any other capital investment, flood warnings need to be viable, that is to say the benefits must outweigh the costs. It is absolutely clear that for an impact such as that experienced during Hurricane Katrina, the forecasting would always play a critical role in ensuring the safe evacuation of citizens. For lesser events, in fast reacting catchments and for surface water flood, the situation is not so clear-cut, as the time between identifying the potential problem and it occurring is relatively short, and therefore the opportunity is limited.

2.6. Conclusion

The lessons learned from historic flood defence and water impoundment structures are invaluable to design, construction and maintenance engineers. The problems of the past, if understood, should enable us to refine our design codes and their practical application. Whilst it is easy for us to dismiss many of the failures, as they were triggered by extreme natural events that were outside the control of man, this would be to miss the key learning point – that of the human impact of these incidents. The cost in both financial and human terms has been great, and it does not stop when the water recedes, as the pain, suffering and mental trauma continue for years. This fact should drive us on as engineers to better the performance of our structures, so that we can provide as good a service to society as is possible.

REFERENCES

Amey (2008) *Broadway Underpass – Westlink, Belfast. Independent Report into the Flooding Incident on 16 August 2008*. Report IR1/2008. Available at: http://www.roadsni.gov.uk/broadwayunderpassfloodingreport.pdf (accessed 2 February 2012).

Atlantic Oceanographic and Meteorological Laboratory (2011) Available at: http://www.aoml.noaa.gov/hrd/tcfaq/AA1.html (accessed 2 February 2012).

BBC (2007) Hundreds flee over dam burst fear. BBC News Online, 26 June. Available at http://news.bbc.co.uk/1/hi/england/south_yorkshire/6239782.stm (accessed 2 February 2012).

Building Futures/Institution of Civil Engineers (2010) Facing up to Rising Sea Levels: Retreat? Defend? Attack? Available at: http://www.ice.org.uk/News-Public-Affairs/ICE-News/Facing-up-to-rising-sea-levels–retreat,-defend,-a (accessed 2 February 2012).

Cabinet Office (2008) *Pitt Review – Lessons Learned from the 2007 Floods*. UK. Available at: http://webarchive.nationalarchives.gov.uk/20100807034701/http://archive.cabinetoffice.gov.uk/pittreview/thepittreview.html (accessed 2 February 2012).

Chanson H and Felder S (2010) Energy dissipation on embankment dam stepped spillways, overflow stepped weirs and masonry stepped spillways. *Proceedings of the 17th Congress of IAHR Asia Pacific Division, IAHR-APD, Auckland*. Available at: http://espace.library.uq.edu.au/view/UQ:198080 (accessed 2 February 2012).

CNN (2009) Court: Army Corps of Engineers liable for Katrina flooding. 19 November. Available at: http://edition.cnn.com/2009/US/11/18/louisiana.katrina.lawsuit (accessed 2 February 2012).

Defra (2009) *Frequently Asked Questions for Reservoir owners in England and Wales*, Version 2. Available at: http://archive.defra.gov.uk/environment/flooding/documents/reservoir/seminar-faq.pdf (accessed 2 February 2012).

Environment Agency (2007) *Case Study: 2007 Summer Floods, Reservoir Safety – Learning from Ulley*. Available at http://www.environment-agency.gov.uk/static/documents/Research/reservoirscasestudy_1917484.pdf (accessed 2 February 2012).

Environment Agency (2010) *The Cost of the Summer 2007 Floods in England*. Flood and Coastal Erosion Risk Management Research and Development Programme, Project SC070039/R1. Available at: http://publications.environment-agency.gov.uk/PDF/SCHO1109BRJA-E-E.pdf (accessed 2 February 2012).

Foresight (2004) *Future Flooding*, Vols 1 and 2. Available at: http://www.foresight.gov.uk/OurWork/CompletedProjects/Flood/Unknown/Projectsummary.asp (accessed 2 February 2012).

HM Treasury (2006) Stern Review on the Economics of Climate Change. Available at: http://www.webcitation.org/5nCeyEYJr (accessed 2 February 2012).

ICE (Institution of Civil Engineers) (2009) *State of the Nation: Defending Critical Infrastructure*. Available at: http://www.ice.org.uk/Information-resources/Document-Library/State-of-the-Nation–Defending-Critical-Infrastruc (accessed 2 February 2012).

Kamphuis WJ (2000) *Introduction to Coastal Engineering and Management*. World Scientific Publishing, Singapore.

Kayen B, Collins B and Gibbons H (2006) *USGS Scientists Investigate New Orleans levees Broken by Hurricane Katrina*. Available at: http://www.soundwaves.usgs.gov/2006/01/index.html (accessed 2 February 2012).

Knabb RD, Rhome JR and Brown DP (2005) *Tropical Cyclone Report: Hurricane Katrina: 23–30 August 2005*. National Hurricane Center, Miami, FL, USA. Available at: http://www.nhc.noaa.gov/pdf/TCR-AL122005_Katrina.pdf (accessed 2 February 2012).

Litman T (2006) Lessons from Katrina and Rita: what major disasters can teach transportation planners. *Journal of Transportation Engineering* **132**: 11–18.

Mason PJ and Hinks JL (2008) Security of stepped masonry spillways. Lessons from Ulley Dam. *Dams and Reservoirs* **18(1)**: 5–8.

Met Office (2007) *Flooding – Summer 2007*. Available at: http://www.metoffice.gov.uk/about-us/who/how/case-studies/summer-2007 (accessed 2 February 2012).

Met Office (2011) *'Awful August' – Floods 2008*. Available at: http://www.metoffice.gov.uk/about-us/who/how/case-studies/floods-2008 (accessed 2 February 2012).

New Civil Engineer (2007a) Scour threat to Ulley Dam. *New Civil Engineer* 28 June. Available at: http://www.nce.co.uk/scour-threat-to-ulley-dam/72113.article (accessed 2 February 2012).

New Civil Engineer (2007b) Liability concerns could trigger Ulley reservoir demolition. *New Civil Engineer*, 19 July. Available at: http://www.nce.co.uk/liability-concerns-could-trigger-ulley-reservoir-demolition/70426.article (accessed 2 February 2012).

Northern Ireland Assembly (2008) Planning for flooding and the impact of climate change. Briefing Note 106/08. Available at http://archive.niassembly.gov.uk/io/research/2008/10608.pdf (accessed 2 February 2012).

Northern Ireland Executive (2008) *New Broadway Underpass opens in Belfast*. Available at: http://www.northernireland.gov.uk/index/media-centre/news-departments/news-drd/news-drd-july-2008/news-drd-010708-new-broadway-underpass.htm (accessed 2 February 2012).

Rivers Agency (2009a) Minutes of 147th Meeting of the Drainage Council for Northern Ireland, 11 June 2009, Armagh. Available at: http://www.dardni.gov.uk/riversagency/minutes_of_the_147th_meeting_of_the_drainage_council.doc (accessed 2 February 2012).

Rivers Agency (2009b) 08-253C//Meadowside, Post Project Evaluation. Final Report rev B. Available at: http://www.dardni.gov.uk/riversagency/meadowside-tech-rep-revb.pdf (accessed 2 February 2012).

Rivers Agency (2011) *Surface Water Flood Management Roles and Responsibilities Report*. Available at: http://www.dardni.gov.uk/riversagency/index/ra-publications/ra_technical_documents/rivers_agency_publications_swfmrr_report.htm (accessed 2 February 2012).

Rotherham Metropolitan Borough Council (2010) Experts visit Ulley. News feed, 2 February. Available at: http://www.Rotherham.gov.uk/news/article/70dam (accessed 2 February 2012).

TerraDaily (2009) US oil production reels from Hurricane Katrina. 29 August 2009. Available at: http://www.terradaily.com/reports/US_Oil_Production_Reels_From_Hurricane_Katrina.html.

Trial transcript (2009) *Robinson et al* v. *USA et al*, 12 May 2009. Available at: http://www.scribd.com/doc/19268116/Trial-Transcript-20090512-AM (accessed 2 February 2012).

Ulleyweb (2007) Available at: http://www.ulleyweb.co.uk/flood.htm (accessed 2 February 2012).

US Department of Energy (2005) *Hurricane Katrina Situation Report #11*. Office of Electricity Delivery and Energy Reliability, 30 August 2005. Available at: http://www.oe.netl.doe.gov/docs/katrina/katrina_083005_1600.pdf (accessed 2 February 2012).

US District Court (2009) 'In RE Katrina Canal Breaches Consolidated Litigation', Findings of Fact and Conclusions of Law. Civil Action No. 05-4182. SECTION "K"(2). United States District Court, Eastern District of Louisiana, Filed 18 November 2009, Available at: http://www.laed.uscourts.gov/CanalCases/Orders/19415.pdf (accessed 2 February 2012).

US Government (2006a) *The Federal Response to Hurricane Katrina: Lessons Learned*, p. 1. Available at: http://www.asksam.com/ebooks/releases.asp?file=Katrina.ask (accessed 2 February 2012).

US Government (2006b) *The Federal Response to Hurricane Katrina: Lessons Learned*, p. 62. Available at: http://www.asksam.com/ebooks/releases.asp?file=Katrina.ask (accessed 2 February 2012).

US House of Representatives (2006) *A Failure of Initiative. Final Report of the Select Bipartisan Committee to Investigate the Preparation for and Response to Hurricane Katrina*, p. 7. Available at: http://katrina.house.gov/full_katrina_report.htm (accessed 2 February 2012).

Walker JP (2008) The discontinuance of Boltby reservoir, North Yorkshire, UK. *Dams and Reservoirs* **18(1)**: 17–21.

FURTHER READING

Environment Agency (2010) *Guidance for the Design and Maintenance of Stepped Masonry Spillways*. Environment Agency, Rotherham, UK.

Institution of Civil Engineers (1996) *Floods and Reservoir Safety*, 3rd edn. Thomas Telford, London, UK.

Institution of Civil Engineers (2002) *Flood Risk Management*. Thomas Telford, London, UK.

Flood Risk
ISBN: 978-0-7277-4156-1

http://dx.doi.org/10.1680/fr.41561.043

Chapter 3
Visual inspections and surveys, remote sensing techniques and sensors for detailed analyses of flood defence infrastructure

Michael J. Wallis,* Gavin Long† and Martin I. Smith†
* HR Wallingford, Wallingford, UK
† The University of Nottingham, Nottingham, UK

This paper discusses many of the tools and techniques that can assist in the assessment of the condition and performance of flood defence infrastructure. Methods range from simple, non-intrusive observation and inspection, through to complex investigative and destructive testing. Modern technology in the form of remote measurement techniques (using both passive and active systems), global positioning systems, new integrated sensor technology, and data recording and communication functionality have broadened the capability of assessments and refined the level of precision and accuracy in recent years.

3.1. Introduction
3.1.1 Background
The assessment of the condition of flood defence infrastructure is an essential function in flood defence asset management. Many countries adopt a tri-level approach to such assessment. The first level involves the inspection or application of limited survey methods, and is aimed at providing basic information about such assets to inform managers about potential weak spots and deteriorating or damaged defences. A second, or middle, level of assessment aims to assess the potential damage or deterioration of 'level-one' identified defences in more detail and to 'trigger' either further analysis or remedial action. A third, higher level of assessment is often targeted at 'high-risk' or 'level-two' identified assets that protect large numbers of people and property or vulnerable critical infrastructure and services – where detailed analysis of condition is required to ensure that a full understanding of the condition and level of performance of such assets can be ascertained.

In the UK, relevant authorities such as the Environment Agency, local authorities and internal drainage boards are responsible for the inspection, monitoring and maintenance of flood defences in their respective areas. Around 70 000 flood defence assets are managed by the Environment Agency. This amounts to approximately 35 000 km of linear flood defences (flood embankments, walls, etc.) that need to be periodically inspected and assessed by qualified staff (Halcrow, 2004).

Without effective condition assessments it is impossible to determine the likely performance of flood defence assets within a catchment, reach or flood defence system. A key requirement in moving towards better risk-based asset management is the need to understand the structural and operational performance of individual assets – as well as of systems and processes. The importance of good-quality data cannot be overestimated, including direct access to basic parameters such as the location, condition and the standard of protection an asset affords, as well as more accurate and usable information on probability of failure, dominant failure modes and critical uncertainties (Sayers *et al.*, 2011). Monitoring and assessment is, therefore, a key component in the day-to-day management of any flood defence system, and for effective decision-making regarding its maintenance, adaptation/modification and investment.

3.1.2 Classification of assessment techniques

The wide variety of types of flood defence assets and the materials from which they are constructed means that there is not one 'analytical' methodology that can be applied to flood defence asset condition assessment. It is important, therefore, that the methodologies applied are framed within a consistent approach to their assessment, and that these techniques collect data that is relevant to a risk-based analysis of their condition. Such 'data-gathering' methodologies range from visual inspection and survey using simple techniques, through to complex testing, sampling and remote sensing/measurement methods.

Data-gathering methods can be classified into two basic groups.

- Non-destructive testing techniques (such as visual inspection and remote sensing/measurement, for example, Light Detection and Ranging (LiDAR)).
- Destructive (or intrusive) testing techniques (e.g. coring).

(In this context 'destructive (or intrusive) testing' is taken to mean any techniques that require the penetration into or removal of material from the asset for examination or analysis, or for the installation of monitoring instruments.)

Due to their application and utilisation in flood defence assessment a more logical classification of data-gathering methods is

- visual inspection and surface survey
- remote sensing/measurement techniques
- intrusive techniques, point sensors and arrays.

This classification has been adopted here to describe the methods and techniques in more detail.

3.1.3 Absolute and relative assessment

The assessment of flood defence asset properties such as geometry, material condition and any associated defects can be done on an absolute or relative basis. Absolute assessment, such as traditional surveying, requires the acquisition of the actual value of the property being assessed, such as crest height or crack location. Relative assessment, such as change detection (e.g. crack width), provides an assessment of some asset property relative to another property or in comparison with a previous value for that property.

Absolute values are defined in terms of a datum. For example, in the UK absolute crest height would normally be measured as height above Ordnance Datum Newlyn in metres.

Relative values measure changes in quantities/features. This is often a previous, historical, value recorded for the same feature, enabling the degree of change to be assessed without absolute assessment of the feature itself. In some monitoring applications the degree of change for a given asset feature is all that is necessary. An example of a monitoring method that can only assess relatively (without additional data) is interferometric synthetic aperture radar (InSAR). InSAR data alone cannot provide an assessment of elevation, but comparison between two InSAR datasets of the same feature or area can detect any change to a very high level of precision and accuracy.

Absolute assessment is often preferred to relative assessment. However, for some monitoring applications a relative assessment of change is sufficient. Other factors such as cost, accuracy or ease of use may also make a relative assessment more appropriate for a specific monitoring application.

3.1.4 Accuracy and precision of assessment

In colloquial terms, 'accuracy' and 'precision' are often used interchangeably, and there is indeed a close relationship between the two concepts. However, it is important to distinguish between accuracy and precision in terms of quantitative assessment. It is also important to highlight in general terms the degree to which accuracy and precision are required for the various types and spatial levels of infrastructure asset monitoring.

Accuracy can be defined as the closeness of a measured quantity to the actual true value of that quantity. As stated above, one of the main objectives of asset monitoring is to assess the current condition of the asset and its likely performance. The accuracy of this assessment is obviously of critical importance. An inaccurate assessment of asset condition makes it much more difficult to manage that asset efficiently and ensure its continued performance to specification. Low accuracy typically makes absolute assessment methods of limited use.

Precision is a measure of the variation in results obtained when repeating or reproducing a quantitative assessment. The greater the dispersion of the results from an assessment, the lower the precision of that result. In the context of infrastructure asset monitoring, it is important to be as precise as is possible, and where imprecision does exist it should be defined explicitly. High levels of imprecision are a particular concern when employing relative assessment, because the comparison between two measurements may not reflect the change in the value of the assessed feature but instead may reflect the imprecision of the assessment method. Conversely, a value may be very precise but not very accurate. A precise but inaccurate method could be countered by applying a correction factor assuming the source of inaccuracy can be identified.

'Statistical noise' is another term sometimes used to define the imprecision of a measurement. More specifically, it typically refers to a variation in results that is unexpected or unexplained. A relevant example of noise would be the effect of vegetative cover on the results obtained using remote measurement methods such as LiDAR or photogrammetry. In both instances the measured surface elevation obtained will vary according to the state of the vegetation and the degree to which it obscures the terrain surface.

3.2. Visual inspection and surface survey

3.2.1 Visual inspection

Visual inspection is the simplest and most commonly used method of non-destructive examination (NDE) testing. It is usually carried out within a tri-level approach of asset assessment and monitoring. Visual inspection is usually the bottom level or tier. It is comparatively cheap and easy, and therefore is practical to conduct on a regular basis. This lower tier of assessment only aims to provide broad screening, and therefore needs to be supplemented for high risk/complex assets by methods that are better but more expensive and time consuming.

Asset inspectors are the eyes and ears of the competent authority 'on the ground'. The visual inspection methodology, in addition to facilitating data collection and recording on asset condition, must also enable the inspector to 'flag' information or 'trigger' other actions that might be important or required for various reasons. This would primarily concern the need for more detailed inspection to ascertain the state of the defence asset. In addition, these might relate, for example, to public or operational safety, recreational use, environmental protection and conservation, access rights or commercial use. Actions will vary depending on the type of flood defence asset being assessed (e.g. sheet-piled wall, embankment, channel, gravity wall). They may also vary depending on the location or environment of the asset (e.g. coastal, rural, urban, fluvial – including combinations of these).

The main objectives of these inspections are to (Long and Mawdesley, 2011)

- record a snapshot of the asset condition
- provide accurate and repeatable assessments of the asset condition
- be a practical solution, balancing assessment accuracy against practicality (the visual inspection method should provide a quick assessment of condition without requiring expert (e.g. qualified engineer) knowledge)
- produce an assessment of condition for all types of defence assets
- identify weak assets (and transitions) prior to their failure
- act as a decision support tool for asset management
- aid in the prioritisation of maintenance options or new capital schemes.

Before an inspection is undertaken, the inspector should obtain all relevant data and information on the site to be assessed. Types of information that should be sought include

- previous written and photographic inspection records of the flood defence
- local geography and asset location
- topographic information on the asset
- standard of protection afforded by the asset (i.e. the performance specification).

Sufficient information should be sought in order to enable a comparison with the inspection to be carried out, to identify any changes to the asset occurring over time, and to highlight any points of weakness. Occasionally, some basic instruments (such as levelling tools) are used to enhance or add value to the inspection, but this is not common practice in the UK.

One limitation of visual inspections is that they can only assess the details of the surface of an asset. In some special cases, a visual inspection can be used to infer something about the subsurface conditions, but any internal change to condition will only be noted if it also manifests in a change in the surface layer or structure of the defence. Such inspections are also restricted where the asset extends below the waterline in a river for example, or is covered by the tide at the coast. This can render particular surface components of defences (e.g. the toe, or the revetment of the structure) 'unobservable' by the inspector.

A typical visual inspection or NDE does not require any quantitative assessment of the asset, and instead relies upon a qualitative assessment based on the training, judgement and experience of the inspector. A set of visual features or indicators is assessed visually, and the results are recorded and combined to produce an overall score for an asset measured on some predefined, abstract scale representing asset condition. Examples of such scales are Condition Grade (Environment Agency), Condition Index (US Army Corps of Engineers) or Soil Slope Hazard Index (Network Rail).

Both the strengths and weaknesses of visual inspection or NDE have been well documented (Long *et al.*, 2006). Some key points of relevance can be summarised as follows.

- Visual inspection can provide a quick assessment of asset condition without requiring the input (and cost) of a professional consultant or measured survey.
- Assessments produced are subjective judgements of surface condition only. However, visual findings at the surface may be indicative of subsurface condition.
- Visual inspection does not generally include the quantitative assessment of features. For example, in flood defence monitoring it is not efficient and accurate enough for assessing asset geometry, either for individual assets or entire asset systems.
- Typically, visual inspection only classifies condition into a small set of possibilities. For example, current visual inspection of flood defences in the UK classifies condition into only five potential states (condition grades) for an asset.

Figure 3.1 The Environment Agency condition assessment manual and a typical condition guidance page
Contains Environment Agency information © Environment Agency and database right

In most of England and Wales, for flood defences such inspections have been carried out by the Environment Agency since its inception in 1996. Prior to this, the National Rivers Authority carried out any visual inspection of flood defence assets, although the process then was less formalised. Local (maritime) authorities are responsible for inspecting the majority of coastal defences and urban drainage assets in their areas. The current process of visual inspection involves the inspector assigning a grade of 1–5 (very good – very poor) to each element of an individual asset guided by a condition assessment manual (Figure 3.1). From these (sub-)gradings, an overall condition grade is calculated using a weighted combination method. Elements are identified (or checked) and coded according to type and material via a simple diagram and table on a recording form. Overall condition grades can also be assessed directly using a manual override.

This visual assessment of the condition of defences is important, both in terms of using the information for management and reporting purposes, and in terms of communication with the public and other government departments and agencies.

The significance of the information for management purposes relates to the following.

- The assessment of the state of the defence, which affects the condition of the related asset system as a whole and hence the present flood risk. This is an important process for understanding the particular state of both defence structures and the asset system as a whole. Although a snapshot of condition at a particular point in time, repeated assessment or monitoring of this condition records changes over time.

- The decisions that may need to be taken on the management of that defence. Comparison of actual condition with minimum or trigger conditions or use of systems analysis can help to suggest interventions to prevent unwanted deterioration in specific structures or in the level of performance of the defence system.

The significance of the condition of the defences in terms of communication with others has been in the use of condition grade as a metric against which the performance of the Environment Agency's flood risk management function has been monitored. (In particular, the National Audit Office has used it in this way and has sometimes been critical of the failure to meet targets, although differences in interpretation have meant that National Audit Office saw targets as minimum conditions below which defences should not fall, whereas others have seen them as aspirational targets to achieve.)

3.2.2 Performance-based inspection

To make the best use of visual inspection it must relate in some way to more than just the 'superficial' appearance of the flood defence – it must relate to aspects that reflect the structural integrity, reliability and performance of the structure. An approach that is also common in other industries is based on failure modes (i.e. the ways in which an asset might fail). There are a number of potential modes of failure for any flood defence structure, most of which can be represented by physical process or statistical models. Performance models can be used to determine the likelihood of the type or 'mode' of failure occurring given a set of data relating to the flood defence system being analysed.

Understanding these failure modes (and any linkage between them) is important for condition and performance assessment, for two main reasons.

- To allow indicators of these failure modes to be included in condition assessments.
- To ensure that the right process-based models are applied when analysing the performance of defences.

A major challenge here, and a constraint of this approach, is the fact that many of these 'performance features' associated with incipient failure in a particular mode are the result of subsurface processes within the structure, which may not be observable. Such features are thus omitted from the visual inspection, or only included via some 'proxy' observable feature. Consistency of the method, wherever possible, is another factor considered in the identification of appropriate performance features. The final choice of performance features is, therefore, a balance between what it is possible to assess visually and the failure modes that need to be considered. Ideally, the selected set of performance features should have some link to all relevant failure modes, even if only indirectly.

Some failure modes and performance features for a typical linear flood defence, such as an embankment or sloping seawall, are listed in Table 3.1.

Table 3.1 Typical failure modes of a flood embankment, and indicators of failure or 'performance features'
Data taken from Long *et al.*, 2011 courtesy of FRMRC

Failure mode	Description	Performance parameters	Visual indicators
Non-structural failure (overtopping)	Flooding occurs without breach of the defence due to the water level exceeding the crest height	Water level Crest height	None
Overflowing/ overtopping leading to erosion of the outer slope	Water running down the outer slope leads to degradation of the surface protection and eventual erosion of the outer slope over time. Eventually leads to a breach of defence	Crest height Grass quality Slope angle	Rutting of crest Crest height below standard of protection Vegetation on outer slope
Slope instability	Geotechnical weaknesses initiate slipping or sliding of either slope. Range of potential causes, such as poor design, third-party damage, etc. Subcategories include circular slipping, sliding, deep-seated failure, etc.	Crack width Slip distance Slope angle Slip width Slip height Slip circle radius	Cracking, slumping or uplift evident Slope movement Animal burrowing Third-party damage to slope or toe
Piping	A pathway for water to pass through or under the embankment forms (e.g. due to poor soil condition and/or vermin infestation). Water seeps into the asset, washing out fill material. Eventual creation of a piping channel from the inner to the outer side of the embankment, and can cause explosive blowout and breach	Embankment width Soil coefficients Seepage length Water level difference Creep ratio	Signs of seepage Presence of washed-out fines Animal burrowing Altered vegetation on bank
Backfill washout	The rear of the embankment or the ground behind the embankment is washed away, therefore weakening the structure. It can be caused by a number of sources, such as excessive seepage through or behind the embankment, caused by weak soils and high water levels, or deep cracks in the embankment. The presence of foreign objects such as drains or pipes carrying water could lead to washout of the backfill	Internal void dimensions Slump dimensions Crack length Crack width Crack depth	Erosion/slumping of outer slope, crest or ground behind the asset. The presence of washed-out fines in or around the outer slope. The presence of foreign objects (drains/ pipes)

3.2.3 Performance features

The assessment of performance features requires an inspector to perform a detailed observation of the flood defence to assign condition (and an associated confidence score) to each performance feature in turn.

Performance features (derived from failure event tree analysis) represent the features of an asset that are to be inspected in order to produce the condition index or 'condition grade'. Performance features must possess a number of attributes. These are listed below in order of importance.

- Related to failure mode – They must be related to at least one failure mode, and therefore to the performance of the asset. For a performance-based visual inspection, the primary objective is to inspect visual indicators of likely asset performance. If the visual condition of a feature plays no role in asset performance, then it should not be inspected.
- Visible – The feature to be inspected must be easily assessed on a visual inspection. For example, many of the performance models have geotechnical parameters that cannot be observed on a visual inspection. Their assessment requires there to be a visible feature on the asset surface that directly or indirectly relates to their current value.
- Gradable – In addition to being visually identifiable, it must be possible to visually grade the current condition of the feature. There must be sufficient visual indictors related to the performance feature in order to assign the range of condition values associated with that feature.
- Mutually exclusive – Ideally, performance features should be mutually exclusive. There should be no chance of mistaking the condition of one for that of another. However, this attribute can be relaxed to some degree in order to satisfy the above three, more important, attributes if necessary.

A performance feature may relate to a number of failure modes or be associated with a single failure mode. Performance features that are uniquely associated with a single failure mode are important in the differentiation of most likely failure mode, and usually will have a high contribution to that failure mode. An example of this is the class of performance feature that relates to deformation of the asset structure or cross-section. These features are directly related to a specific failure mode, and represent visual signs of asset deformation caused by that failure mode. It is possible that not all failure modes will have a specific deformation type feature associated with them. Some failure modes may not produce singular forms of deformation that are visually identifiable. An example to illustrate this is the structural collapse failure mode for vertical wall structures. Structural collapse can occur due to a failure in resistance of the wall caused through general time- and load-based deterioration of the structure. It may not show any specific antecedent deformations of the wall and/or surroundings to distinguish it from other failure modes, and therefore there is no obvious deformation of the structure and/or its surroundings relevant to structural collapse.

Performance features can apply to a single element of the asset, and therefore be repeated for each element, where relevant, or can apply to the whole asset. The decision on whether a performance feature should be applied to individual elements or to the whole asset depends on the level of detail of inspection required. It is also determined by whether the location of the performance feature across the asset cross-section has any differential impact on performance. Also, the combination of a given performance feature with different elements of the asset may be linked to different failure modes. An example of this is the combination of vegetation coverage with inner or outer slopes of an embankment, linking to different failure modes. The performance feature 'vegetation coverage' might apply to each slope and the crest of the embankment individually, as the condition of the vegetation coverage within these different elements would represent different performance issues. It could also be possible that the vegetation coverage is only of interest, from the viewpoint of the condition indexing process, when applied to a certain element. In this example, the vegetation condition of the outer slope is important, because it is a visual indicator of overflow/overtopping leading to failure by erosion.

3.2.4 Visual inspection guidance and training

Guidance and training for inspectors is vital to ensure accuracy and consistency of the method across different operational conditions. Inspectors need to be competent in asset assessment. This requires a skills and knowledge base plus tailored training and guidance. The skills required include basic numeracy and literacy, together with a basic (possibly intuitive) understanding of engineering and physical processes. An inspector must be capable of accessing and locating the correct asset 'in the field', which requires spatial awareness and an appreciation of geo-referencing and map reading. An inspector will need to possess an understanding of the reasons for such assessments and how the operational processes and procedures surrounding asset assessment interact with each other. An inspector needs to know how to recognise the different asset types, and must understand 'performance' and how assets may deteriorate and fail. Computer literacy is required for entering collected data into portable computers (laptops or similar) and uploading to the asset database. Knowledge will also be required on what guidance on asset assessment is available, and from where it can be sourced.

A number of mechanisms and measures can be used to deliver guidance and training to operational staff, including

- condition grade descriptions and images (see Figure 3.1)
- flowcharts – a structured mechanism for assigning condition grade
- checklists
- accredited training courses in visual inspection (including an understanding of failure modes)
- online learning and guidance
- field-based tools to aid decision-making and recording (e.g. electronic guides on laptops and hand-held computers).

3.3. Walkover survey

Another form of non-intrusive investigation (or NDE) is the walkover survey. Walkover surveys are sometimes undertaken in addition to or in combination with visual inspections. The main difference between the two is that visual inspections do not generally involve the use of any measurement devices or recording of geometry or other parameters in the assessment. Subjective judgement of structure geometry or other performance features is not sufficient for accurate assessment of performance. A walkover survey, on the other hand, involves the use of various measurements, either in order to fill in missing information about the asset (geometry, extent, crest height, etc.), or to determine if there has been any change in such parameters over time. Walkover surveys can be seen as quantifying the performance features that can only be observed or assessed qualitatively in a visual inspection. Further surveys may then be undertaken to monitor these changes. Remote sensing/measurement and/or embedded sensor methods (discussed below) may also then be employed to monitor assets, where this is deemed necessary.

Walkover surveys are the domain of qualified engineers, surveyors and specialist geomorphologists expert in conducting such surveys, and typically employed by civil engineering consultancies. Various approaches and methods utilise a range of tools and techniques to collect data on specific parameters of the asset. As with visual inspection, walkover surveys are limited to the measurement of surface features such as cracking, slumping or rotation and uplift. One key advantage of the walkover survey over the visual inspection, however, is the ability to determine asset geometry and crest height with, for example, the use of modern real-time kinematic (RTK) global positioning system (GPS) surveying techniques.

Such accurate assessments of these parameters are necessary for many of the performance models of flood defences, and this emphasises the importance of quantitative assessment. A detailed survey can enable changes in topography to be quantified by means of measuring tools and photography. Repeated measurements over time enable asset managers to confirm whether any geotechnical instability was an isolated incident or an ongoing process. This is useful information for determining the correct form of intervention or deciding whether an intrusive geotechnical investigation might be required.

Table 3.2 lists major visual indicators for an embankment that can be quantified by measurement. It also provides physical dimensions for slight, minor and major occurrences of each. However, it should be noted that the severity of defects strongly depends on the local situation. The values given are, therefore, only illustrative. In terms of the dimensions shown, x represents the value parallel to the embankment crest line, y represents the value horizontally perpendicular to the crest line, and z is the change in elevation associated with the feature (Long *et al.*, 2011).

3.3.1 Measuring techniques

The method used to measure and record quantifiable data in a walkover survey depends predominantly on the type of defence, the material it is constructed of and the performance feature to be measured. Some key parameters, especially if localised in extent, can

Table 3.2 Surface features related to damage to and failure of an embankment
Data taken from Long *et al.*, 2011 courtesy of FRMRC

Visual indicator	Location	Description	Dimensions: m		
			Slight	Minor	Major
Rutting	Crest/slope	Wearing of crest or slope due to traffic (human or livestock)	*x*: 0.1–0.3 *y*: 0.2–0.5 *z*: 0.05–0.1	*x*: 0.3–0.6 *y*: 0.5–1.0 *z*: 0.1–0.3	*x*: 0.6 + *y*: 1.0 + *z*: 0.3 +
Circular slip	Slope (either)	Semicircular cracking and lowering of slope section	*x*: 0.1–0.3 *y*: 0.05–0.1 *z*: 0.01–0.05 SA: 1–5°	*x*: 0.3–1.0 *y*: 0.1–0.5 *z*: 0.05–0.2 SA: 5–10°	*x*: 1.0 + *y*: 0.5 + *z*: 0.2 + SA: 10°
Vermin holes	Slope (either)	Holes in slope caused by vermin. Slight = vole/rat size; minor = rabbit size; major = badger/fox size	*x*: 0.05–0.1 *y*: 0.05–0.1 *z*: 0.05–0.2	*x*: 0.1–0.3 *y*: 0.1–0.5 *z*: 0.1–0.3	*x*: 0.3–0.6 *y*: 0.3–0.6 *z*: 0.3–0.6
Slumping	Toe/crest	Depression at toe or crest. If at toe there may also be movement of the slope above the slump, leading to a change in the slope angle	*x*: 0.1–0.2 *y*: 0.1–0.3 *z*: 0.02–0.05 SA: 1–5°	*x*: 0.2–1.0 *y*: 0.2–1.0 *z*: 0.05–0.4 SA: 5–10°	*x*: 1.0 + *y*: 1.0 + *z*: 0.4 + SA: 10° +
Heaving	Toe/crest	Uplift at toe or crest caused by geotechnical or hydraulic issues. There may be slumping behind a heave point	*x*: 0.1–0.2 *y*: 0.1–0.3 *z*: 0.02–0.05	*x*: 0.2–0.6 *y*: 0.2–0.6 *z*: 0.05–0.2	*x*: 0.5 + *y*: 0.5 + *z*: 0.2 +
Cracking and fissuring	All	Presence of openings in bank and potential movement or erosion at crack or fissure points	*x*: 0.01–0.05 *y*: 0.001–0.01 *z*: negligible	*x*: 0.05–0.5 *y*: 0.01–0.2 *z*: 0.001–0.1	*x*: 0.5 + *y*: 0.2 + *z*: 0.1 +

SA, slope angle; *x*, value parallel to the embankment crest line; *y*, value horizontally perpendicular to the crest line; *z*, change in elevation associated with the feature

Figure 3.2 (a) Callipers being used to measure surface cracking of an embankment, and (b) a variety of photographic tapes and rulers
Courtesy of ELAS Brno, v.o.s.

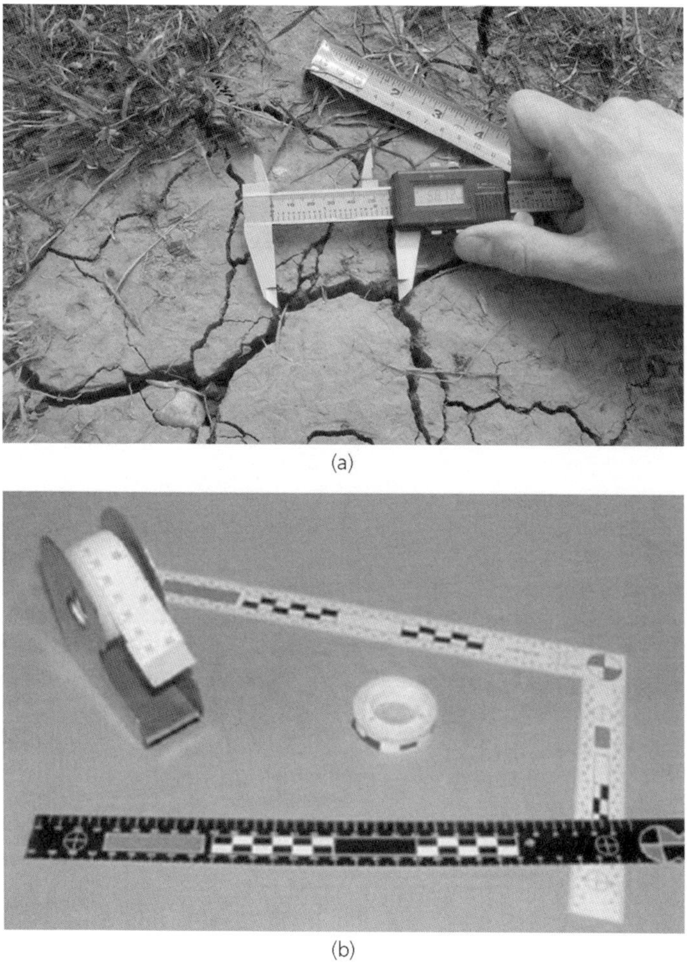

only be measured effectively and with sufficient accuracy and precision by asset-level assessment methods. Simple devices such as photographic rulers and tapes, inclinometers, callipers and gauges can be used to measure various parameters (Figure 3.2), as well as more sophisticated GPSs – used generally for determining position and the structure geometry.

Basic measurement and levelling tools can be used to quantify performance features such as cracking, rutting, slumping and crest and slope angle to a sufficient degree of accuracy. Markers can be used for measuring changes in an asset surface. For example crack extent can be marked on a wall so that subsequent inspections can record any progression or

Figure 3.3 An integrated measuring system of an RTK GPS, digital compass, camera and surveying wheel used for embankment crest profiling
From Long *et al.* (2011) courtesy of FRMRC

expansion. Tell-tale markers and simple gauges can also be placed onto the surface of an asset to monitor movement over time.

Crest height in particular, is a critical measure of likely asset performance for raised defences. It represents the standard of protection (SoP) of the flood defence and cannot be solely or accurately assessed by visual inspection. Figure 3.3 shows a sophisticated combination of tools, including an RTK GPS, digital compass and camera mounted on a surveyor's wheel for crest profiling, and Figure 3.4 shows an example of the plotted data from such a survey.

RTK GPS is a well established method of surveying. A roaming GPS receiver is used to survey a site. The instrumentation can be installed on a vehicle, carried in a backpack or mounted on a pole. Differential corrections from a fixed-point GPS station (reference station, or network of stations) are used to reduce the positional errors to 1–2 cm (using dual-frequency, survey-grade receivers with open sky). A GPS antenna accuracy of around 10 cm in elevation can be obtained using a simple backpack-mounted system. Real-time communication between the roaming unit and the fixed station or network is necessary. This is usually achieved through the use of mobile phone (GSM/3G) networks or a radio transmitter/receiver. The technology is already used by some Environment Agency Asset System Management teams to profile crest height.

Figure 3.4 An example of a crest profile recorded by RTK GPS
From Long *et al.* (2011) courtesy of FRMRC

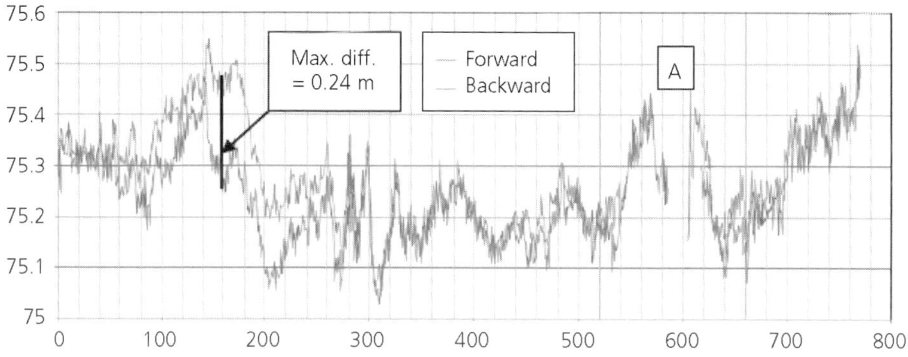

This approach has been adopted in beach profiling surveys (often conducted by quad bike) on the south coast of the UK. For embankments subject to regular grass mowing, the mounting of a kinematic GPS on mowing equipment that fully traverses the embankment would enable collection of asset profiles for the entire asset. This offers significant efficiency savings, as asset geometry data could be collected without the need for additional time and resources over and above the already scheduled maintenance activity.

Other remote sensing techniques can also be used to determine crest elevation and/or asset topography.

3.4. Remote sensing/measurement approaches

Remote measurement involves the assessment of some properties on the surface or within the structure of an object or entity without the need for intrusive methods or, in some cases, even without close physical contact with the entity. Here 'remote sensing' refers to methods and techniques for assessing flood defence assets or the surrounding terrain from a mobile or fixed platform some distance from the asset. These range from space satellites and aircraft, to terrestrial vehicles, fixed ground positions or surface deployed sensors. Often the term 'remote sensing' is used when satellite-based systems are being considered.

Remote measurement methods can be broadly categorised into passive or active systems. Active systems transmit some form of energy and then record the reflection or back-scatter from the target. Examples of active remote measurement systems include LiDAR, radar and ultrasonics. Passive systems detect existing sources of energy that are naturally present, such as light or infrared (IR) radiation. Photogrammetry or near-infrared photography are examples of passive systems.

It should be noted that, in the context of the tiered assessment approach mentioned earlier, remote measurements gathered by satellite and some aerial systems would be in the lower tier, and possibly even lower than visual inspection. This is because these approaches are better suited to system- and catchment-level survey, where they can be done cheaply for a large area. The main benefit is that they can be used for low-consequence systems, firstly to ascertain that the defence is there, and secondly, where the survey is inconclusive, to suggest which areas visual inspection should be targeted at. Aerial surveys can also be included with higher tier (accuracy) techniques when appropriate methods are adopted.

Remote sensing approaches relevant to flood and coastal defence surveys include

- light detecting and ranging (LiDAR)
- photography (which includes digital imagery) and photogrammetry
  - ground-based fixed-point photography and videography
  - near-IR photogrammetry
  - multi/hyperspectral imaging
- radar (synthetic aperture radar (SAR), interferometric SAR (InSAR), advanced SAR (ASAR)), including ground-penetrating radar (GPR)
- electrical resistivity tomography (ERT)
- sonar.

Brief descriptions of these techniques follow below, but it is beyond the scope of this chapter to describe these techniques in great detail. Further information can be found in Long *et al.* (2011).

3.4.1 Light detection and ranging (LiDAR)

LiDAR enables rapid and accurate assessment of the ground topography and large natural and man-made structures. Its use is commonplace in the production of digital surface models (DSMs) or digital terrain models (DTMs) for flood mapping and inundation modelling. It is also used to survey man-made and natural features for a variety of applications (e.g. slope monitoring in mining and quarrying, structural health monitoring and even meteorological surveying of the atmosphere).

LiDAR surveys can be conducted in three main ways, by

- static ground-based survey, or terrestrial laser scanning (TLS)
- mobile ground-based survey, or mobile laser scanning (MLS)
- aerial survey, by fixed wing aircraft or helicopter.

The basic principle of LiDAR is that the laser unit emits pulses directed at the ground or structure. The receiver detects the multiple reflectances of these pulses which (in the case of aerial LiDAR) are processed to produce a ‘cloud’ of three-dimensional coordinated points. Normally, from a ‘point cloud’ produced by aerial survey a regular gridded DTM can then be generated by post-processing the data. Generally, high-resolution DTM data (<1 m grid size) is necessary for assessing linear flood defence asset geometry. Lower resolution data can be useful in detecting large-scale trends and surface features, but lack detail and accuracy for individual assets. Significantly, a resolution of 0.25 m or less is required to detect small surface deformations related to asset failure modes. To detect surface slumping or heave with a dimension of 1 × 1 m on an embankment, for example, would require a DTM with a grid size of 0.5 m or less. Small assets presenting a slender cross-section such as vertical walls require sufficiently low grid size for accurate assessment of crest height. However, their hard, reflective surface can provide a more precise assessment of elevation in comparison to a grass-covered embankment. This increased precision also applies to most revetment materials. On the other hand, the nature of these asset types means that their structural tolerances are typically smaller, increasing the requirements for precision and accuracy of monitoring.

Static ground-based surveys (TLS) can provide more accurate results than aerial or terrestrial mobile surveys – however, they take a long time to cover a large area. They are best used to accurately monitor small-scale, safety-critical structures such as, for example, critical flood embankments and dams.

Vehicle-mounted LiDAR surveys (MLS) offer the greatest benefits to asset monitoring due to their accuracy and coverage. As distance from the scanner to the target is a key factor in the precision of the technique (Anderson *et al.*, 2005), the proximity of the platform and its survey path determine the trade-off between accuracy and coverage in the results obtained. The main drawback of MLS is the requirement for vehicular access – a major constraint for many flood defences where there is no existing access or pathway for the vehicle to use.

Both terrestrial and mobile laser scanning have potential uses in crest height measurement, but generally have practical limitations. TLS from a tripod can achieve high levels of detail and accuracy but, due to the nature of the geometry of most asset crests, is not an appropriate method. It would also be very slow when only a crest height is required. For these reasons, airborne LiDAR surveying has advantages and is normally the preferred option. The grid size of the DTM produced can be an issue for embankments with narrow crests (Figure 3.5). Generally, the grid size should be less than crest width (ideally it should be half the crest width or less) if the crest height is to be accurately assessed. With grid sizes equal to or greater than the crest width, the interpolation of the point cloud to the 'gridded' DTM will potentially include slope elevation values in the grids representing the crest, thereby reducing the elevation value calculated for the embankment crest. It is now possible to get grid sizes as low as 25 cm or even 12.5 cm if necessary. Such high-resolution DTMs should mitigate the problems of false readings of crest height for all but the most slender of embankments.

Vegetation cover (and other surface obstructions) can affect elevation accuracy of LiDAR. LiDAR is capable of penetrating light vegetation, but significant cover such as trees or shrubs can produce misleading assessments of surface elevation. This is not generally a major issue in flood defence embankment profiling, where the management practice is to limit vegetation near to and on embankments wherever possible. However, this is not always the case, and vegetation cover should be considered when planning to utilise LiDAR for crest profiling. Terrestrial-based surveying alternatives

Figure 3.5 Effect of LiDAR resolution for aerial monitoring of gravity walls: (left) oblique photography © Google Earth, (centre) 2 m gridded, (right) 0.25 m gridded
Contains Environment Agency information © Environment Agency and database right

may be required for those areas where vegetative cover is a significant obstruction of the bank crest from the air.

Low altitude aerial surveys are typically employed in asset monitoring applications. Helicopter-mounted systems such as FLI-MAP have been used in a number of similar surveys in the UK (e.g. TE2100, Network Rail mapping of the east and west coast mainlines), and are particularly suited to mapping narrow corridors such as highways, railways, fluvial channels or coastlines.

Standard LiDAR is reflected by water, and therefore cannot detect asset geometry below water level. LiDAR using a blue–green laser can penetrate water, but is currently too costly for regular or practical use in flood defence monitoring.

3.4.2 Photography, digital imagery and photogrammetry

Various forms of photography/digital imagery and photogrammetry can be used for flood defence asset monitoring, including

- ground-based, fixed-point photography and videography
- aerial photography
- near-IR photogrammetry
- multi/hyperspectral imaging.

A standard video or still camera that takes regular images of an asset or group of assets from a fixed location is a method that is increasing in popularity. It reduces the need for inspection staff to visit a site until there is a definite need for intervention. The degree of assessment provided by a fixed camera will be determined by the quality of the camera sensor and the fixed point chosen. A fixed camera capable of some rotation and optical zoom can increase its utility, although both these options can significantly increase power consumption. As the camera position is fixed and known, photogrammetric techniques can be employed to enable quantitative assessment of the images obtained through single- or multi-camera methods.

Photogrammetry, or more accurately stereo photogrammetry in this context, is the practice of determining the geometric properties of an object from measurements on photographs or (digital) images. In terms of infrastructure asset monitoring, the principal form of photogrammetry used is aerial images taken from an aircraft. Like LiDAR, aerial photogrammetry ideally needs a resolution of 5–10 cm ground sample distance (GSD) to be sufficient to identify performance features for asset monitoring purposes. Terrestrially taken photographs from a fixed reference point or ground vehicle are also capable of producing quantitative assessments of asset geometry and resolution of surface performance features.

In some cases, photogrammetry can exceed the accuracy of LiDAR. The resolution of slender features such as vertical walls is a particular example where the technique can be of benefit. Oblique aerial photogrammetry can be useful in assessing the condition of perpendicular elements such as the inner and outer faces of vertical walls.

Detailed analysis of photogrammetry and its capabilities can be found in Smith *et al.* (2006), but some key points regarding the use of aerial photogrammetry of relevance are as follows.

- Camera, GPS and inertial measurement unit (IMU) are often used to produce orthorectified imagery (imagery with the same geometric characteristics as a map) of the terrain.
- Wider coverage per swathe than LiDAR can be achieved due to the camera geometry of a photogrammetric survey.
- High-resolution photography from fixed-wing aircraft can produce quantitative assessment of terrain and assets to a similar level of accuracy as LiDAR.
- With manual photograph/image interpretation, photogrammetry can pick up features not detectable by LiDAR (such as vertical walls with a slender cross-profile).
- Photogrammetry requires a clear view of the target surface – the capture of ‘normal’ photographs on film and digital imagery can be limited by atmospheric conditions such as low cloud, fog or heavy rain, and by general light levels.

In principle, near-IR photogrammetry works in an identical way to standard photogrammetry, except that the sensor or film used is more sensitive to the IR section of the electromagnetic spectrum. The ‘near’ refers to the fact that the sensor or film only works passively and cannot detect far into the IR band of the spectrum, as in the case with thermal imaging/thermography. Near-IR photogrammetry has a singular property that makes it potentially useful for flood defence asset investigation, i.e. vegetation and water are highlighted due to their reflectance to the near-IR band of the spectrum. Near-IR photogrammetry has long been used to monitor crop growth and type. In terms of flood defence asset monitoring, the highlighting of vegetation means that bare sections of earth stand out in comparison. This is useful when assessing grass cover on earth embankments. It can also highlight differences in vegetation type and structure, which is indicative of the phreatic surface or even potential failure (e.g. the presence of water-loving plants where seepage is occurring).

Multi/hyperspectral imaging includes systems that capture data at a range of wavelengths across the electromagnetic spectrum. It can include wavelengths that lie outside the visual spectrum, including near-IR and deeper into the IR region. The difference between multi- and hyperspectral imaging is that the latter is more discriminatory in the individual bands of the spectrum it analyses.

These types of imaging systems are typically mounted on space satellites or aircraft. The CASI imaging system is an example of this type of remote measurement equipment, and is already being used by the Geomatics Group in the Environment Agency in the UK.

Multi- and hyperspectral imaging systems are also commonly employed in monitoring the extent of agricultural vegetation, and would be of potential use in assessing the presence of vegetation on flood defences. Their increased ability to monitor specific bands of the spectrum raises the potential of identifying specific vegetation types. Of most relevance to earth embankments, measurements from this system could be used as a proxy for subsurface conditions, such as the presence of saturated ground or specific soil types.

3.4.3 Radar (SAR, InSAR, ASAR), including ground-penetrating radar (GPR)

Radar is another active remote measurement system, but it is not at present widely used for flood defence asset investigation. There are a number of potential methods and techniques for remote measurement using radar, including SAR and its variants (e.g. ASAR and InSAR).

Radar-based remote measurement is not capable of providing absolute positional values unless it is integrated with other positional datasets (e.g. data from GPS stations). However, it is capable of detecting change between two radar datasets with a high level of accuracy. It is not affected by cloud cover, and is therefore suited to space-based or high-altitude surveying. It can, therefore, provide very wide area coverage and is ideal for monitoring terrain over very large areas. It is suited to detecting large-scale land movements associated with mining subsidence or natural processes. It therefore has the potential to be useful at the regional level and as part of long-term flood risk planning.

Specific details regarding the nature and operation of these technologies can be found in Smith *et al.* (2006). InSAR is the method most suited to asset monitoring applications due to its aptitude and accuracy in the detection of change or movement of a target over time.

InSAR can survey the earth surface and generate topographical information. It can be used to produce a digital elevation model and, like the other system-level technologies, can utilise a variety of space-based, airborne or terrestrial platforms. A key feature of InSAR is that it is not affected by cloud cover or darkness. The standard shorter wavelength systems (C band and X band) cannot penetrate vegetation, and therefore produces elevation as the top of any vegetative cover. Longer wavelength systems (P band and L band) can penetrate vegetative cover to give a bare-earth elevation of the ground features. These systems may, therefore, be of greater use in surveying flood defence assets, although the longer wavelengths are limited by their resolution, and therefore produce less detailed data.

GPR is a non-intrusive geophysical investigative technique that is capable of detecting subsurface features down to a depth of approximately 15 m. However, its penetrative capability is reduced, sometimes to only a few centimetres, in the clayey, saturated soils that are typical in flood defence embankments in the UK. That said, where penetration is not limited, GPR is effective at identifying, for example, subsurface voids in embankments or beside walls. However, GPR can often generate false-positive results and requires specialist operation and interpretation of the results. Further research into GPR technology is being undertaken.

There is a range of other geophysical methods that can be used for subsurface investigation (e.g. gravity gradiometry, aeromagnetic surveying and seismic reflection), but these are intended for analysing rock and soil types at much greater depths than is relevant to flood defence monitoring. They do not offer the same accuracy as GPR at shallow depths, and are primarily used in oil and gas exploration.

3.4.4 Electrical resistivity tomography (ERT)

Electrical resistivity tomography (ERT) (also known as electrical resistivity imaging (ERI)) is a geophysical method that measures the electrical resistance of soils at various points from the surface. The results can be used to infer the type and nature of the soil conditions due to the difference in electrical resistance between soil types. For example, clay has a very low resistance (1–20 Ωm), whereas gravel has a relatively high resistance (300–2000 Ωm).

Moisture content greatly affects the resistance value of the material, and the repeated use and calibration of this method can be used to assess the permeability of the embankment and its moisture content. The permeability of the soil within or under an embankment is an important physical property used in many of the performance models for embankments, particularly those relating to seepage, piping and internal erosion. Differences in permeability between the embankment fill and the foundation soil can lead to differential hydraulic loading and potential failure due to the increased pore water pressure in the soil.

3.4.5 Sonar

Sonar and ultrasound scanners mounted on or towed behind various types of water craft can be used to measure underwater topography or bathymetry. Water-based surveying can be useful at both the system and the asset level. At the asset level, boat-based monitoring equipment can be used to produce a view of the submerged aspects of assets (e.g. the submerged toe of structures) over a subreach or reach.

Sonar technology is used currently at the Environment Agency in the UK and by the Port of London Authority (PLA), and the channel profiles generated can prove useful in asset management. Tidal regions are most commonly associated with the use of sonar-based surveying of fluvial and estuarial channels. The erosive effect of regular incoming and outgoing tides makes toe scour at flood defence structures a major concern. Sonar is of most use where observation of the toe of an asset is not possible through a standard visual inspection due to difficulty of access and/or adverse hydraulic conditions (water clarity and/or depth over the toe).

Figure 3.6 shows a remote-controlled boat used by the Environment Agency in the UK, containing an Acoustic Doppler Current Profiler for measuring river velocity, discharge and cross-section profiles.

Modern, accurate georeferencing of data is now enabling the combination of above-water laser-scanned datasets and photographic images to be 'joined' to underwater swath bathymetric plots to provide seamless below/above water mapping.

3.4.6 Ultrasonic/pulsed eddy current testing

Ultrasonic testing is a technique used to measure the thickness of steel-sheet piled walls – a common structure type in many riverine and coastal flood defences. It can measure wall thickness to an accuracy of about 1 mm in clean, unobscured conditions. Another method, 'pulsed eddy current testing', is a similar method that is regularly used (mainly in ports and harbours) and has comparative levels of accuracy to ultrasound. Both methods require specialist equipment and expertise to deploy and operate. Neither technique is used regularly at present for flood defence asset assessment.

3.5. Intrusive techniques, sensors and arrays for detailed assessment, diagnosis and monitoring

A wide variety of parameters useful for assessment and performance diagnosis can be sampled and measured in more specialist detailed studies by techniques and sensors installed in and on flood defence structures (both non-destructive and destructive evaluations). Some intrusive methods have been in use for a long time, and include physical techniques such as coring or trenching for sampling soils, Schmidt hammers for testing the strength of concrete structures, settlement gauges to detect consolidation and settlement in embankments, and strain gauges to measure tension in sheet pile wall ground ties. Other techniques are generally limited to research and development projects or highly critical defence studies, although some are now commercially available and being used more widely. The majority of these are electronic sensors that can be used, for example, to detect water tension, temperature, inclination, self potential, acceleration and hydraulic head. Techniques and sensor types include

Figure 3.6 (a) Radio-controlled survey boat/instrument platform used by the Environment Agency, UK, for profiling river bathymetry and measuring water velocities and discharge. (b) ADCP velocity contour plot of the river channel cross-section
Contains Environment Agency information © Environment Agency and database right

(a)

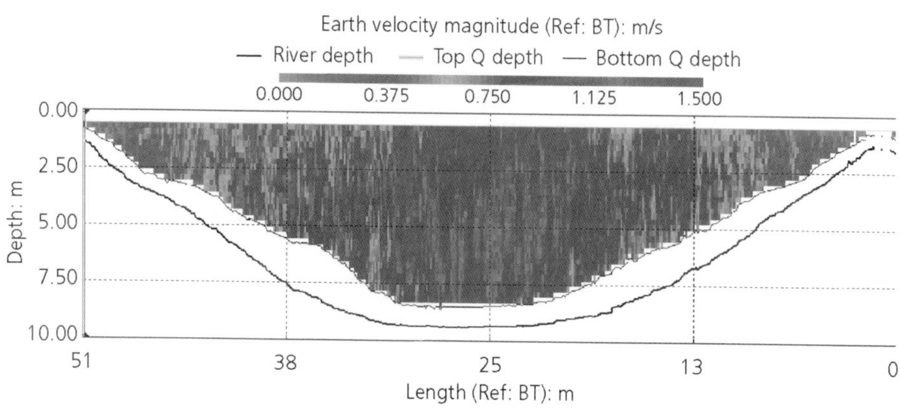

(b)

- point sensors
  - accelerometers
  - inclinometers
  - potentiometers
  - extensometers and positional sensors (total station)
  - moisture sensors (pore water pressure transducers), piezometers
- integrated sensor arrays (micro-electromechanical systems (MEMS) and shape acceleration arrays (SAAs))
- sensor-enabled geotextiles (optic fibre).

Various types of subsurface sensor (e.g. for motion and moisture detection) can be installed within the surface of a gravity wall or embankment. Motion sensors such as accelerometers or inclinometers can detect any movement in the wall associated with sliding, slipping or overturning. They can also be placed in the earth at the toe of the wall to detect toe scour.

Moisture sensors can be used to detect signs of seepage within or under a gravity wall. The main limitations associated with subsurface sensors are: their installation in an existing structure is invasive; their power requirements can be an issue; and their number and placement within the structure constrains their effectiveness. However, for new-build walls in critical locations, sensors are possibly the most effective and efficient means of subsurface monitoring.

Sensor-enabled geotextiles incorporating fibre optic technology are able to detect soil strain by distributed light back-scattering. SAAs are able to measure three-directional soil deformation and/or pore pressure. MEMS modules are able to detect local tilt, pore pressure and temperature (Figure 3.7), the latter as a proxy method for detecting water flow. MEMS modules can be placed in an embankment and measure, without

Figure 3.7 An example of an integrated/multi-sensor system ready for installation in an embankment
Reproduced by kind permission of Alert Solutions

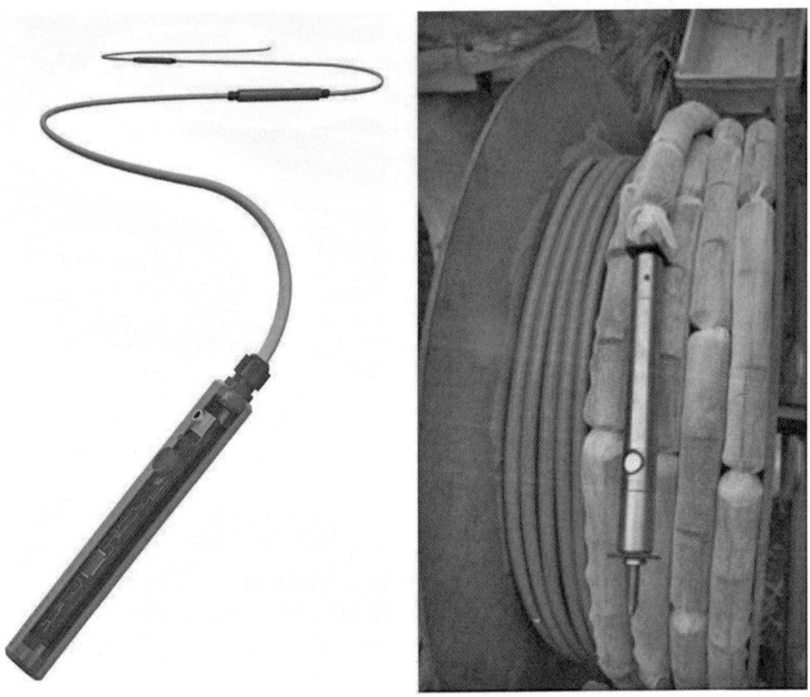

human intervention, the water pressure, temperature and movement, to very high accuracy. The measurements take place in real time and at frequent intervals, which means that the effects of events such as a local heavy rainstorm can be monitored, and the data can be read remotely using mobile data networks (Langius *et al.*, 2009).

The benefit of many of these new techniques is that, once the sensors are installed, they can supply data for continuous monitoring of structure stability and performance, and the technology allows much longer stretches of embankment to be monitored at low cost compared with individual point sensors.

3.6. Conclusions

This chapter has described in brief many of the methods available for assessing and inspecting flood defence infrastructure – both NDE and intrusive techniques. One difficulty is in deciding whether NDE methods are in fact applicable, and can or will provide sufficient data or information on a particular asset, material or terrain. Some methods may even be prohibitively expensive, or may not provide a high enough level of resolution for detailed asset survey or assessment. Table 3.3 lists the benefits and limitations of various surveying and NDE methods that are often used for such detailed assessments.

3.7. Future developments

Remote sensing/measurement technology has improved greatly since its introduction, and is likely to continue to improve. Image and data resolutions that are now standard which would have been thought to be very high just a decade ago. The accuracy of GPS and IMUs required for mobile surveys has also improved significantly.

Another area that has shown great advances in the last decade is the data processing of remote measurement data. There is still scope for further improvement, particularly in terms of automation and feature-extraction algorithms. It is in this area in particular that further advances would be of great benefit from a flood defence monitoring perspective.

As alluded to earlier, continuous monitoring using electronic sensors, and also fixed cameras, is an area where technological advances could offer major benefits and opportunities. Current technology is limited or expensive to deploy in most instances. Improvements and associated cost reductions will make the use of continuous monitoring systems a much more viable method for flood defence monitoring, in the same way that it is currently used for monitoring dams and reservoirs. Advances in wireless sensor networks in terms of their accuracy, power consumption, robustness and reduced cost are likely to make them suitable for flood defence management. However, power usage and supply is still a limiting factor in their use.

Radio frequency identification (RFID) tags is another technology that (although not discussed here) could find a role in asset monitoring, assuming that their cost and accuracy improve. A large number of such tags could be installed within assets, and scanned periodically to detect internal movement and deformation. Currently, this

Table 3.3 Benefits and limitations of various surveying and NDE methods
Data taken from Long *et al.*, 2011 courtesy of FRMRC

Method/technique	Benefits	Limitations	Applicable assets
Kinematic global positioning system (GPS)	■ **Can survey geometry to a very high level of accuracy** ■ Backpack mounted version can be operated by a single operative and produce crest profiles over large assets quickly ■ GPS mounted in mowing equipment has the potential to profile the entire asset surface without additional resource requirements ■ Can be used to profile large faults, e.g. slip circles in embankment	■ Equipment is relatively expensive ■ Requires connection to GSM or two receivers for kinematic corrections ■ Requires a clear view of satellites ■ Some data processing and analysis may be required ■ Requires continuous access to transverse crest; vertical walls may be problematic in this regard	Embankments Gravity walls Revetments
Ground-penetrating radar (GPR)	■ **Can detect subsurface performance indicators** ■ Identification of voids caused by geotechnical issues or vermin ■ High penetration and accuracy in concrete, asphalt and masonry ■ Non-destructive and does not require extensive setup and calibration	■ Limited accuracy and penetration in heavy, clayey soils ■ Expensive and requires training to employ in the field ■ Data produced requires expert interpretation ■ Can produce false-positive results ■ Cannot be easily set up and operated by a single operative	Embankments Gravity walls Revetments Backfill material
Electrical resistivity tomography (ERT)	■ **Can detect subsurface performance indicators** ■ Identification of voids caused by geotechnical issues or vermin ■ Highlights areas of excessive moisture within the subsurface which could indicate seepage or piping ■ More suited than GPR to soil conditions typically in flood defences and their foundations	■ Expensive and requires training to employ in the field ■ Data produced requires expert interpretation ■ Accuracy of results for flood defence monitoring requires further research ■ Requires multiple trained operatives to set up and calibrate	Embankments Backfill material

Measurement of key performance indicators	■ Low cost and requires minimal equipment and training ■ **Acquires quantitative data on performance features quickly** ■ Can easily be integrated with photography to produce images with quantitative data	■ Can be difficult to measure features accurately on embankments due to changes in vegetation ■ Without georeferencing it may be difficult to ensure that repeated measurements assess identical features	All
Use of levelling devices	■ Simple, easy to use device ■ **Quickly assesses wall level or slope angle** ■ Low cost ■ Logging of data enables detection of change over time	■ Only measures level or angle over a small area ■ Multiple measurements required for large assets ■ For change detection, fixed reference points may be required for accurate assessment	Vertical walls, Revetments (contiguous only)
Fixed-point photography	■ **Simple and fairly accurate method for acquiring comparative imagery for an asset** ■ Can be carried out with minimal training and equipment ■ Can acquire quantitative data with additional measurement	■ Requires a fixed reference point to be set up ■ Multiple points needed for full coverage of an asset ■ A tripod is required to ensure a fixed viewpoint is accurately used for each image	All
Georeferenced photography	■ Image location is recorded ■ **Staff can locate feature easily on repeated inspection/remedial work**	■ Ordinary hand held GPS only accurate to around 3 m ■ cannot extract quantitative measurements from imagery	All
Ultrasonic testing (UT)/pulsed eddy current (PEC)	■ **Can measure thickness of sheet pile** ■ detects subsurface indicators of corrosion and accelerated low water corrosion ■ Non-invasive and non-destructive	■ Expense of equipment ■ Requires processing and interpretation by a highly trained operative ■ UT requires a clean, smooth surface	Sheet pile walls

technology is too limited for useful application in flood defence monitoring, but it is likely that these limitations will reduce over time.

For detailed inspection work there is a range of technological advances that could assist in quantitative assessment by inspection staff. Initial advances in this area are already evident, and are likely to develop in the next few years. Mobile devices such as smart phones already offer the ability to perform a number of functions of potential relevance. For example, a theodolite application (or App) is now available for particular mobile phones. This App allows the user to perform basic surveying operations and take georeferenced imagery. The high growth and penetration in the smart-phone market makes it likely that the hardware functionality and software available for these types of mobile device will continue to develop and improve in the near future (Long *et al.*, 2011).

Acknowledgements

Much of the research reported in this paper was conducted as part of the work of the Flood Risk Management Research Consortium (FRMRC). The FRMRC is supported by Grant EP/F020511/1 from the Engineering and Physical Sciences Research Council (EPSRC), in partnership with the Department for Environment, Food and Rural Affairs (Defra)/Environment Agency Joint Research Programme on Flood and Coastal Defence, United Kingdom Water Industry Research (UKWIR), the Office of Public Works (OPW) Dublin, and the Northern Ireland Rivers Agency (DARDNI). This financial support is gratefully acknowledged. The authors are also grateful to the Environment Agency for providing LiDAR data and to the Ordnance Survey for providing Mastermap data. The authors gratefully acknowledge the work undertaken at The University of Nottingham. Thanks go to Michael Mawdesley and Dr Ahmad Taha (formerly), of The University of Nottingham, and Dr Geoff Baxter, of the Environment Agency, and to those who have supplied the photographs and images.

REFERENCES

Anderson ES, Thompson JA and Austin RE (2005) LiDAR density and linear interpolator effects on elevation estimates. *International Journal of Remote Measurement* **26(18)**: 3889–3900.

Halcrow (2004) *NADNAC – National Assessment of Defence Needs and Costs.* Halcrow in association with HR Wallingford, 2004.

Langius E, van der Waaij B, Bastiaans A *et al.* (2009) *Robust Monitoring. Added Value of Monitoring Sensor Streams in Dike Monitoring Systems.* FloodControl 2015 Report 2009.02.01.1. Available at: http://urbanflood.eu/Documents/FloodControl%202015%20report%202009.02.01.1%20Robust%20Monitoring.pdf (accessed 2 Februsry 2012).

Long G and Mawdesley MJ (2011) In *Flood Risk Science and Management* (Pender G and Faulkner H (eds)). Wiley-Blackwell, London.

Long G, Mawdesley M and Simm J (2006) *Improved Approaches to Condition Assessment – Volume 1: Performance Based Visual Inspection of Flood Defence Assets.* Flood Risk Management Research Consortium (FRMRC) Research Report UR10. Available at: http://web.sbe.hw.ac.uk/frmrc/downloads/UR10%20Vol1%20signed%20off.pdf (accessed 2 February 2012).

Long G, Mawdesley M and Simm J (2011) *Improved Approaches to Condition Assessment – Volume 2: Detailed Technical Report*. FRMRC Research Report UR10. Available at: http://web.sbe.hw.ac.uk/frmrc/downloads/UR10%20Vol2.pdf (accessed 2 February 2012).

Sayers P, Wallis M, Simm J, Baxter G and Andryszewski T (2011) Towards the next generation of risk-based asset management tools. In *Flood Risk Science and Management* (Pender G and Faulkner H (eds)). Wiley, London.

Smith M, Edwards E, Priestnall G and Bates P (2006) *Exploitation of new data types to create digital surface models for flood inundation models*. FRMRC Research Report UR3. Flood Risk Management Research Consortium. Available at: http://www.floodrisk.org.uk/images/stories/Phase1/UR3%20signed%20off.pdf (accessed 2 February 2012).

FURTHER READING

Buijs F, Simm JD, Wallis MJ and Sayers PB (2007) *Performance and Reliability of Flood and Coastal Defences*. Joint Defra/Environment Agency Flood and Coastal Erosion Risk Management R&D Programme. Department for Environment, Food and Rural Affairs. London, UK. Technical Reports FD2318 TR1 and TR2.

Defra/Environment Agency (2004) *Use of Non-destructive Testing within Flood and Coastal Defence*. Environment Agency, Bristol, UK. R&D Technical Report W5A-059/2/TR.

Defra/Environment Agency (2009) *PAMS (Performance-based Asset Management System) – Phase 2 Outcome Summary Report*. Environment Agency, Bristol, UK. Environment Agency Flood and Coastal Erosion Risk Management Research and Development Programme Report SC040018/R1.

Deltares, TNO, IBM. (2009) Robust monitoring. Added value of sensor streams in dike monitoring systems. Flood Control 2015 program. Available at: http://www.floodcontrol2015.com (accessed 2 February 2012).

Feurer D, Bailly J-S, Puech C *et al.* (2008) Very-high-resolution mapping of river-immersed topography by remote sensing. *Progress in Physical Geography* **32(4)**: 403–419.

Koelewijn AR, Gijs JCM and van Hoffmans MA (2004) Lessons learned from a full-scale dyke failure test. *Proceedings of the 5th International Conference on Case Histories in Geotechnical Engineering*, New York, USA.

Rajabalinejad M, van Gelder PHAJM and Vrijling JK (2008) Probabilistic finite elements with dynamic limit bounds; a case study: 17th Street Flood Wall, New Orleans. *Proceedings of the 6th International Conference on Case Histories in Geotechnical Engineering*, Arlington, VA, USA.

TenCate (2011) DeoDetect Solution White Paper. The Future of Geomonitoring Performance. Available at: http://www.tencategeodetect.com (accessed 2 February 2012).

Urban Flood (2010). Monitoring, rich browsers and cloud service technologies for an online EWS hosting platform. Work Package 6. Groningen, Netherlands. Available at: http://www.urbanflood.eu (accessed 2 February 2012).

Van MA, Zwanenburg C, Koelewijn AR and van Lottum H (2011) Evaluation of full scale levee stability tests at Booneschans and corresponding centrifuge tests. *Proceedings of the 17th International Conference on Soil Mechanics and Geotechnical Engineering*, Alexandria, Egypt.

USEFUL WEBSITES

Alert Solutions: http://www.alertsolutions.nl (accessed 2 February 2012).
Flood Control: http://www.floodcontrol2015.com (accessed 2 February 2012).
Flood Risk Management Research Consortium: http://www.floodrisk.org.uk (accessed 2 February 2012).
TenCate: http://www.tencategeodetect.com (accessed 2 February 2012).
Urban Flood: http://www.urbanflood.eu (accessed 2 February 2012).

Flood Risk
ISBN: 978-0-7277-4156-1

http://dx.doi.org/10.1680/fr.41561.073

Chapter 4
Supporting flood risk management through better infrastructure design and management

Paul B. Sayers,* Jaap Flikweert† and Andreas Kortenhaus‡
* Sayers and Partners, Oxfordshire, UK
† Royal Haskoning, Peterborough, UK
‡ Technische Universität Braunschweig, Braunschweig, Germany

As any flood defence asset manager will acknowledge, ensuring acceptable performance of flood defence assets and asset systems is a considerable challenge. The wide variety in asset types and forms and, uniquely to flood and erosion risk management, the interaction between each asset and its physical surrounding (including other assets) further complicates the task. Within this context, the concepts of risk and performance provide the asset manager with a consistent framework to integrate short to longer term actions to maintain, repair, improve or replace assets appropriately alongside non-structural measures, while avoiding unnecessary expenditure. In particular, they help identify the critical components of an asset system and enable effort in further data collation or physical intervention to be appropriately targeted. This chapter explores some of the challenges as well as some of the tools and techniques available to assist the asset manager in making informed decisions, from the requirement for further data collection and analysis through to actions to repair, renovate, replace or, indeed, remove assets.

4.1. Introduction

Good flood risk management is based on evidence – evidence of the risks of action and inaction – and the transparent recognition and management of uncertainty. Much of the developed and developing world has already invested heavily in flood defence assets, and engineered structures are likely to continue to play an important role as part of a broader portfolio approach to flood risk management in the future. This chapter provides an overview of some of the challenges faced, as well as some of the emerging tools that can help the asset manager make risk-informed investment decisions.

This chapter explores the state of the art in assessing the performance of individual assets and asset systems. It provides a discussion of reliability analysis and system-based risk analysis tools. It also provides a forward look towards the practical application and more detailed discussion of specific asset types in later chapters.

4.2. Asset management within the context of flood risk management

Flood risk management, as opposed to traditional flood defence or flood control paradigms, aims to utilise limited resources of time, social effort, environmental capital and money to deliver multiple benefits (Figure 4.1). There is also a recognition that structural responses (including the ongoing management of existing assets) will continue to play a major role in flood risk management into the future (e.g. Evans *et al.*, 2004a,b; United States Army Corps of Engineers (USACE), 2008).

In meeting these goals the modern flood risk manager no longer relies on engineered flood defences alone but utilises a range of other measures and instruments to deliver the desired outcomes. This paradigm shift, away from traditional engineering design and safety standards, and to a risk management approach, has a profound influence on the way flood management is considered and implemented. Fully understanding the importance of this change is not straightforward, but is a prerequisite to delivering good flood risk management, and in particular the appropriate use of infrastructure (defined here as measures to reduce the chance of flooding as opposed to the consequences of flooding).

4.3. The challenge of asset management in a resource constrained world

Asset management is not a simple 'construct and maintain' process, but exists as a continuous process of data gathering, analysis, planning, action and review. This cyclic process has long been recognised in manufacturing and process industries, and is starting to be more formally embedded in many flood risk management organisations, including for example the Environment Agency (Figure 4.2).

Figure 4.1 The primary purpose of flood risk management (Sayers *et al.*, 2011)

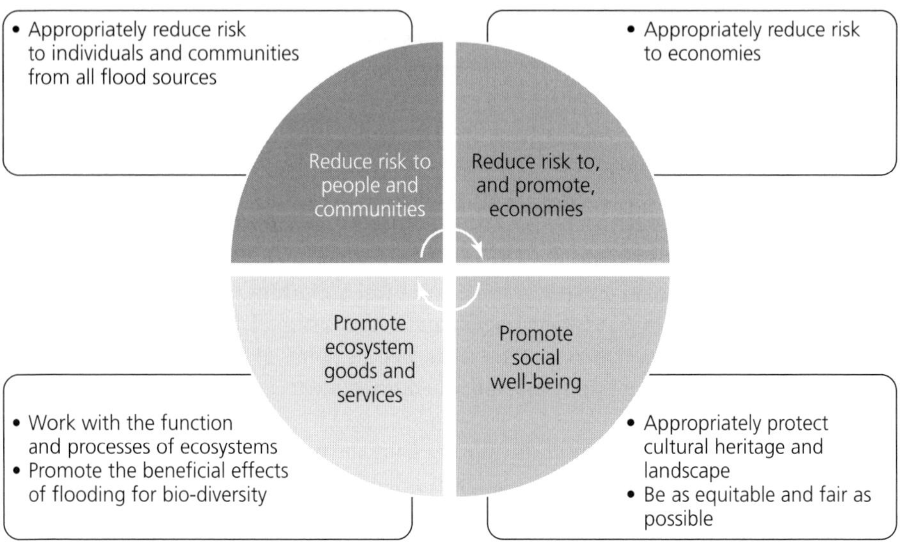

Figure 4.2 The cyclic process of asset management followed within the Environment Agency, UK

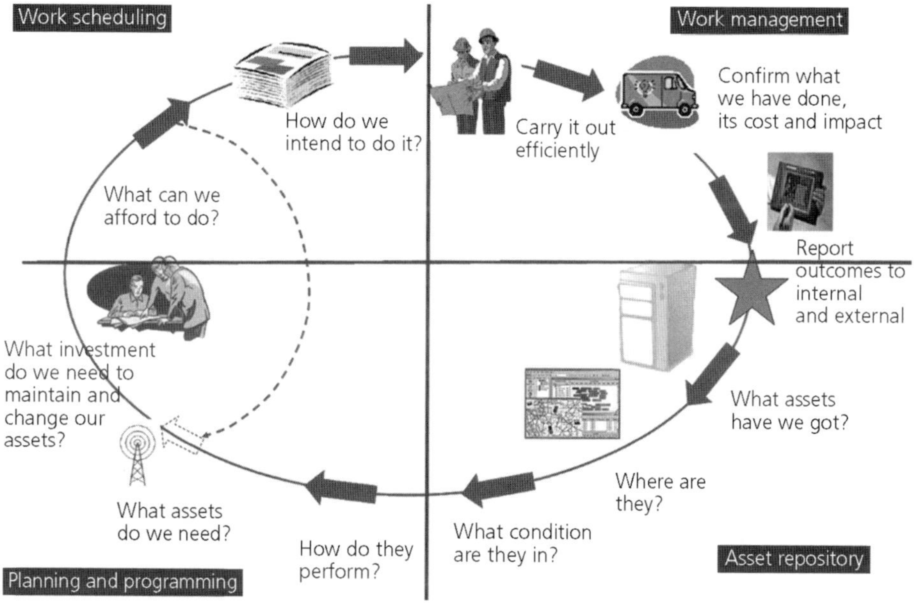

Whole-life considerations are at the heart of this process, linking actions from inception through to demolition/removal. Implementing the approach outlined in Figure 4.2, however, presents a number of practical challenges, including the following.

- *Applying the concepts of infrastructure asset management to flood defences* – Infrastructure asset management is a mature discipline with well-developed methods and tools, but it is not straightforward to apply these to flood defence assets that interact closely with the natural environment. The close coupling of the concepts of performance and risk in the context of flood defences (where a key measure of performance is the risk reduction achieved) further complicates the task. This introduces more complexity into the determination of performance, and makes the use of the asset management tools used elsewhere, which are often based on standard procedures and statistics reliability, difficult to apply meaningfully.
- *Understanding the role infrastructure has as part of a wider portfolio of responses* – Increasingly, flood risk management is recognised as a wide-ranging approach that implies a portfolio of measures and instruments (both structural and non-structural) to manage risk appropriately (e.g. Sayers *et al.*, 2002a, 2011). This need to utilise infrastructure appropriately as part of a wider response to managing flood risk places new demands on asset managers to become more proactive and integrated with others.
- *Incomplete understanding of the existing asset base* – Many of our towns and cities prone to flooding are already 'protected' by some form of structural defences.

Often these have been constructed over many years, with changing design and construction practice and functional requirements. The physical dimensions and engineering properties of these existing assets are often unknown or poorly resolved. In recent years, many countries have devoted significant effort to improving data and marshalling it into structured, accessible, databases (see e.g. Buijs *et al.*, 2005a; National Committee on Levee Safety (NCLS), 2009). It would, however, be impractical to seek to maintain comprehensive data on all assets. Typically effort is devoted to providing a minimum level of data (often considered as the location, type, notional standard provided and associated condition), with further data gathered only when required. An incomplete understanding of the existing asset base will, therefore, always exist (regardless of the effort directed at data collection).

- *Incomplete understanding of structural/operational performance* – Assets are often a complex composite of structural components with spatially varying materials, profile, operational rules, etc. The physical processes that lead to failure are equally complex and often poorly understood in detail (e.g. internal erosion and associated piping failure), and can be costly to analyse without significant gains in knowledge. The performance of an asset will also vary over time due to deterioration, a process that will be influenced by maintenance, fatigue (due to on-demand usage) and climate change (e.g. accelerated desiccation and associated fine fissuring of soils (Dyer *et al.*, 2009)).
- *Variability of impact* – The impact of failure can vary markedly from one asset to another and can change depending on the time of day or year when the failure occurs (e.g. in summer when tourists are camping in the flood plain, or during the rush hour when the roads downstream of a dam are congested with traffic). Not all assets are therefore equally important, and hence there is no requirement for them to have a common standard or condition. The impacts can also vary over a longer time scale as the land use within the flood plain or downstream valley changes (through increased development, a change in the demographic or the inhabitants, or simply through a change in the awareness of the flood risk). Many examples exist where the construction of structural defences has promoted the development of the protected flood plain, radically altering the potential consequences of a failure and perhaps undermining the adequacy of the design standards originally used (e.g. the continued development of the Thames flood plains and the extensive flood plain development in Sacramento, USA). Such development often takes place despite planning regulations such as PPS 25 that, in England, seek to limit residual risk (Communities and Local Government, 2010).
- *Affordability* – Budgets are limited, and it is common to have insufficient resources (of time, money, social and environmental capital) to undertake, maintain, periodically inspect, and properly operate all ‘desirable’ works. For example, in the USA it has been estimated that $2.2 trillion would be needed to raise all linear defences (levees) to the desired standard and condition (Stockton, 2009). Historically, funds have frequently been made available for the initial construction, but in only a few cases have the funds necessary for maintenance and inspection been provided. This separation of capital and revenue funding streams continues to persist today, and continues to undermine good, whole-life

asset management. This is especially true when the funding responsibility is devolved to local communities (rather than national or regional governments) or to commercial partnerships, where long-term funding can be difficult to secure.

- *The need to balance different interests* – Flood defence assets seldom have the single objective of reducing the chance of flooding. Visual impact (material and profile choice, working with nature, etc.), amenity (beach management activities, etc.), ecosystem services (wetland creation and protection, maintaining sediment connectivity, etc.), and transport and navigation are all common functions that flood defence assets must also support. Balancing these different, and often conflicting, interests presents a major challenge to the asset manager, and demands an open and transparent dialogue as to the trade-offs being made. Truly integrated actions are often undermined by separate funding streams and by differing time horizons and priorities. This fundamental constraint is starting to be recognised, and policies to promote multiple-function and cost-shared projects are starting to emerge (see e.g. Defra, 2011).
- *Decision complexity* – The invariable complexity of an asset system and the flood plains they protect make expert and engineering judgement difficult to apply. For example, within an asset system of 100 or more assets protecting a heterogeneous flood plain, identifying those assets that contribute most to the residual risk can be very difficult. Given the imperative to utilise limited resources to best effect, this often leaves asset managers with a rational doubt as to which action to take and when.

4.4. Adopting a hierarchical approach to decision making

Asset management involves a vast range of asset types. A flood defence *asset* can be described as any feature that is actively managed to reduce the chance of flooding (as opposed to the associated consequences), and as such this includes a wide variety of individual structures and activities that act together to form infinitely diverse asset systems composed of

- linear assets (above ground) – for example, a raised defence (levee or dyke) through to major dam structures
- linear assets (below ground) – for example, urban drainage networks
- interface assets (linking above- and below-ground systems) – for example, culverts, gulleys, manholes
- point assets – for example, a pump, gate or culvert trash screens
- watercourses and channels – which can include the vegetation and sediment within a channel and flood plain
- coastlines – for example, a groyne, beach or backshore
- defence systems – that recognise multiple assets combine to reduce the chance of flooding to a given part of the flood plain; this can include multiple lines of linear defences (secondary dykes, etc.) and different assets working in combination (pumps, linear assets, etc.).

A nested approach, where policies set the direction for the type of approaches used and the 'on-the-ground' realities inform policy, is a prerequisite to good management (see e.g.

Figure 4.3 The management of infrastructure assets takes place across a range of scales of time and space – from the identification of national needs and priorities, through to specific remedial actions for individual structures

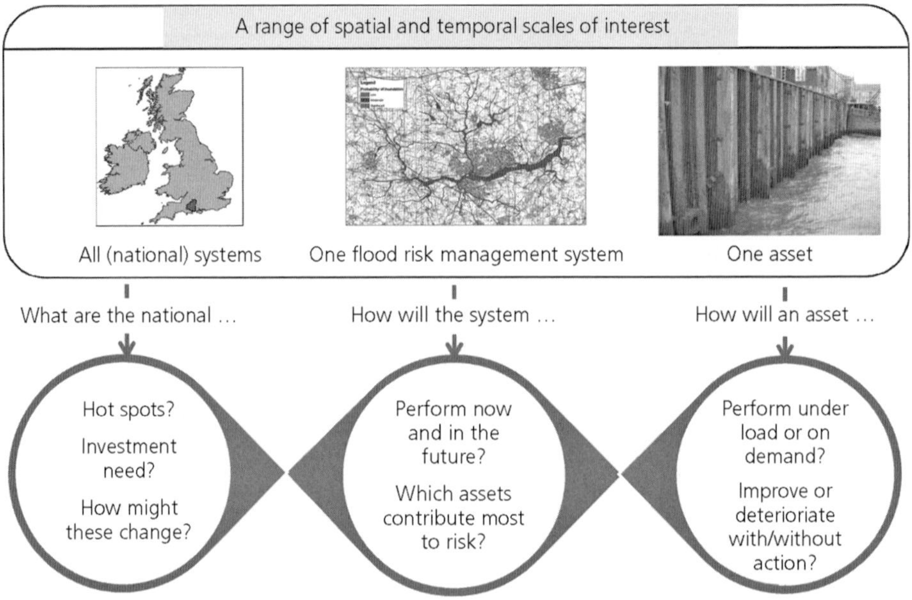

Sayers *et al.*, 2010). Within this context, infrastructure assets are managed across a range of spatial scales – from a single asset to the national allocation of funding – and across temporal scales – from short actions to long-term investment planning. Across these multiple scales the questions and decisions vary in nature, and so does the nature of the supporting evidence (Figure 4.3).

This approach is now starting to become a reality in practice (e.g. the national long-term investment strategy through to local system asset management planning undertaken by the Environment Agency (Environment Agency, 2009a)). Although these are positive steps, it is likely to be some time before policies, strategies and actions are routinely integrated.

4.5. Common issues faced when assessing the performance of flood defence infrastructure

Many common issues are faced by asset managers as they attempt to manage an ageing and extensive asset base and appropriately integrate new engineered structures, including the following.

- *A need for better evidence on the condition and performance of individual assets.* In England and Wales, the Environment Agency has stated that it will have succeeded in its asset management role when it knows exactly 'what assets we have; where they are; what standard of protection they provide; how they were constructed; their current engineering integrity; and, how they work together to provide a flood defence system' (Tim Kersley, Head of Asset Management, Environment Agency, 2008). Similar, seemingly basic, requirements can be seen to exist around the world and across sectoral disciplines (within rail, road, etc.) and are a central thrust of the USACE National Levee Safety Program (USACE, 2008).

- *Better decision making – how, where and when to invest.* All asset managers seek to make *good* investment decisions – decisions that minimise whole-life costs and maximise environmental gain, while ensuring that communities are appropriately protected from flooding both now and in the future. Such decisions will reflect a set of common characteristics, including
  - *robustness*: ensuring that the strategy performs well in the context of a wide range of possible futures
  - *flexibility*: determining whether future choices are constrained by previous choices, or alternative actions can be taken at a future date with limited additional cost
  - *adaptability*: embedding the capacity to adapt as the reality of the future unfolds (e.g. Can an asset be raised/widened, etc. with minimal cost?).
- *A need to deal with uncertainty better and more explicitly.* In addition to the severe uncertainty regarding the future climate and demographic conditions within which an asset will perform, uncertainty in the data and models used to assess risk is unavoidable. Handling this type of uncertainty is fundamental to the progressive nature of a hierarchical approach to risk assessment. Without an understanding of the nature of the uncertainty at each stage it is impossible to determine when the analysis and data used are 'sufficiently' credible in terms of the decision being made.

4.6. Data and tools to support a better understanding of risk and performance

To make informed choices asset managers must have access to evidence that

- *is transparent and auditable* – recognising the need for asset managers through to the public to be able to challenge the evidence, and the justification for decisions
- *reflects the performance of the whole system* – recognising that the protection afforded to a given person, property or other valued feature in the flood plain reflects the performance of the asset system as a whole under a wide range of loads (and not the performance of an individual asset during a single design storm).

To be efficient, the tools and techniques that are starting to emerge must be as follows.

- *Amenable to progressive refinement in order to meet the demands of the decision at hand* – the nature of the decisions to be made will vary from national allocation of resources to local specific intervention actions, and the supporting analysis must allow for progressive refinement of the data and analysis to reflect the demands of the decision at hand (being *just* sufficient to ensure a robust choice; defined as one that further refinement would not alter).

- *Based on the principle of collect once and use many times* – it should be possible to reuse data through the hierarchy of decisions – both bottom-up and top-down. Creating this value-added chain of data use and reuse is central to development of efficient modelling tools and relies on the uncertainties associated with the data being recorded and, where appropriate, reduced through the analysis. In many countries, national databases that provide a hub for all asset data are now becoming well established to aid this process (Figure 4.4).
- *Simple and practical to apply* – many good modelling approaches have failed due to perceived, and/or actual, complexity. Modelling approaches must utilise expert judgement where appropriate, and be practical to operate within acceptable timescales while having realistic data and skill demands.

4.6.1 System risk analysis tools – developing a whole system understanding

Structured approaches for dealing with whole systems of infrastructure assets, rather than individual structures and defences, are becoming embedded in practice (see e.g. Sayers and Meadowcroft, 2005). A key aspect of these whole-system tools is the structured description of the system components. Within the UK, the source–pathway–receptor terminology (used widely within environmental assessment; Department of the Environment, Transport and the Regions (DETR), 2000) has in recent years been adopted within flood risk management (Sayers *et al.*, 2002a). In this model, consideration is given to the extreme climatic conditions (the *sources* that initiate a flood), through the response in the form of the hydrological, hydraulic and structural behaviour of the rivers, coasts and control infrastructure (including breach, blockage, failure to open or close, etc. – the *pathways* that link the source to the receptors), to the individuals, properties and other features in the flood plain that suffer the consequences (the *receptors*) (Figure 4.5).

Within this framework, the asset manager is focused on managing the pathway of flooding, and in this context the river channel, flood plain surfaces and topography, near-shore morphology and natural backshore features are all legitimate parts of the asset system alongside the man-made infrastructure. The performance of these assets modifies the probability of flooding and its nature (the depth, velocity, debris content, etc.). The action taken may influence either the ultimate limit state failure (e.g. a breach or mechanical failure) or a serviceability failure (e.g. overflow or overtopping of the crest of an embankment, or the flow capacity of the pump being exceeded).

Two primary issues are, therefore, of concern in understanding the performance of a flood infrastructure, including the following.

- *What volume of water is likely to enter the flood plain, where is it likely to enter, and how is it likely to flow in?* (Note: This could include multiple breach or asset failure locations.)
  - Two situations must be considered, i.e. whether the asset
    - fails and structurally degrades (i.e. an ultimate limit state failure such as a breach of a linear asset, or the blockage or inability to operate a point asset)

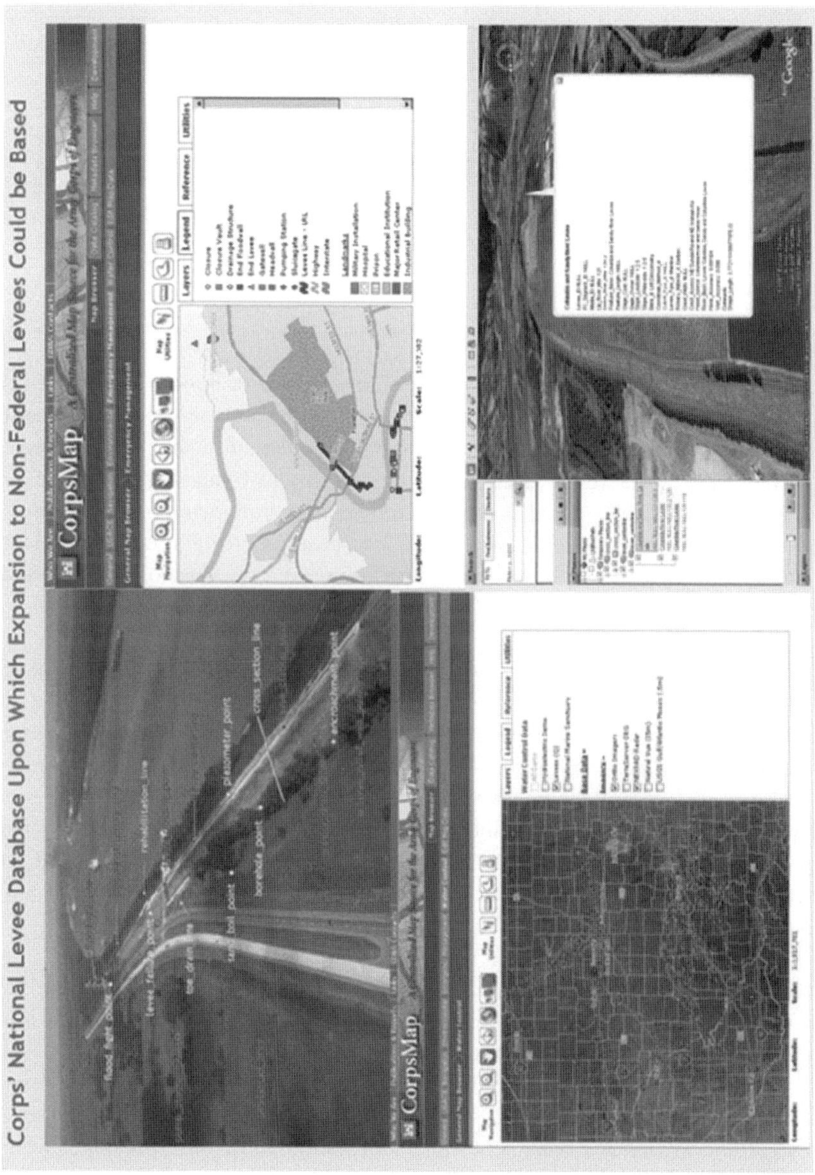

Figure 4.4 Example of a national levee database under development by the US Army Corps of Engineers (USACE, 2006) in the public domain. Similar tools exist in the UK (through the National Flood and Coastal Defence Database), The Netherlands (through the dyke safety programme) and elsewhere

Figure 4.5 The source–pathway–receptor terminology provides a useful framework for describing the flooding system and the influence of the infrastructure assets (Sayers *et al.*, 2004)
Contains Environment Agency information © Environment Agency and database right

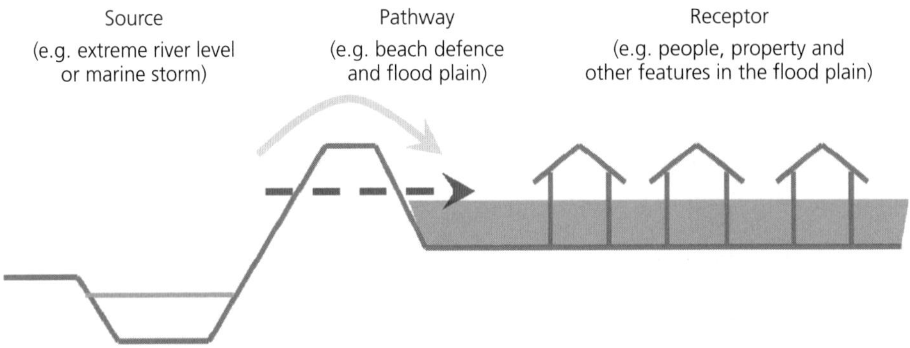

- remains structurally intact but fails to prevent flood water entering the flood plain (i.e. a serviceability limit state, such as overtopping, through periodic wave action or overflowing, as the still-water level exceeds the crest of a linear asset, or the surcharging or bypassing of a point asset).
- *What is the probability of either an ultimate limit state or serviceability limit state failure under given load or on demand?* For example, for a given marine storm or river flow/water level, how likely is the failure of a given embankment, or how likely is a pump or barrier to fail when required to pump or close?

The significance of the failure will depend on the associated consequences. The contribution of an asset to the residual risk will, therefore, reflect both the chance of failure and the associated consequences should failure occur (given the performance of the other assets in the system at the time of failure). Only through consideration of the system of assets, their probability of failure (and hence the possible *system states* that may exist within a system of assets that act to protect a given flood plain – no asset failures, one or more asset failure) and the consequences associated with each possible system state can risk be calculated and attributed to individual assets (e.g. Gouldby *et al.*, 2010).

Understanding the performance of the intervening system of infrastructure assets is, therefore, critical, and dominates the understanding of the probability of flooding in the majority of occupied flood plains (as they are typically protected by a levee system and related gates, barriers and pumps). Within risk analysis models, the reliability of individual structures and systems of assets must, therefore, be represented. Within England and Wales, the Risk Assessment for Flood and Coastal Defence for Strategic Planning (RASP) (Sayers *et al.*, 2004) approach provides a framework for system risk analysis; a framework that is now embedded in many Environment Agency strategic fluvial and coastal risk analysis and investment tools (e.g. Hall *et al.*, 2003; Sayers and Meadowcroft, 2005; McGahey *et al.*, 2007; Gouldby *et al.*, 2010). The RASP framework of analysis is not unique but shares an important similarity with all other system models, namely that it attempts to represent all important components of the flood risk system at a resolution and accuracy appropriate to the decisions they support (Figure 4.6).

4.6.2 Understanding the performance of a single asset – the chance of failure (reliability)

To understand in detail the performance of a single infrastructure asset under load or on-demand can be a major undertaking. Often, such an analysis will involve geotechnical, structural and hydraulic considerations, models and data. If, however, the particular asset has a limited role in managing risk, or the management decision is clear in the absence of detailed analysis, such detailed investigations are not required.

Hierarchical frameworks of reliability analysis that enable more progressive detail and data to be used and uncertainties to be reduced (where possible) have started to emerge. The basis for all levels of analysis – from the most simple to this most detailed – is an understanding of the failure process and modes. For example, the initial analysis, such as potential failure mode analysis (PFMA) or failure mode and effects analysis (FMEA), through to the full detailed reliability analysis, all start with an understanding of how failure develops through the asset. Typically, two approaches are used to provide a framework of thinking.

- *Fault tree analysis* (first proposed by Watson (1961) and revised by many authors since) – here a top-down, deductive, framework of thinking is adopted, where the processes that may have led to a hypothesised undesirable event, such as a breach, are deduced.
- *Event tree analysis* (first applied to the dam industry in the context of a risk assessment by Whitman (1984)) provides a bottom-up, inductive, framework of thinking where initiating processes are hypothesised, such as piping, and the ensuing processes of failure explored.

Although fault trees and event trees are infinitely extendable, perhaps the fault tree analysis is most convenient in the context of an hierarchical risk analysis (Figure 4.7). The skill of the asset manager is in ensuring that the tree remains as simple as possible, but not too simple, while capturing the most significant failure modes and process.

To assess the reliability of an asset, the failure modes must be described and their correlations known (or inferred) as set out in either a fault tree or event tree. Each failure process and failure mode within the fault or event tree must be described in quantified terms and the threshold at which failure is assumed to occur must be known (a so-called 'limit state equation'). The process of analysis is summarised in Table 4.1. There are various software tools available to support the fault tree and reliability analysis elements of this process, and some of these are summarised in Table 4.2 together with some key characteristic features and the link to where more information can be found. In general, however, in order to establish the response variable as a probability distribution, some method of integration of the input probability distributions is required. Where the distributions are continuous, Monte Carlo simulation techniques are often used to sample the input probability distributions.

Figure 4.6 The components of an analysis of a flood risk system, including the performance of the intervening system of assets
Sayers *et al.*, 2011

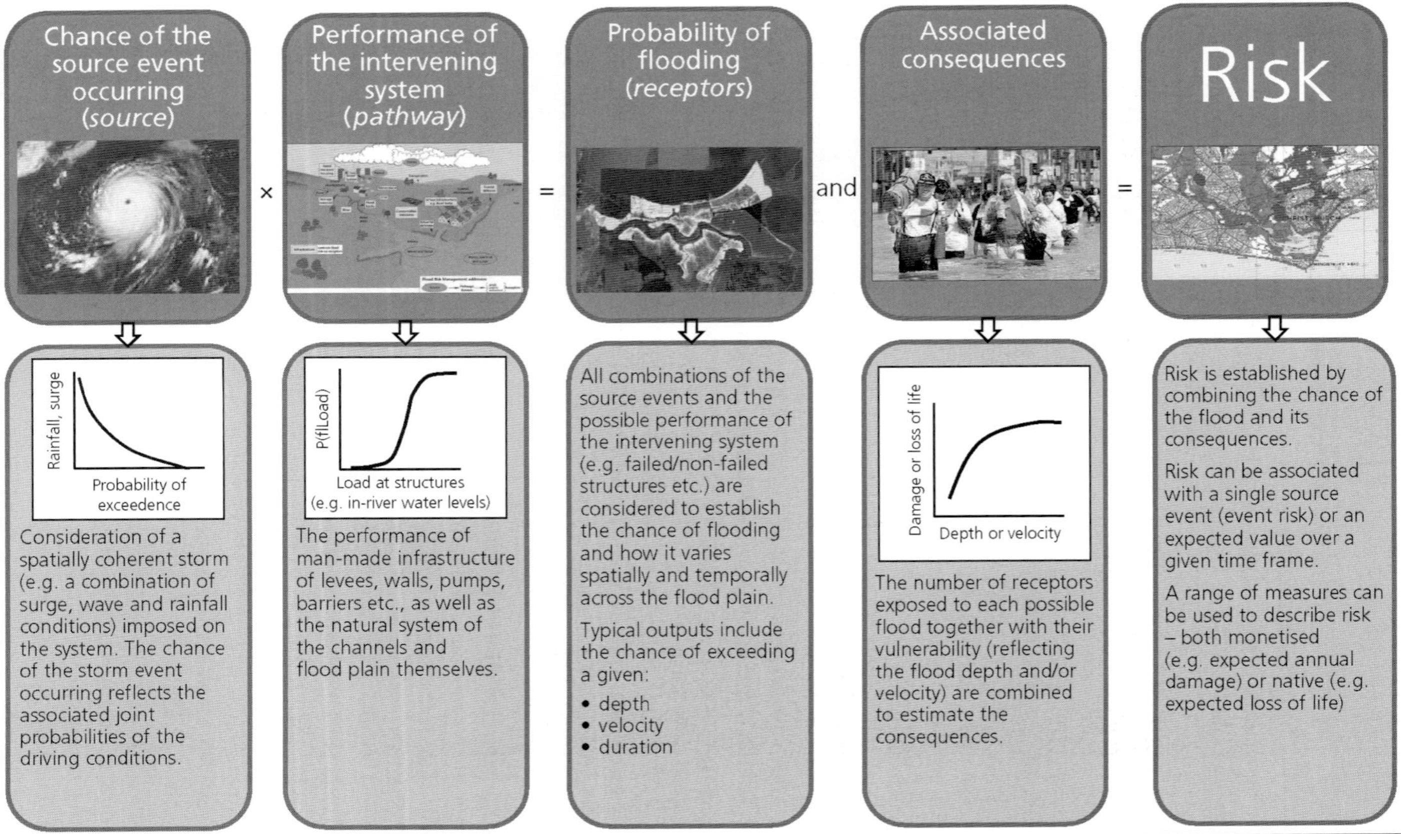

Figure 4.7 An example fault tree for a generic mass concrete vertical wall, showing varying levels of detail associated with different failure modes and processes (see Allsop *et al.* (2007) for a wide range of generic fault trees and associated limit state equations)

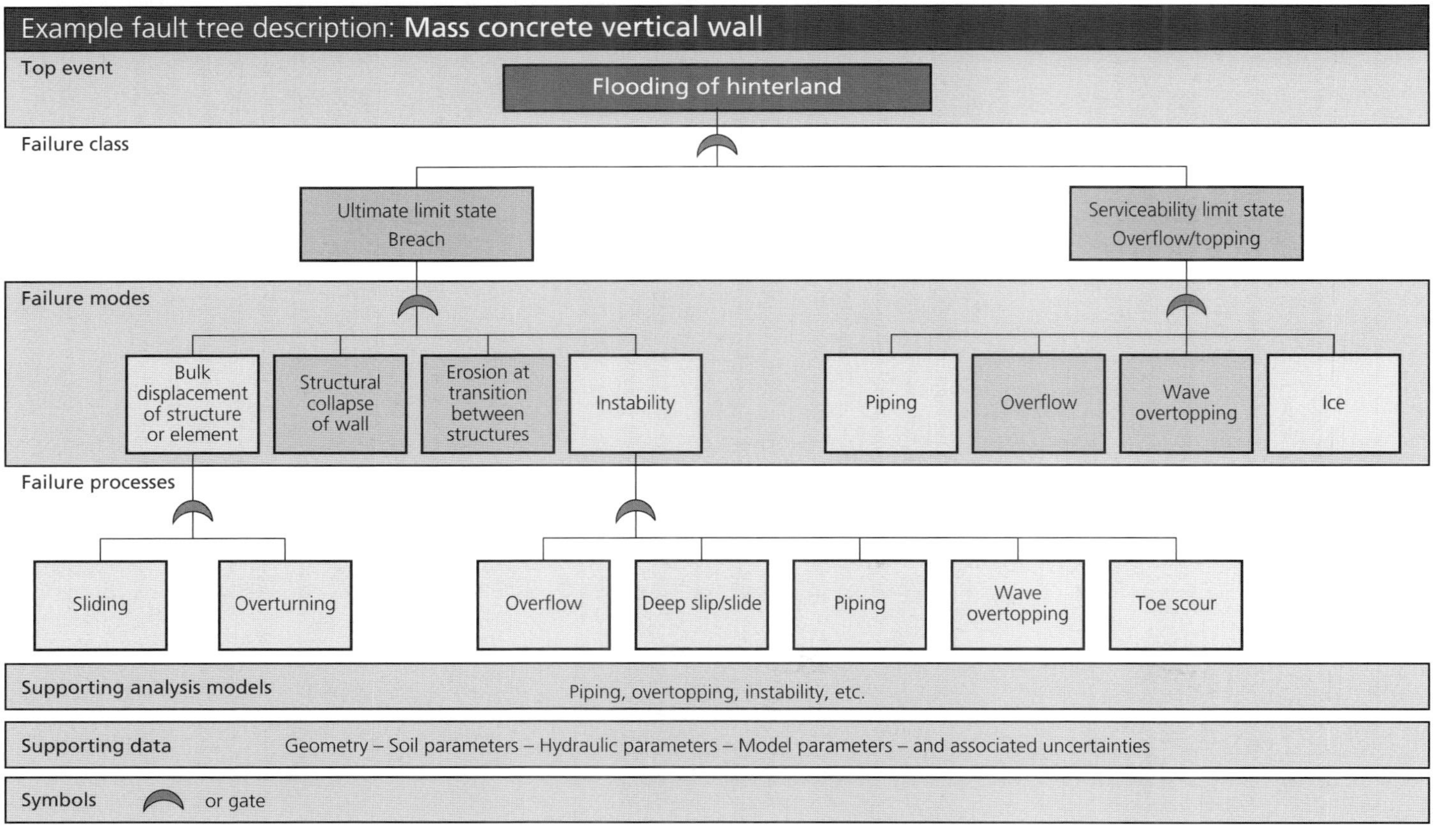

Table 4.1 Basic steps in the analysis of infrastructure reliability
Adapted from Simm *et al.*, 2008

Step	Description
1. Define asset function	A flood defence asset rarely acts solely to protect from flooding, often also functioning as a valuable environmental habitat, navigation or amenity asset. Understanding the multi-functionality of the asset is an important precursor to understanding how to manage it
2. Establish incident loading	An asset may be subject to a range of loading conditions – joint wave and water levels, marginal high or low water levels, groundwater levels, or perhaps a combination
3. Identify failure modes	The *failure mechanisms* (processes that can lead to ultimate failure) and the *failure modes* (processes that define ultimate failure) also need to be described. To avoid unnecessary effort, conventional deterministic approaches can be helpful to eliminate unrealistic failure mechanisms (i.e. individual events with a relatively low probability in comparison with the likely overall reliability of the asset). Research into failure mechanisms continues to be vital to understand asset performance better (e.g. Allsop *et al.*, 2007; Dyer *et al.*, 2009, Sentenac *et al.*, 2009)
4. Prepare a fault tree	Fault trees provide a useful visual, and formal, encapsulation of the failure mechanisms and their relationship to the failure modes
5. Identify/establish appropriate limit state equations	An appropriate model needs to be selected to represent each failure mechanism/mode. In many cases, empirical relationships will exist, and these can be easily translated into the form of a limit state equation (used in the reliability analysis – see below). In some cases, the failure mechanisms are complex (e.g. slip failure) and demand the use of more sophisticated models (e.g. traditional slope stability analysis or finite-element model). It is possible to link such models within the reliability analysis (Vrouwenvelder, 2001a,b; Lassing *et al.*, 2003), but this is often difficult and can incur an unacceptable runtime overhead. Emulation of these more complex models (e.g. through artificial neutral networks) provides an efficient and effective means of enabling such complete mechanisms to be incorporated in the reliability analysis
6. Document uncertainty in model variables and parameters	The engineering parameters, and the empirical variables, within the limit state equations will not be perfectly understood. Describing the uncertainty within these relationships and the supporting data on the asset of interest is an important task. In describing the uncertainty it is important that this process is comprehensive (ignoring uncertainty at this stage is to assume the data are perfectly known). Two groups of uncertainties can typically be distinguished (USACE, 1999; Sayers *et al.*, 2002b): ■ *natural variability* (aleatory uncertainty) – uncertainties that stem from known (or observable) populations and therefore represent randomness in samples

Table 4.1 Continued

Step	Description
	■ *knowledge uncertainty* (epistemic uncertainty) – uncertainties that come from basic lack of knowledge of fundamental or measureable phenomena Perhaps most critically, it is important to record the assumptions made regarding the uncertainty in the variables and parameters and the associated supporting evidence for these choices. This provides a vehicle for peer review and audit (Hall and Solomatine, 2008)
7. Undertake reliability analysis and display results	Once the above inputs have been established, the reliability analyses can be undertaken. For each hydraulic loading condition a series of simulations (across the uncertainty bands for each input parameter) is resolved. Failure arises in a particular case when the combinations of parameter values in the limit state function Z yield a value for Z that is less than or equal to zero. The probability of failure for that given loading condition is then the number of times when the simulation gives Z as less than or equal to zero divided by the total number of simulations. Repeat for all hydraulic loads (Kortenhaus *et al.*, 2002; Lassing *et al.*, 2003; Simm *et al.*, 2008; Gelder *et al.*, 2008)
8. Display results	Present the results of interest (annual probability of failure, fragility curve, etc.)

This approach avoids analytical integration, which can be complex or even impossible. There is a range of commercially available software tools and packages that can facilitate this process.

4.6.3 Expressing the results of a reliability analysis

The results of a reliability analysis can be expressed in a number of ways. The most convenient for both expert review and validation, as well as for onward use within a system risk model such as RASP (Sayers and Meadowcroft, 2005), is perhaps a fragility curve or a fragility surface. A fragility curve is a means of displaying the probability of failure for a given loading condition. The Environment Agency has developed a set of generic fragility curves, covering all basic types of coastal and fluvial linear defences, for application in broad-scale risk models (see e.g. Hall *et al.*, 2003; Sayers *et al.*, 2004; Gouldby *et al.*, 2010). Only where required are these high-level curves refined using more detailed analysis. The form of the fragility curve remains unaltered regardless of the level of detail; only the degree of certainty that is assigned changes. A comparison of the fragility curves resulting from a high level and a more detailed analysis are shown in Figure 4.8.

4.6.4 Accounting for deterioration

All assets are subject to deterioration. Deterioration of relevance to a flood risk manager can include: lowering of the defence crest through settlement (increasing the chance of

Table 4.2 Overview of fault tree analysis software packages (www.faulttree.net)

Name	Features
CAFTA	■ Full graph editing of fault trees ■ Thousands of failure events ■ Creation of new gates by pointing and clicking ■ Database editor, cut set generator ■ Document-quality graphics of fault trees ■ Calculation of probabilities for all gates in a fault tree without cut sets
Care	■ Handles fault and event trees to calculate probability and rate of events ■ Multiple events (common causes) ■ Minimal cut sets calculation with probabilities ■ Quick and accurate calculation of probability and rates for all events ■ Conditional effects probability may be given under OR gates
ITEM	■ Cut, copy and paste fault tree symbols ■ Creation and editing using drag and drop techniques ■ Multiple document interface ■ Library facilities for storing common data ■ Customisable paging facilities ■ Fully customisable reports, graphs and charts
FaultTree +	■ Fault and event tree construction ■ Full minimal cut set analysis ■ Common cause failure analysis ■ Post-processing facilities for accurate upper bound calculations ■ Uncertainty analyses allowing confidence levels ■ Time-dependent analysis providing intermediate values
FAULTTrease	■ Graphical user interface with WYSIWYG display ■ Transfer menu allows the fault tree to be split into several branches ■ Sstatistics can be probabilities, frequencies, or multipliers ■ Direct evaluation for fast computation of statistics and cut sets ■ Hierarchical references for numbers ■ Find dialog box searches for text in the tree by label, reference or tag
OpenFTA	■ Advanced user interface to create and edit fault trees ■ Powerful algorithms for determining minimal cuts and failures, including Monte Carlo simulation ■ No limits (other than physical) on the numbers of trees and events, or on the numbers and depths of transferred-in trees
Fault Tree/ Event Tree	■ Fast minimum cut set (MCS) calculation algorithm ■ Exact solutions ■ Auto-aligned graphical layout ■ Several importance measures ■ Common cause failures ■ Generate fault tree analysis from failure mode and effects analysis

Table 4.2 Continued

Name	Features
BlockSim FTI	■ Provides a full array of system reliability, maintainability and availability analysis options ■ Facility to work with fault trees or reliability block diagrams (RBDs) in the same environment ■ Automatically convert any fault tree to an RBD ■ Support for all major fault tree analysis gates and events
TTREE	■ Simple and logical tree input method ■ Clear outputs providing cut set ranking by unavailability and failure rate ■ Clear graphics that can be easily imported into a word processing package ■ Technis Beta + method for taking account of common cause failure ■ Facilities for one-shot probability such as human error

overtopping at lower water levels); animal infestation (increasing the chance of piping and the probability of a breach); siltation of a watercourse or debris blockage of a culvert (reducing the conveyance capacity of the channel); etc.

The consideration of deterioration in *design* typically leads to two types of design criteria.

- Minimising deterioration by the choice of materials and structure types.
- Taking deterioration into account by considering the expected design life and the need for (and ease of) inspection and repair.

An example of the choice of materials would be the use of imported high-quality rock for a revetment rather than locally available poor-quality stone that would break down quickly under hydraulic forces. An example of allowing for deterioration would be increasing the thickness of steel in a sheet pile wall to allow for corrosion over the life of the structure (30–50 years).

In flood risk analysis, understanding deterioration is an essential element of asset management, and is crucial for assessing whether or not it is worth extra initial investment to prolong the life or reduce the maintenance interval of an asset. In recent years, a series of R&D projects has been undertaken to help understand the process of deterioration – from more detailed process-based models (Buijs *et al.*, 2005b) through to more expert-judgement-led deterioration curves (Figure 4.9) (Sayers *et al.*, 2001; Environment Agency, 2009b). Although improving, the level of understanding remains basic, and will be an important area of research going forward.

4.6.5 Understanding the performance of a single asset

Understanding the chance of failure is, of course, only part of the story. The implications of failure, in terms of the increased flow into the flood plain, are equally as important as the performance of an individual asset. This includes understanding the following.

Figure 4.8 Risk models require a probabilistic representation of levee performance. (a) High-level fragility curves developed for all linear structures in England and Wales (Environment Agency, 2004). (b) An example from a more detailed reliability analysis in the Thames (Environment Agency, 2008). (c) A fragility surface for a coastal defence along the seafront of Towyn, North Wales (Environment Agency, 2004)

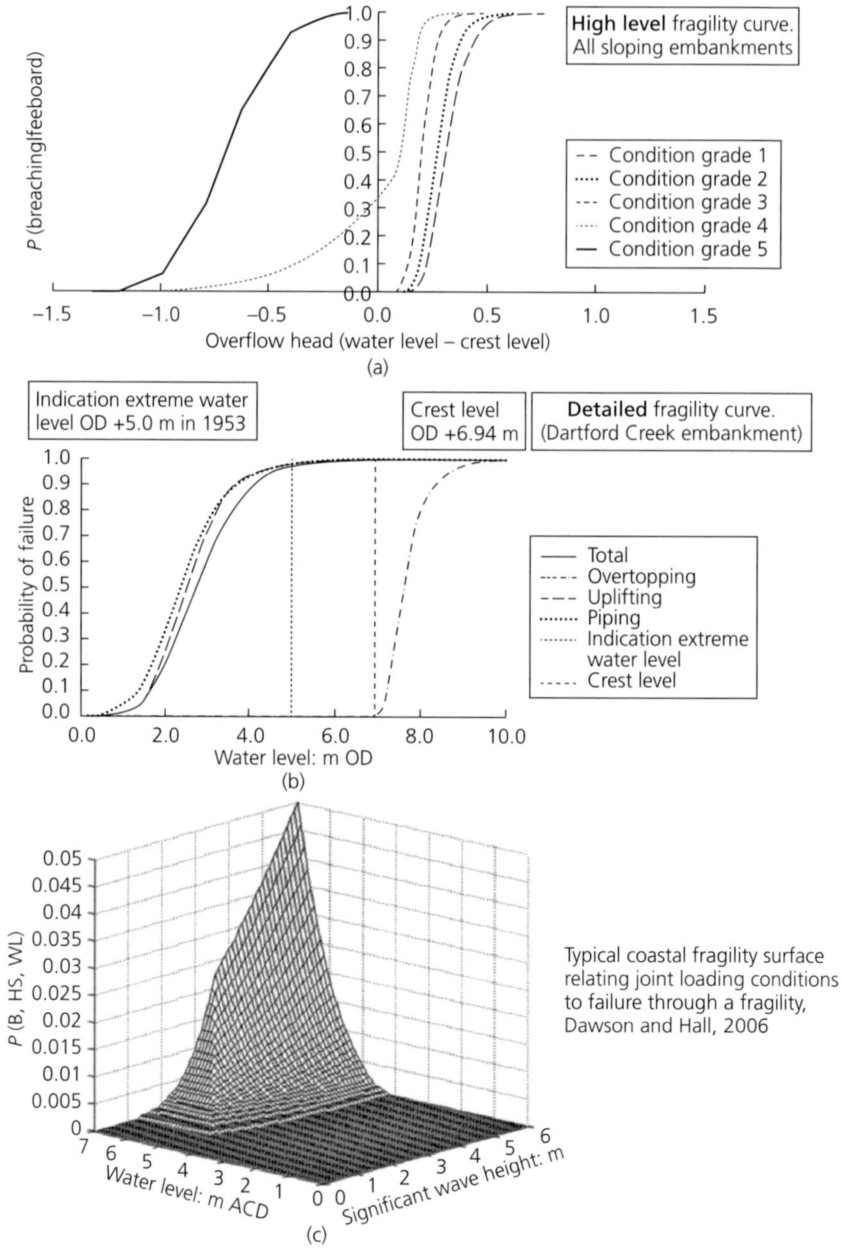

Figure 4.9 High-level deterioration curves for a narrow, turf-covered fluvial embankment under the assumptions of business as usual, enhanced maintenance and decreased maintenance (Environment Agency, 2009c)

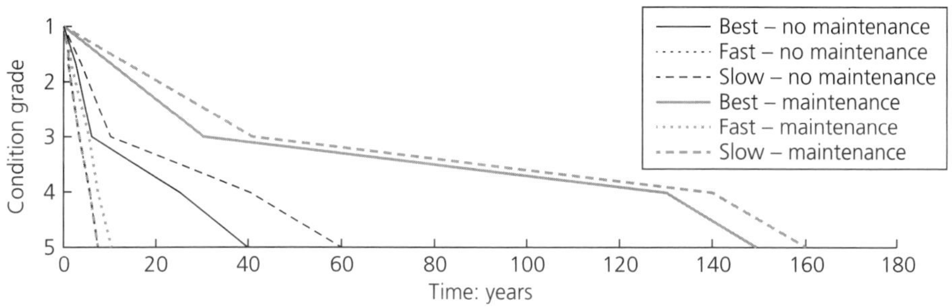

- *The breach growth and inflow* – Understanding breach is important, not only to improve the ability to calculate volume of water entering the flood plain, but also, and most importantly, to understand the velocity and rate of rise in flood waters and how these develop around the breach, and the associated risk to life. Various research projects have been directed at improving our understanding of breach and developing breach models (Mohamed, 2002; D'Eliso *et al.*, 2006; Gee, 2010). Such models represent the state of the art, but also demand information on various geotechnical parameters (which are often simply not known). As in the assessment of reliability, simplified methods are starting to emerge that support higher level risk analysis. For example, through the Flood Risk Management Research Consortium (FRMRC), effort is being devoted to the development of rapid and simplified breach models (Work Package 4.4 led by Oxford University (Van Damm) with support from HR Wallingford (Mark Morris).
- *Overtopping* – Wave-driven overtopping often dominates coastal flooding and is often highly sensitive to changes in beach levels and subsidence of the seawall crest. In recent years the approaches to coastal overtopping have been consolidated through the EurOtop manuals (see Further reading).
- *Blockage of point structures* – Blockage of culverts, bridges and other point assets by debris – both anthropogenic and natural – can lead rapidly to significant flows entering the flood plain as the watercourse attempts to find a new route. Through the FRMRC, effort is being devoted to updating long-standing guidance on how to assess the potential recruitment of debris and the degree of blockage. Although early within the research, promising predictive capability is emerging (see Chapter 9).

4.7. Using uncertainty and sensitivity analysis as aids to decisions

Uncertainty and sensitivity analysis are closely related, but are not the same. Uncertainty seeks to enable the decision-maker to understand better the confidence within the evidence presented and the choices taken. Sensitivity analysis seeks to highlight to the decision-maker those aspects of the analysis to which the evidence presented, and the choices

Figure 4.10 (a) The estimated annual damage (90th percentile). (b) The confidence in the estimate of risk expressed by plotting the value of 2 × variance in the risk estimate (Environment Agency, 2009c)

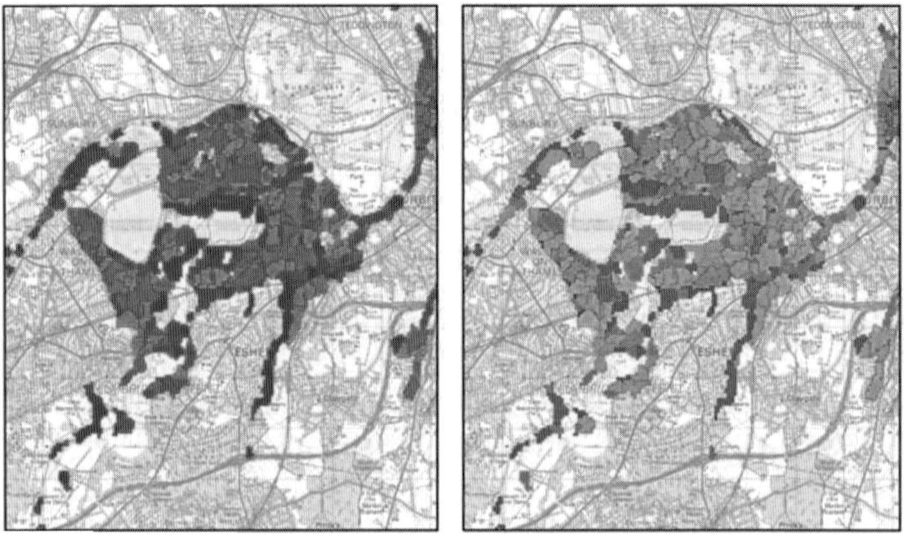

90th percentile (a)

2 × variance (b)

being made, are most sensitive. Both of these aspects are important within good practice management of flood infrastructure (and all other forms of infrastructure).

Uncertainty and sensitivity analysis can add considerable value to the risk analysis process; for example, by highlighting those areas of the flood plain where there is more or less confidence in the assessment of flood risk (Figure 4.10).

Such an analysis provides the decision-maker with an understanding of which uncertainties are most important in terms of their contribution to risk. For example, a crest level of one levee may be poorly known but, due to the circumstances of that levee, refining the knowledge regarding its crest level would do little to reduce the variance in the estimate of risk. Together, formal assessment of risk and uncertainty provides a significant insight to target data collection and further engineering investigations.

4.7.1 Determining the impact of uncertainty on decisions and targeting improvements in data and analysis

Formal sensitivity testing is a powerful tool to support the asset manager in determining when it is appropriate to spend additional effort in reducing uncertainties and where to target this effort (i.e. which parameters and models to refine). For example, is it better to invest in research, perhaps to improve the representation of the flood physics within the model components (e.g. the representation of breach size or flood propagation), or data collection, or perhaps to improve topography or crest level data – which would reduce the uncertainty more? Two of the basic approaches to sensitivity analysis that are in use include the following.

- *Sensitivity testing to assess the impact of uncertainty* – This typically involves examining a number of expert-defined scenarios without attaching probabilities to them, and determining by how much key variables can change before an alternative preferred option is identified. There will then follow some judgement of the likelihood of that change actually being applicable. Sensitivity testing in this way usually involves varying the selected parameter over a plausible range, with other parameters held at their 'best estimate' value. Although limited in scope, this approach is practical and transparent. It can also be credible, if done well, in enabling key variables within the analysis of risk to be identified and the associated uncertainty either reduced or managed. (It is often appropriate to conduct some sensitivity tests before embarking on the more thorough simulation methods discussed below.)
- *Simulation approaches to assessing the impact of uncertainty* – The approach involves representing uncertainties by, for example, probability distributions. These probability distributions are then combined to provide a probability distribution of the response variable (e.g. probability of dam failure and associated consequences), which incorporates the uncertainties in the parameters, variables and model relationships. Where uncertainties are expressed as confidence intervals, as opposed to probability distributions, it is necessary to make an assumption regarding the type of probability distribution to be used in the simulation. If there are many different types of uncertainty, involving many different parameters and variables, this approach can become complex. This is particularly so where there are dependencies between separate parameters and variables. To avoid overcomplicating the process, it is worth considering the sensitivity of the response variable to each of the parameters, together with the associated uncertainty. If a parameter has a narrow confidence interval (small uncertainty) and has a minor effect on the response, it is feasible to consider it as known. In addition, it may be necessary to consider the different sources of uncertainty as separate elements, and to structure the analysis to calculate specific uncertainty sources before combining these analyses in an overall simulation (Gouldby *et al.*, 2010).

4.7.2 Understanding the results – not all risks are equal

The resulting risk can be expressed and viewed in a number of ways, and the way in which the risk is constructed will influence the way its importance is perceived and the ability of the asset manager to respond to it. For example, risk can be expressed as follows.

- *The consequences that are expected to occur within a given time frame* – This reflects the average risk. Typically, the expected annual damage (EAD) is used as a convenient measure of the average damage in a given year. Alternatively, the expected lifetime damage may be used, reflecting the damage that is expected to occur, say to a house, over an average lifetime. Although the 'expected' damage is a useful term when looking to compare the efficiency of various management options (e.g. within a benefit–cost analysis), it provides only a limited view of the risk (e.g. failing to differentiate between high consequence/low probability and low consequence/high probability contributions). As such, although useful when exploring the efficiency of asset management investments over the medium to longer term, the tolerability of the residual risks is obscured.

- *The consequences that are expected to occur during a given storm event* – reflecting the consequences that would be expected (damage, loss of life, etc.) in the event of storm of a given return period (e.g. measured as the return period of the rainfall or flow in the river). To determine the event risk, the integration is done over all possible states of the intervening pathways (including the performance/reliability of the levees, pumps, barriers, etc.) and the performance of the non-structural measures (flood warning, etc.), but not across all return period storms (as in case expected risk). This provides the asset manager with a useful and important insight into the profile of the risk as the storm event increases in severity (Figure 4.11).
- *Significance of the risk* – Typically, risk is calculated through a simple process of multiplying probabilities and consequences. However, although this is a simple and pragmatic calculation, it does not always reflect the 'true' significance of the risk faced. To understand the significance of the risk, the magnitude of the consequences must be explicitly considered in order to differentiate 'risks' of the same numerical value. For example, a risk with the same product value may result from a high-probability/low-consequence or a low-probability/high-consequence combination. The significance of the second risk may be considered much higher, and would be treated differently in developing an appropriate management response. Such approaches require a common currency of risk to be used. Two approaches are typically used: monetarisation of all impacts (direct and indirect); and maintaining impacts in their native parameters before combining them through a process of weighting. Both approaches are problematic and continue to be an area of active research.

In summary, perhaps most important for asset managers is that the risk (either expected annual damages or expected event damages) can be attributed to individual components within the infrastructure system (Figure 4.12) and the components of an individual asset (Figure 4.13).

4.8. Conclusions

Risk analysis tools that incorporate the performance of infrastructure assets are now becoming available to support decision-making (McGahey and Sayers, 2008). These tools and techniques provide a step change in the 'richness' of the evidence provided to decision-makers at all levels (across national, flood system and individual asset levels). In particular, they provide the following.

- *An improved understanding of the role that an individual asset plays within a larger asset system* – by highlighting those assets, within a system of assets, which contribute most to residual risk.

Figure 4.11 Example of a risk profile for the Thames Estuary. (a) Potential economic damage arising from damage to the Teddington weir right bank (West Ham/Royal Docks flood area) – the risk increases with storm return period (so-called 'event risk'). (b) The expected annual damage (£). Reproduced by permission of Ordnance Survey on behalf of HMSO © Crown Copyright 2011. All rights reserved. Ordnance Survey License Number 100052127

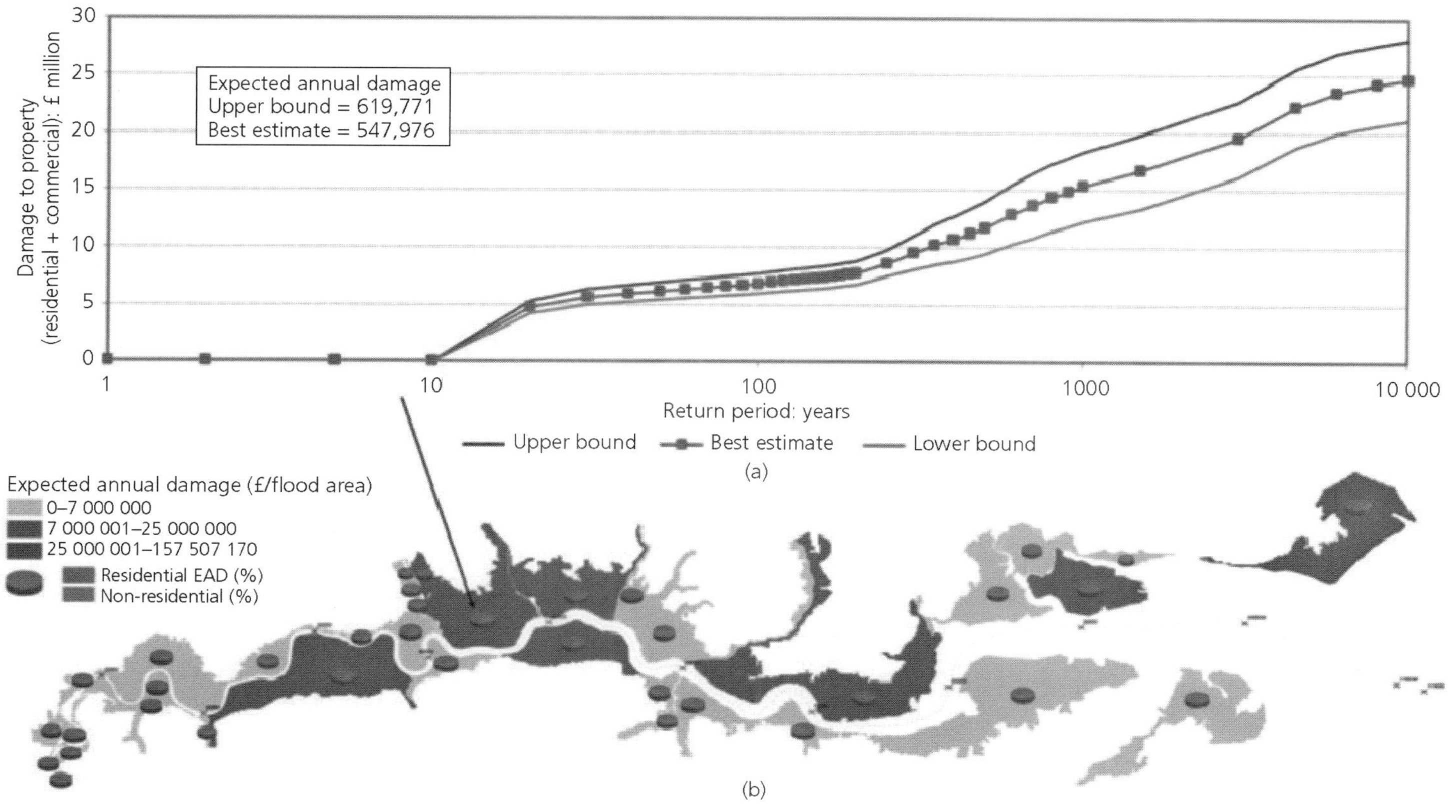

Figure 4.12 The expected annual damages attributed to individual defence assets with an asset system (as well as spatially within the flood plain). Reproduced by permission of Ordnance Survey on behalf of HMSO © Crown Copyright 2011. All rights reserved. Ordnance Survey License Number 100052127

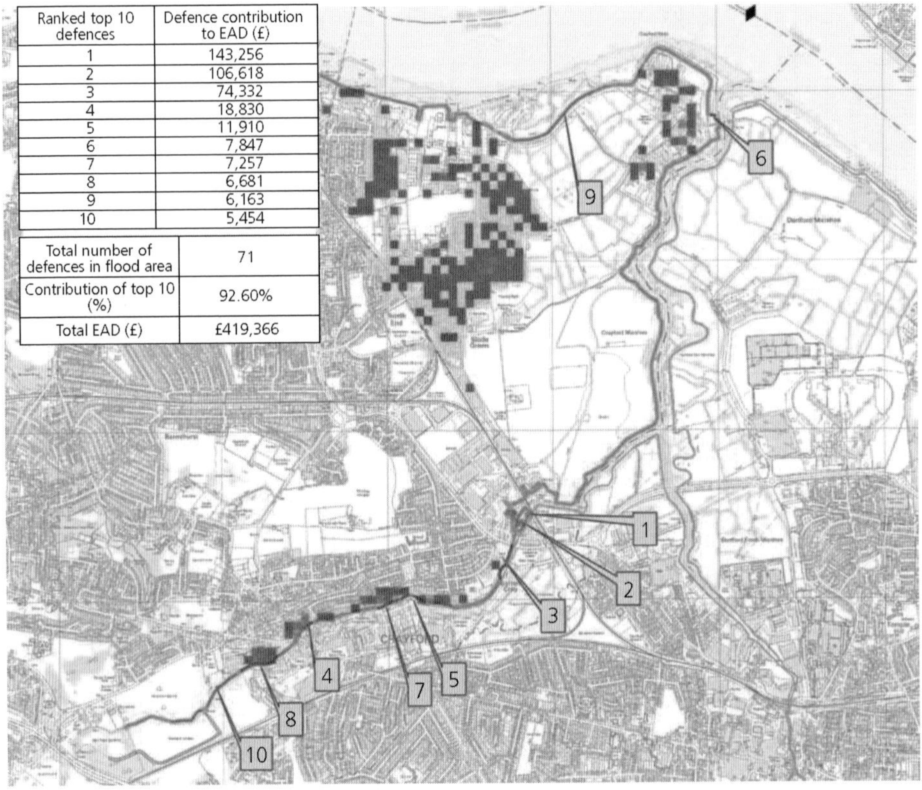

Ranked top 10 defences	Defence contribution to EAD (£)
1	143,256
2	106,618
3	74,332
4	18,830
5	11,910
6	7,847
7	7,257
8	6,681
9	6,163
10	5,454
Total number of defences in flood area	71
Contribution of top 10 (%)	92.60%
Total EAD (£)	£419,366

- *A better understanding of the impact of uncertainty within the estimated risk* – by highlighting those assets that contribute most to the uncertainty in the estimates of risk, the asset manager is able to prioritise the need for further data collection or engineering investigations on a common basis alongside structural measures.
- *The ability for progressive refinement of the analysis detail* – Attributing risk to an asset, and an associated understanding of the critical contributors to the uncertainty in that estimate, enables the decision-maker to target further analysis or data collection as appropriate for the decision at hand. Although a tiered analysis is a well-recognised concept in support of risk analysis, until recently it has been very difficult to achieve a hierarchical process whereby data and models evolve (rather than change) from one tier to the next.
- *Support to develop optimal investment strategies* – Asset managers face difficult choices: (i) *Where* to act? (ii) *When* to action, now or later? and (iii) *How* to act,

Figure 4.13 Combining risk and uncertainty analysis enables the identification of critical defences (see Figure 4.3) and which physical components within an asset are most critical (either requiring physical attention or further study) (Environment Agency, 2010)
Contains Environment Agency information © Environment Agency and database right

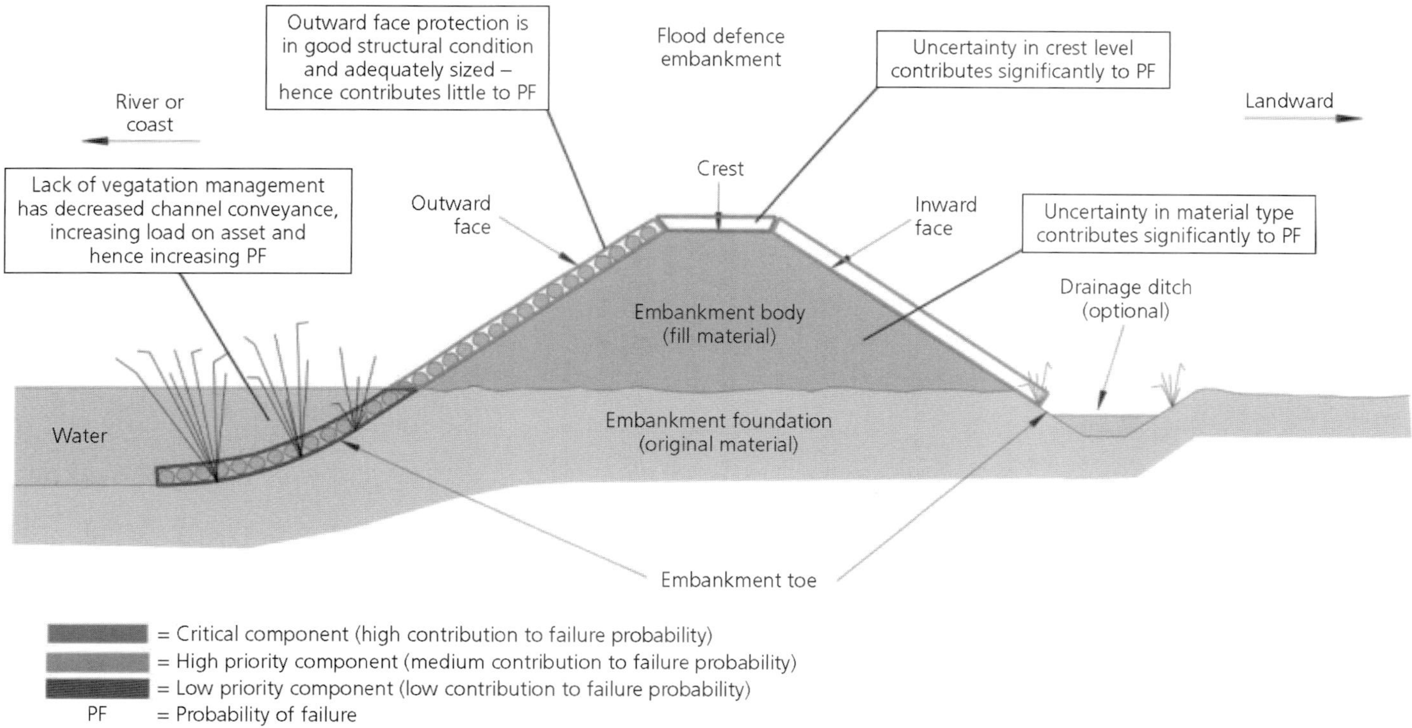

collect more data, undertake more analysis or intervene? Increasingly, it is not possible, or acceptable, to answer these questions on the basis of intuition.

- *A need to overcome practical challenges through expert input* – The need for simplifications and assumptions often limits the level of detail at which the tools are practically applicable. As the level of detail increases it becomes necessary for the asset manager to take ownership of the analysis and the actions proposed. This is a critical step, and enables local knowledge and judgement to be incorporated. The skill sets required, however, are now quite different from those traditionally used, requiring the asset manager to understand both the practical engineering issues as well as the analysis process and risk assessment. Developing these skills continues to be a barrier to wider uptake.

Acknowledgements

This paper draws upon a number of projects from the Environment Agency and Defra joint research programme, the European funded FLOODsite Project and the EPSRC funded Flood risk Management Research Consortium (Grant No. EP/FP202511/1). The contribution from these organisations is gratefully acknowledged.

REFERENCES

Allsop W, Kortenhaus A and Morris M (2007) *Failure Mechanisms for Flood Defence Structures*. FLOODsite project report T04_06_01. Available at: http://hikm.ihe.nl/floodsite/data/Task4/pdf/failmechs.pdf (accessed 2 February 2012).

Buijs F, Simm J, Wallis M and Sayers P (2005a) *Performance and Reliability of Flood and Coastal Defences*. R&D Technical Report FD2318/TR 1. Defra, London, UK. Available at: http://www.rasp-project.net/documents_info/TR1%20defra.pdf (accessed 2 February 2012).

Buijs FA, Hall JW, Sayers PB and Van Noortwijk JM (2005b) Time-dependent reliability analysis of flood defences using gamma processes. *Proceedings of ICOSSAR*. Rome, Italy.

Communities and Local Government (2010) *Planning Policy Statement 25: Development and Flood Risk*. HMSO, London, UK.

Dawson R and Hall J (2006) Adaptive importance sampling for risk analysis of complex infratructures systems. *Proceedings of the Royal Society of London: Mathematical, Physical and Engineering Sciences* **462(2075)**: 3343–3362.

Defra (2011) *Flood and Coastal Resilience Partnership Funding – An Introductory Guide*. Available at: http://archive.defra.gov.uk/environment/flooding/funding/documents/flood-coastal-resilience-intro-guide.pdf (accessed 2 February 2012).

D'Eliso C, Oumeraci H and Kortenhaus A (2006) Breaching of coastal dikes induced by wave overtopping. *Proceedings of the 30th International Conference Coastal Engineering (ICCE)*. ASCE, Vol. 3, San Diego, CA, USA, pp. 2844–2856.

DETR, Environment Agency and Institute of Environmental Health (2000) *Guidelines for Environmental Risk Assessment and Management*. The Stationery Office, London, UK.

Dyer M, Utili S and Zielinski M (2009) Field survey of desiccation fissuring of flood embankments. *Proceedings of the ICE – Water Management* **162(3)**: 221–232.

Environment Agency (2009a) *Investing for the Future: Flood and Coastal Erosion Risk Management – A Long Term Investment Strategy*. Environment Agency, Bristol, UK.

Environment Agency (2009b) *Guidance on Determining Asset Deterioration and the Use of Condition Grade Deterioration Curves*. Product code SCHO0509BQAT-E-P. Available at:

http://publications.environment-agency.gov.uk/PDF/SCHO0509BQAT-E-E.pdf (accessed 2 February 2012).

Environment Agency (2009c) *Assessment and Measurement of Asset Deterioration including Whole Life Costing*. Environment Agency, Bristol, UK. Science Report SC060078/SR.

Environment Agency (2010) PAMS *Performance-based Asset Management System – Phase 2: Outcome Summary Report*. Environment Agency, Bristol, UK. Flood and Coastal Defence R&D Programme Science Report SC040018/SR1.

Evans E, Ashley R, Hall J *et al.* (2004a) *Foresight Future Flooding, Scientific Summary: Volume 1: Future Risks and their Drivers*. Office of Science and Technology, London, UK.

Evans E, Ashley R, Hall J *et al.* (2004b) *Foresight Future Flooding, Scientific Summary: Volume 2: Managing Future Risks*. Office of Science and Technology, London, UK.

Gee M (2010) Use of breach process models to estimate Hec-Ras dam breach. *2nd Joint Federal Interagency Conference*. Las Vegas, NV, USA.

Gelder P, Buijs F, Horst W *et al.* (2008) Reliability analysis of flood defence structures and systems. In *Floodrisk08, Oxford*. Taylor & Francis, London, UK.

Gouldby B, Sayers PB, Panzeri M and Lanyon J (2010) The development and application of efficient methods for the forward propagation of epistemic uncertainty and sensitivity analysis within complex broad-scale flood risk system models. *Canadian Journal of Civil Engineering* **37(7)**: 955–967.

Hall JW and Solomatine D (2008) A framework for uncertainty analysis in flood risk management. *Journal of River Basin Management* **6(2)**: 85–98.

Hall J, Dawson R, Sayers PB, Rosu C, Chatterton J and Deakin R (2003) A methodology for national-scale flood risk assessment. *Proceedings of the ICE – Water and Maritime Engineering* **156(3)**: 235–247.

Kortenhaus A, Oumeraci H, Weißmann R and Richwien W (2002) Failure model and fault tree analysis for sea and estuary dikes. *Proceedings of the 28th International Conference Coastal Engineering (ICCE), ASCE*, Cardiff, UK, Vol. 2, pp. 2386–2398.

Lassing BL, Vrouwenvelder ACWM and Waarts PH (2003) Reliability analysis of flood defense systems in the Netherlands. *Proc. Conf. ESREL* **2**: 1005–1013.

McGahey CM and Sayers PB (2008) Long term planning – robust strategic decision making in the face of gross uncertainty (tools and application to the Thames). In *Flood Risk Management: Research and Practice* (Samuels P (ed.)). Taylor & Francis, London, UK.

McGahey CM, Sayers PB, Evans E and Surendran S (2007) Modelling and Decision Support Framework 2: a risk-based approach to catchment, estuary and coastal flood management planning. *Proceedings of the 1st Environment Agency Conference of River and Coastal Management*. Thomas Telford, London, UK.

Mohamed M (2002) Embankment breach formation and modelling methods. PhD thesis. The Open University, Milton Keynes, UK.

National Committee on Levee Safety (NCLS) (2009) *Draft: Recommendations for a National Levee Safety Program. A Report to Congress from the National Committee on Levee Safety*. Available at: http://www.leveesafety.org/docs/NCLS-Recommendation-Report_012009_DRAFT.pdf (accessed 2 February 2012).

Sayers PB and Meadowcroft IC (2005) RASP – a hierarchy of risk-based methods and their application. *40th Defra Flood and Coastal Management Conference*. York, UK.

Sayers P, Crossman M, Gouldby B *et al.* (2001) Integrating reliable flood modelling with probabilistic analysis in practice. *Proceedings of the Ministry of Agriculture Fisheries and Food Conference for Coastal and River Engineers*, Keele University: Defra, London, UK.

Sayers P, Hall J and Meadowcroft I (2002a) Towards risk-based flood hazard management in the UK. *ICE Journal of Civil Engineering* **150(5)**: 36–42.

Sayers PB, Meadowcroft IC, Hall JW and Gouldby BP (2002b) *Risk, Performance and Uncertainty in Flood and Coastal Defence – A Review*. Defra/Environment Agency, London/Bristol, UK. Scientific Report 587.

Sayers P, Meadowcroft I, Hall J, Dawson R, Chatterton J and Rosu C (2004) *RASP – Risk Assessment for Flood and Coastal Defence Strategic Planning*. Environment Agency, Bristol, UK. HR Wallingford Report SR603. Available at: http://publications.environment-agency.gov.uk/PDF/SCHO0305BIQN-E-E.pdf (accessed 2 February 2012).

Sayers P, Tarrant O, Gouldby B, Kavanagh D and Panzeri M (2006) Thames Estuary – establishing a robust flood system model to support engineering investment decisions. *Proceedings of the 41st Defra Conference of River and Coastal Management*. Defra, London, UK.

Sayers PB, Gouldby BP, Benwell D, Hassan M and Panzeri MC (2008) *Thames Estuary 2100 – Interventions and Assets Report*. Environment Agency, Bristol, UK.

Sayers P, Wallis M, Simm J, Baxter G and Andryszewski T (2010) Towards the next generation of risk-based asset management tools. In *Flood Risk Science and Management* (Pender G (ed.)). Blackwell, London, UK.

Sayers P, Galloway G, Penning-Rowsell E *et al.* (2011) Flood Risk Management: A strategic approach. WWF (UK and China) and GIWP, China. Published by Asia Development Bank.

Simm J, Gouldby B, Sayers PB, Flikweert J, Wersching S and Bramley ME (2008) Representing fragility of flood and coastal defences: getting into the detail. In *Floodrisk08, Oxford*. Taylor & Francis, London, UK.

Sentenac P, Zielinski M and Dyer M (2009) *The Influence of the Desiccation Fine Fissuring on the Stability of Flood Embankments*. Flood Risk Management Research Consortium, Swansea, UK. Final Report including Laboratory Studies.

Stockton (2009) Remark during a presentation to the Association of State Floodplain Managers, Orlando, FL, USA. Primary source unknown.

USACE (United States Army Corps of Engineers) (1999) *Risk Analysis for Flood Damage Studies*. USACE.

USACE (2006) National levee database program. Available at: http://www.usace.army.mil/LEVEESAFETY/ACTIVITIES/Pages/act_nldb.aspx (accessed 2 February 2012).

USACE (2008) *A Safe Public and Reduced Economic Losses by Means of Reliable Levees – Part of an Integrated Solution to Flooding*. National Committee on Levee Safety. National Levee Safety Program, USACE. Available at: http://www.leveesafety.org (accessed 2 February 2012).

Vrouwenvelder ACWM, Steenbergen HMGM and Slijkhuis KAH (2001a) *Theoretical Manual of PC-Ring, Part A: Descriptions of Failure Modes* (in Dutch). TNO Building and Construction Research, Delft, The Netherlands. Report No. 98-CON-R1430.

Vrouwenvelder ACWM, Steenbergen HMGM and Slijkhuis KAH (2001b) *Theoretical Manual of PC-Ring, Part B: Statistical Models* (in Dutch). TNO Building and Construction Research, Delft, The Netherlands. Report No. 98-CON-R1431.

Watson (1961) *Launch Control Safety Study*, Section VII, Vol. 1. Bell Labs, Murray Hill, NJ, USA.

Whitman RV (1984) Evaluating calculated risk in geotechnical engineering. *ASCE Journal of Geotechnical Engineering* **110**: 143.

FURTHER READING

See the related chapters in this book for more detailed discussions on specific asset types. More general reading on asset management can be found on the following websites:

Environment Agency – research outputs on sustainable asset management: www.environment-agency.gov.uk (accessed 2 February 2012).

EurOtop – manuals and tools: http://www.overtopping-manual.com/eurotop.pdf (accessed 2 February 2012).

Fault trees: http://www.faulttree.net (accessed 2 February 2012).

Flood Risk Management Research Consortium (FRMRC), funded by the Engineering and Physical Sciences Research Council (EPSRC) – outputs and tools: http://www.frmrc.org.uk (accessed 2 February 2012).

FLOODsite project – outputs and tools from the project, including useful contributions on failure modes (Task 4), breach growth (Task 6) and reliability analysis (Task 7): http://www.floodsite.net (accessed 2 February 2012).

Flood Risk
ISBN: 978-0-7277-4156-1

http://dx.doi.org/10.1680/fr.41561.103

Chapter 5
Performance of coastal flood defences

Andreas Kortenhaus
Technische Universität Braunschweig, Braunschweig, Germany

This chapter describes the performance of coastal flood defences under extreme loading conditions. Special emphasis is on failure mode analyses and available models and tools used to analyse the stress and resistance of various defence types such as breakwaters, sea dikes, seawalls, beaches and dunes. Wherever available, the reader is pointed to simple tools or models and provided with more detailed reading at both design and research level.

5.1. Introduction

Flood defences play a major role in preventing floods from low-lying and flood-prone areas and thus in protecting thousands of people and their belongings. Their major role is to prevent high water levels in rivers, estuaries and along the coastline from entering into the hinterland and thus leading to major catastrophes, including loss of life, injury, disease and major damage. Numerous events in the past decades have shown that flood defences have not always fulfilled this role sufficiently well. River dikes have been overtopped or weakened, sea dikes have been eroded and breached, and dams have been broken, always leading to huge areas being flooded.

Within risk analysis or reliability studies, the role of flood defences has to be understood and formulae or models have to be found to describe their functions. In the past, detailed failure analysis of many flood defences helped us to understand the processes leading to either damage or complete failure of these structures. Each process has been described, and this has improved our understanding of why and how structures have failed. Nevertheless, not all these improvements were used, and then led to better or stronger flood defences, due to insufficient resources, poor maintenance and/or deterioration processes that had not been considered so far. Even unexpected and new processes leading to either damage to or complete failure of structures have occurred in the recent past, so that the understanding of all processes and the performance of flood defence structures will remain a permanent process.

Considering the several thousands of kilometres of flood defences worldwide, the tremendous costs of protecting ourselves from floods and their consequences, including the necessary maintenance costs, have to be considered and taken into account when designing such structures. It is, therefore, a pivotal requirement for flood defence design to optimise the cost–benefit ratio as efficiently as possible; or, in other words, overdesigned flood defences might be sufficiently safe but will be unaffordable, whereas underdesigned structures save a lot of money but might fail.

The performance of flood defences, therefore, requires the consideration of the following points

- different types of flood defence and their principal functions
- processes that lead to failure of these types of flood defence
- uncertainties in the models with which these processes are described, including uncertainties and variations (temporal and spatial) in the input parameters
- present and future challenges (storm surges, sea level rise, etc.)
- knowledge of potential measures to either increase the height of flood defences, their strength, or their resistance capabilities.

In this chapter, these issues are discussed, together with the most recent solutions and answers to many of the most urgent questions related to them. First, a brief overview is given of existing flood defence structures and their key characteristics, and then the failure modes and fault trees for each of these defence types are introduced. It is not the purpose here to provide detailed design equations for each of the failure modes, but to guide the reader to where they can be found and to comment on their practical use (see Section 5.3). Where possible, links are given to existing models and software for calculation of the probabilities of failure for flood defences. The uncertainties in both the models and the input parameters are discussed and, finally, some example cases are presented.

5.2. Defence types

Flood defence structures are earthen or non-earthen structures that prevent a (low-lying) hinterland from being flooded. This is mainly done by: (i) increasing the structure's height to a level higher than the expected water level in rivers or along the coast, and (ii) making sure that the stability of the structure is sufficient to withstand all possible loads. The types of flood defence are manifold, and may be distinguished both by their type of construction (earthen, concrete, etc.) or the type of flood they are built to withstand (flash floods, river floods, coastal floods, etc.).

The construction of flood defences along rivers depends on various considerations and 'boundary conditions', such as the availability of material, space, maximum allowable height of defences, costs, tourism-related aspects, political issues, multi-purpose functionality, etc.

River embankments are widely used along river courses where space and material are more readily available than in cities or areas with not much room for the rivers. Vertical walls are often used when space is scarce and buildings and other assets near the river have to be protected. Barriers are used for water-level control and to prevent major flood waves from entering vulnerable areas (e.g. the Thames barrier, which protects the City of London from major floods resulting from storm surges in the North Sea). Mobile flood defences are increasingly used along river basins in areas that cannot be protected by permanent (and usually too high) flood defence structures, especially in areas where there are large differences between normal and extreme water levels of rivers. River gates have very similar functions to barriers, and are mainly used to control water levels and discharges in rivers. Finally, domestic flood protection measures are the last possibility for houses and other private properties to protect against flooding after other flood defences have failed.

An overview of some key shoreline structures is given in Figure 5.1. These comprise

- seawalls
- beaches and cliffs
- dunes
- breakwaters
- point structures, such as sluices, outlets, etc.

Sea dikes are very common structures in The Netherlands and Germany, where they protect the very long and flat, low-lying coastal areas, and where usually a lot of space and earthen material is available. Dunes are often natural barriers against the open sea, and are maintained as flood defences by planting special types of grass and thus avoiding sand erosions. Dunes are common in sandy areas on some of the North Sea islands, and also along the Baltic Sea coast. Other natural coasts with beaches or cliffs can be found quite frequently worldwide, mostly at locations where erosion or coastal recession is not a major problem but may take place to some extent. Revetments and sea walls are hard structures that are used to protect more vulnerable areas, such as cities near the coastline, where any landward erosion would lead to extensive costs. Breakwaters are generally used for harbour protection, very often with significant loads acting on them and with high costs in case of failure. Harbour protection walls have similar roles but usually experience lower loads and may be significantly longer than breakwaters. Sluices, outlets, barriers and other single-point structures usually serve very specific purposes, and are single points on coastlines. They often involve high investment costs, and are usually confined to relatively short lengths, but may play a significant role in coastal defence.

The European FLOODsite project has classified four general types of flood defence structure (Allsop *et al.*, 2007)

- foreshores, dunes and banks
- embankments and revetments
- walls
- point structures.

This classification is useful for different types of flood defence and different types of flood, and is therefore used in this chapter.

5.3. Failure modes and standard fault trees

This section discusses failure modes and failure mode analyses as observed from failures of flood defences and similar coastal structures in the past (Section 5.3.1). In addition,

Figure 5.1 Coastal flood defences at North Sea and Baltic Sea dikes. (a) Sea dike near St. Peter-Ording, Germany. (b) Revetment and seawall at Baltrum, Germany. (c) Sluice near Ribe, Denmark. (d) Storm surge barrier at Eider, Germany (Wiki Commons, Dr Torsten Henning). (e) Outlet of Konge Å near Gredstedbro, Denmark. (f) Flood protection wall, Hamburg, Germany. (g) Dunes on Sylt island, Germany. (h) Vertical breakwater at Granville, France

a brief introduction is given of fault tree analysis and the background principles (Section 5.3.2).

5.3.1 Failure modes and analyses

A failure of a flood defence structure may be defined as a process taking place at the structure that will change its current condition in such a way that it leads to flooding of the hinterland behind it. Therefore, waves overtopping a sea dike may represent a failure of the sea dike, even if there is no damage to the dike (erosion, grass removal, etc.). On the other hand, failure of a dike can also mean a full dike breach with 100 m of the dike being washed away. Generally, a total failure of a flood defence is not caused by one single mechanism but by several processes following each other in time.

All processes for the different flood defences result from different types of loading acting on the structures. Three different types of loading may be distinguished (Allsop *et al.*, 2007).

- Water level difference across a structure.
- Wave loading.
- Lateral flow velocities.

This is, of course, a very simplified view of the complex loads acting on a flood defence structure. In reality, there will most often be a combination of the aforementioned loads, and various other factors will play a role. These factors should be considered when designing a coastal flood defence in detail, but will not be taken into account in the description of a performance-based design given in this chapter.

- Differential loading – rapid changes in water-level differences may affect the pore pressure distribution in soils and hence the stress on defence structures.
- Vortex shedding – vertical vortices may result from sharp corners and can induce vertical (uplift) forces on the structure (elements).
- Tension cracks – initial cracking may decrease the stability of the soil of an earthen embankment or dike.
- Surcharge loadings – placing material on banks or dikes may cause additional loading and stress on the structure.
- Presence of trees or shrubs – trees or shrubs may affect the stability of a structure, although, on the other hand, the roots may increase the soil strength.
- Vandalism – removal of stones from revetments or grass from a dike may result in erosion problems.
- Burrowing animals – may cause piping or erosion failures.
- Debris/ice impact – may be significant and should be considered if relevant.
- Transitions – may cause excessive erosions.

Limit state equations may be identified from analyses of failures that have occurred in the past in which the sequence of damages and failures to a flood defence may be identified and analysed. This will help in setting up all failure modes relevant for the specific coastal flood defence. Failure analyses can be achieved by the following methods (see e.g. Stewart and Melchers, 1997)

- preliminary hazard analysis (PHA)
- failure modes and effect analysis (FMEA)
- failure mode, effect and criticality analysis (FMECA)
- hazard and operability analysis (HAZOP).

Hazard analyses have been performed for some flood defence structures. In the case of dams, much work has been published in reviews and/or books (Johnson and Illes, 1976; Singh, 1996; Morris and Hassan, 2002, Nagy, 2004, and Johnson and Matalo, 2004). In the case of sea and river dikes, hazard and failure analyses have been performed (e.g. Richwien, 1995; Oumeraci *et al.*, 2001b; Schüttrumpf and Oumeraci, 2004). In the Netherlands, many failures of river dikes have been reported (TAW, 1995).

For vertical breakwaters failures have been analysed (Oumeraci, 1994) which have later been used for fault tree and probabilistic analysis of those breakwaters (Oumeraci *et al.*, 2001a). Similar work has been performed for rubble mound breakwaters (Permanent International Association of Navigation Congresses (PIANC), 1992; Burcharth and Liu, 1996) leading to probabilistic design for rubble mound breakwaters (Burcharth, 2001).

5.3.2 Fault trees

A fault tree is formed of events, which are often described by binary (Boolean) variables (the event occurs or not) and related by logical functions, essentially OR and AND. Graphically, these logical functions are represented by Boolean 'gates'. The output of a gate (the event represented immediately above the gate) is exactly equal to the Boolean function of the inputs of the gate (the events represented immediately below the gate). Each input event, in turn, can be the result of a logical function of a set of events, down to the point where all inputs are basic events that cannot be practically analysed any further.

Fault trees are constructed by using inductive logic (backward logic), i.e. by identifying a top event – failure of all or part of the system – and sequentially identifying unions or intersections of events that entirely describe each successive binary variable (Paté-Cornell, 1984).

Finally, the fault tree allows the user to obtain a logical identity between the 'top event' and a set of basic events. Then, on the basis of that identity, one can compute the probability of the top event as a function of the probabilities of the basic events. Phenomena that are not necessarily part of a failure scenario, but increase the probability of failure of the system, can be introduced at that stage. The probability of the top event can be computed conditional on the occurrence of these events at different 'levels' if it is appropriate.

Fault trees are generally used in three steps: (i) construction of the fault tree; (ii) reduction of the fault tree (e.g. by minimum cut sets); and (iii) probability of the top event. Some further details of these steps are given below. Further details of the principal

Figure 5.2 Definitions of the symbols and calculation rules used in fault trees

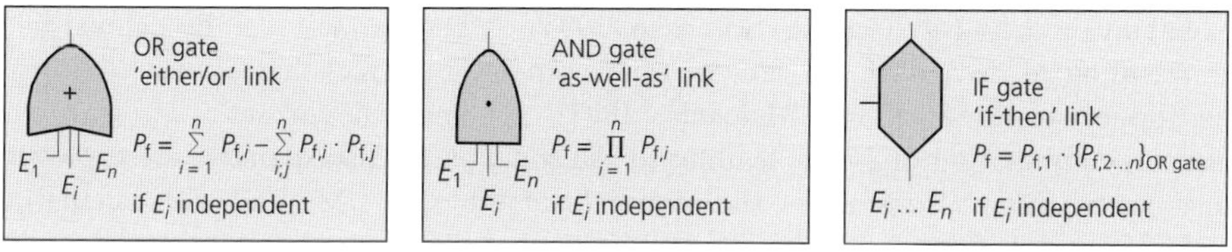

knowledge on fault trees can be found in the literature (Modarres, 1993; Schneider, 1994; Stewart and Melchers, 1997; Barlow, 1998; see also the faultree website, which provides further reading and tutorials for download).

5.3.2.1 Step 1: construction of the fault tree

Fault trees are principally set up in a chronological order of the events. To link the various events, different symbols have been defined (Figure 5.2). Rules how these gates are calculated may be summarised as follows (Andrews and Moss, 1993).

- All incoming events of an AND gate are in parallel, i.e. all events leading to the gate have to be TRUE before the gate opens. The probability of events linked in this way is, therefore, the product of the individual failure probabilities of the events.
- Events leading to an OR gate represent a series system, i.e. only one of the incoming events has to be TRUE to provide access through the gate. In this case, the individual failure probabilities are summed (the negative product of failure probabilities in Figure 5.2 is often ignored, as the product is often negligible compared with the sum of the failure probabilities).
- An IF gate can be treated similarly to an OR gate, i.e. only one of the incoming events has to be TRUE to provide access through the gate. However, in addition, the event E_i in Figure 5.2 must also be to open the gate. The latter requires calculation, as shown for an AND gate.

All calculations are based on the assumption of independent events (failure modes). Assumptions such as fully dependent events or upper and lower bounds for the calculation of the failure probability of the gate require different calculation rules, as shown in Table 5.1.

Table 5.1 Overview of calculation rules for assessing the failure probability of systems

	Parallel system	Series system
Gate	AND	OR
Upper bound	$P_f = \min(P_i)$	$P_f = \prod_{i=1}^{n} P_i$
Completely dependent	$P_f = \min(P_i)$	$P_f = \max(P_i)$
Completely independent	$P_f = \prod_{i=1}^{n} P_i$	$P_f = 1 - \prod_{i=1}^{n}(1 - P_i)$
Lower bound	$P_f = 0$	$P_f = \max(P_i)$

5.3.2.2 Step 2: reduction of the fault tree

A fault tree may be converted to cut sets in order to calculate the overall failure probability more easily (see e.g. Sörensen, 1995). A cut set may also be defined as a combination of events (i.e. an excerpt of the whole fault tree) that always comprises the top event (Stewart and Melchers, 1997). A 'minimal cut set' represents the minimal combination of events leading to the top event. Consequently, the top event only occurs when all events of the minimal cut set occur. Methods for calculating minimal cut sets from fault trees as well as the minimum number of events are also available (Barlow, 1998).

5.3.2.3 Step 3: probability of the top event

The overall probability of failure of the flood defence structure can be calculated using the relevant formulae in Table 5.1 for each gate. If these calculations are performed for each gate of the fault tree, assuming that the probabilities of failure for each individual failure mode are known beforehand, the overall flooding probability can be calculated.

More detailed calculations are required if the correlation between individual failure modes of the fault tree are known and considered. This is, however, not an issue within this chapter, where only the most simple way of calculation (see Table 5.1) will be followed. Tools for setting up and calculating fault trees are commercially available, and some of these are summarised and discussed in Section 5.5.

5.3.2.4 Applications

The application of fault trees to coastal flood defence structures usually involves only a limited number of events; some examples are given in Table 5.2. This table is not complete, but shows the possible use of fault trees for flood defence structures.

5.4. Supporting equations and models

All limit state equations are expressed as the difference between a resistance term R and the stress term S:

$$z = R - S \tag{5.1}$$

resulting in negative z values for failures. Three examples (Figure 5.3) can be used to illustrate the differences between the general deterministic understanding of failure and a probabilistic definition (Kortenhaus *et al.*, 2002).

Figure 5.3(a) shows overtopping over a dike crest. Depending on the allowable amount of overtopping, this is usually acceptable and will not be regarded as failure. Figure 5.3(b) shows an erosion hole of the seaward slope of a sea dike after a severe storm surge. In general, this will be regarded as a failure, because the design of the cover layer of the dike has not been dimensioned properly. In a probabilistic sense, this is not a failure, as flooding did not occur (provided that the overall top event of a probabilistic fault tree is defined to be 'flooding of the hinterland of a dike'). However, the probability of failure can be significantly higher than in the overtopping case. From these observations it is obvious that limit state equations are needed to describe, for example, the erosion process at the seaward side of the dike in a probabilistic sense. Figure 5.3(c) shows a

Table 5.2 Overview of fault tree approaches to coastal structures

Reference	Structure	Failure mode	Remarks
Bakker and Vrijling (1980)	Sea dike	Overflow, wave overtopping, piping, geotechnical failure	First simple fault tree without calculation
CIAD Project Group (1985)	Breakwater	Very detailed	
CUR (1990)	Barrier	Very detailed	
Vrijling (1993), Rijkswaterstaat (1994)	Barrier	Wave overtopping, piping, geotechnical failure	Simple fault tree with calculation of the limit state equation (LSE)
Townend (1994)	Sea wall	Various failure modes	Simple fault tree, no limit state equation, no calculation
Zeidler (1994)	Revetment	Stability, wave overtopping, subsidence, toe erosion	
Van Agthoven *et al.* (1997)	Dike ring	Six failure modes	All failure modes linked by OR gate
Kuijper and Vrijling (1998)	Dike ring		Detailed calculation principles, but no results
Kuijper and Vrijling (1998)	Sluice	Foundation, piping, material, human error	No calculation
Schiereck (1998)	Revetment	Stability, wave overtopping, subsidence, toe erosion	Very simple fault tree for coastal revetments
Vrouwenvelder *et al.* (1999), TAW (2000)	Dike ring	Eight failure modes	Detailed study, all failure modes linked by OR gate
Voortman (2002)	Dike system	Wave overtopping, piping, geotechnical failure	Complex fault tree, but simplified during calculation

complete failure of the dike (from both the deterministic and probabilistic 'viewpoint'), where flooding occurred.

Stress and resistance terms in limit state equations may range from a very simple (empirical) formulation to highly complex and sophisticated modelling, depending on the research already performed in the field, the models available and the accuracy required. Examples of analytical formulations for sea and estuary dikes are provided

Figure 5.3 Examples of 'failures' and the probabilistic definition of failure: (a) wave overtopping event at a sea dike; (b) failure of the seaward slope of a sea dike; (c) breaching of a sea dike during a storm surge; (d) probabilistic definition of failure

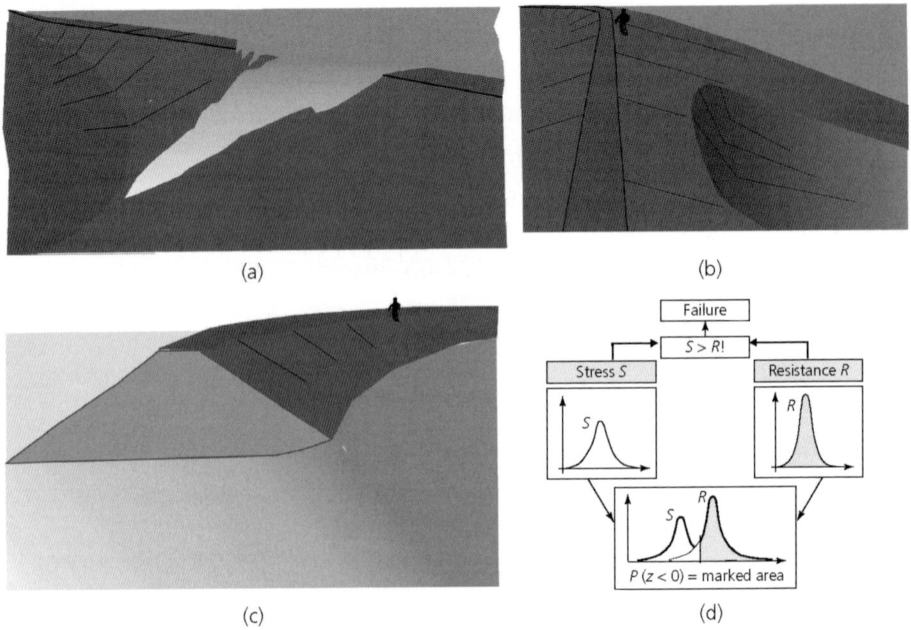

in Table 5.3 (Kortenhaus *et al.*, 2002). It should be noted, however, that most of these models are not formulated as limit state equations (see Equation (5.1)), but still need to be transformed.

A more complete overview of recent limit state equations for different flood defence structures subject to different loading types, including details of the stress and resistance terms, analytical formulation, background literature and other information, has been brought together in the European FLOODsite project (Allsop *et al.*, 2007). A typical description of a failure mode and the corresponding limit state equation is presented in Box 5.1.

Generally, it is also possible to include very sophisticated and complex models in a limit state equation for further use in reliability design. However, this may require extremely long computation times, especially if level 3 calculations (e.g. Monte Carlo simulations) are performed (Rajabalinejad, 2009).

It should be noted that many of the limit state equations provided in Table 5.3 are not expressed as time-dependent functions. This can be seen as a simplification, which in many cases may be tolerable but in other cases is too simplistic (e.g. erosion processes at earthen sea defences very strongly depend on time-dependent parameters, such as

Table 5.3 Overview of limit state equations used for sea and estuary dikes
Kortenhaus *et al.* (2002)

No.	Name	Reference	Comparison	Unit
1	Overflow	Oumeraci *et al.* (1999)	Freeboard	m
2	Overtopping	Schüttrumpf and Oumeraci (2005)	Freeboard	m
3	Breaching	Tuan (2007), Stanczak (2008), D'Eliso *et al.* (2007)	Duration of storm	h
4	Sliding	DIN 1054 (1996)	Forces	kN
5	Stability of revetment	Van der Meer (1998)	Stone diameter	m
6	Wave impacts	Kortenhaus and Buß (2008), Stanczak (2008)	Forces	kPa
7	Uplift of revetment	Klein Breteler and Bezuijen (1998)	Forces	kPa
8	Run-up velocity	Schüttrumpf and Oumeraci (2005)	Velocity	m/s
9	Erosion of grass at seaward slope	Amended after TAW (2000)	Duration of storm	h
10	Erosion of clay at seaward slope	TAW (2000)	Duration of storm	h
11	Cliff erosion	Larson *et al.* (2004)	Duration of storm	h
12	Stability of seaward slope	DIN 4084 (1983)	Forces	kN m
13	Overflow velocity	Schüttrumpf and Oumeraci (2005)	Velocity	m/s
14	Overtopping velocity	Schüttrumpf and Oumeraci (2005)	Velocity	m/s
15	Erosion of grass at shoreward slope	TAW (2000)	Duration of storm	h
16	Erosion of clay at shoreward slope	Rose *et al.* (1983)	Duration of storm	h
17	Infiltration	Weißmann (1999)	Duration of storm	h
18	Failure of cap (Kappensturz)	DIN 4084 (1983)	Forces	kN m
19	Phreatic line at shoreward slope	See Scheuermann and Brauns (2001)	Duration of storm	h
20	Clay uplift	Richwien and Weißmann (1999)	Forces	kN
21	Sliding of clay	Richwien and Weißmann (1999)	Forces	kN
22	Stability of shoreward slope	DIN 4084 (1983)	Forces	kN m
23	Partial breaching shoreward slope	Visser (1995)	Duration of storm	h
24	Piping	Weijers and Sellmeijer (1993)	Hydraulic gradient	–
25	Matrix erosion	De Mello (1975)	Diameter	m

Box 5.1 Uplift of revetment surface

Summary: Uplift of revetment surface/elements under wave drawdown, dealt with by dimensioning armour thickness (and other factors) in relation to the wave height and steepness

Reliability equation

The reliability function is expressed as

$$z = G(\cos\alpha) - F_u$$

where:

G = mass force of revetment element (kN/m)
F_u = uplift force underneath the revetment (kN/m)

Loading equations

Acting uplift force calculated as

$$F_u = \frac{1}{2} p_a b \left(2 - \frac{b \sin\alpha}{z_{-98} + D\cos\alpha}\right)$$

with:

$$p_a = \rho_w g H_s [0.7 + 0.7 \tanh(\xi_d - 2.1)] + \rho_w g D \cos\alpha$$

$$\frac{z_{-98}}{H_s} = 0.7 + 0.7 \tanh(\xi_d - 2.1)$$

Resistance (strength) equations

Mass force of revetment elements

$$G = \gamma_s D b \cos\alpha$$

Parameter definitions

F_u = wave-induced uplift force (kN/m)
G = mass force of revetment elements (kN/m)
D = thickness of revetment elements (m)
b = width of revetment elements (m)
z_{98} = wave rundown being exceeded by 2% of all run-downs (m)
α = slope (°)

Sources of failure mechanism equations/methods

Klein Breteler *et al.* (1998) and Schüttrumpf (2001)

Sources of uncertainties in failure equations/input parameters

For details on uncertainties see Schüttrumpf (2001), Kortenhaus (2003) and Allsop *et al.* (2007)

Remarks

Further formulae are available to calculate the uplift forces under wave attack. The aforementioned formulae include the description of Schüttrumpf (2001) for wave rundown

the quality of the soil material, and therefore need to be considered). Long-term processes, such as the variability of a coast or the degradation of a strength parameter of a flood defence, may significantly influence reliability analyses of flood defences. In addition, the sequencing of storms may play a role in the probabilistic analysis of flood defences, as the previous event may have left the defence in a different condition than usually considered for design. All these temporal aspects need to be considered for future design (see e.g. Buijs, 2007) and may play an important role when considering key processes that lead to failure of flood defences.

5.5. Design and assessment aids

This section seeks to give an overview of the available tools and aids for describing limit state functions and of fault tree approaches, as described in the previous chapters. It is not the intention to provide design recommendations for full probabilistic or reliability based approaches and risk-based tools.

Limit state equations can be derived from Equation (5.1), where both stress and resistance may be taken from deterministic approaches already available. Hence, many of the tools listed here are available from deterministic approaches and are thus well known. If tools do not include any assessment of uncertainties, the reader is referred to Section 5.6 for further details.

An excellent source of information on limit state equations is the FLOODsite report on flood defences (Allsop *et al.*, 2007), which already includes the correct format of the limit state equation, details on stress and resistance, background reading, and hints on how to assess the uncertainties. If more detailed information is required on specific topics, the following sources may provide further information.

- Wave overtopping: the *Wave Overtopping of Sea Defences and Related Structures: Assessment Manual* (Pullen *et al.*, 2007) contains all details on wave overtopping of coastal structures and includes online and downloadable tools for the assessment of overtopping rates.
- The *Coastal Engineering Manual* (US Army Corps of Engineers (USACE), 2002) provides extensive background information on all sorts of coastal engineering and design guidance on various types of flood defences. PC-based tools are provided by the CEDAS package, which may be purchased from Veri-Tech Inc.
- *The Rock Manual* (CIRIA/CUR/CETMEF, 2007) provides design formulae and background information on the planning and design of rock works in hydraulic engineering (for more details see the CIRIA website). Some online tools are available on the Coastal and River Engineering Support System (CRESS) website.
- The *Handbook of Coastal and Ocean Engineering* (Kim, 2010) contains many details on loading of coastal defences and the performance of coastal structures
- International standards are written in a very general form, but should be respected when searching for limit state equations (see e.g. ISO 2007).
- Design guidelines may differ in various countries and very often need to be respected when searching for limit state equations (e.g. for Germany see EAK (2002), DIN 1054 (1996), DIN 4084-100 (1996), DVWK (1986), Bundesantalt für Wasserbau (BAW) (2005), EAU (2004); for the UK see CIRIA/CUR/CETMEF (2007), CIRIA (2010) and many more).
- For many earthen flood defences it may be necessary to have detailed knowledge of the subsoil failure modes and breaching of dikes and embankments. For the latter, some limit state equations and background knowledge are available

(Morris *et al.*, 2009), as are some software models (see the FLOODsite website for *http://www.floodsite.net/html/HR_Breach_Model.htm* for the HR Breach model)

- Numerical tools for describing subsoil failures are available from different sources, (e.g. the PLAXIS package, provided by Plaxis bv, The Netherlands).

The aforementioned list is not complete by far, and there are many other textbooks, design guidelines, scientific literature and further publications that may be extremely useful when specifying the limit state equations for flood defences.

Tools and assessment aids for calculating the probability of a failure are also available, and may differ significantly in complexity, user friendliness and costs. PROBAN (by DNV, Norway), STRUREL (by rcp, Germany), FReET (by Červenka Consulting Ltd, Czech Republic), and @RISK (by Palisade) are some examples of such tools that can be used to assist in calculating the failure probabilities from user-defined limit state equations. Further information is also available from the website of the A. James Clarke School of Engineering, Department of Mechanical Engineering, University of Maryland.

Software tools for fault tree analysis and a lot of background reading is provided by Allen Long on the FaultTree website, also contains a selective overview of fault tree analysis programs. Table 4.2 in Chapter 4 gives an overview of available tools, with some key characteristic features and the link to where more information can be found. Again, further information is available on the website of the A. James Clarke School of Engineering, Department of Mechanical Engineering, University of Maryland.

5.6. Uncertainties

Many processes related to the performance of flood defences along coasts are uncertain. This holds true for both the parameters that are relevant for these processes, such as water level and wave parameters, and the models describing these processes. In any design approach, these uncertainties should be considered.

Uncertainty may be defined as the relative variation in parameters or the error in the model description, and so there is no single value describing this parameter but a range of possible values. Due to the random nature of many of the variables used in coastal engineering, most of the parameters should not be treated deterministically but rather should be treated stochastically. The latter assumes that a parameter x shows different realisations out of a range of possible values. Hence, uncertainty may be defined as a statistical distribution of the parameter. If a normal distribution is assumed, the uncertainty may also be given as relative error, which is expressed mathematically as the coefficient of variation of a certain parameter x:

$$\sigma'_x = \frac{\sigma_x}{\mu_x} \tag{5.2}$$

where σ_x is the standard deviation of the parameter and μ_x is the mean value of that parameter. Although this definition may be regarded as imperfect, it has practical value and is easily applied.

The return period of a parameter is defined as the period of time in which, on average, the parameter occurs again. Therefore, it is the inverse of the probability of occurrence of this parameter. If the return period of a certain wave height is T_R, this means that this specific wave height will only occur once in T_R years on average.

Many parameters used in engineering models are uncertain, and so are the models themselves. The uncertainties in input parameters and models generally fall into specific categories.

- *'Fundamental' or statistical uncertainties*: these are elemental, inherent uncertainties, which are conditioned by random processes of nature and which cannot be diminished (measured data always contain uncertainties of this type).
- *Data uncertainty*: these include measurement errors, inhomogeneity of data, errors during data handling, and non-representative reproduction of measurement due to inadequate temporal and spatial resolution.
- *Model uncertainty*: this is due to inadequate reproduction by the model of the true physical processes in nature.
- *'Human' errors*: these cover all the errors that arise during production, abrasion, maintenance as well as other human mistakes that are not covered by the model. These errors are not considered in the following, due to the fact that, in general, they are specific to the problem at hand, and no universal approaches are available.

Considering uncertainties in a design, therefore, means that all the input parameters are no longer regarded as fixed deterministic parameters but can be any realisation of the specific parameter. This has two consequences. First, the parameters have to be checked to determine whether all realisations of a parameter are really physically sound (e.g. a realisation of a normally distributed wave height can become mathematically negative, which is physically impossible). Second, the parameters have to be checked against realisations of other parameters (e.g., a wave of a certain height can only exist in certain water depths, and not all combinations of wave heights and wave periods can physically exist).

An example of calculating a wave run-up on a sea dike may illustrate these uncertainties and their consequences for design. Figure 5.4 shows waves running up a simple sea dike and a simple equation for calculating the run-up height.

Taking the equation and the values provided in Figure 5.4 the wave run-up height R_u can be calculated deterministically as $R_u = 3.25$ m. If only three parameters are taken as stochastic values (H_s with the standard deviation $\sigma_{Hs} = 0.1$ m; T_p, $\sigma_{Tp} = 0.1$ s; and c1, $\sigma_{c1} = 0.1$), and all are normally distributed, a Monte Carlo simulation would result in the same mean value of $R_u = 3.25$ m but with a standard deviation for the run-up height of $\sigma_{Ru} = 0.23$ m. As the run-up height is then also normally distributed, 90% of all possible run-up heights fall between 2.87 m and 3.62 m, spanning an interval of about 80 cm. This indicates that even small uncertainties might lead to considerable consequences and should not be ignored during design. Therefore, probabilistic analyses

Figure 5.4 Examples of a simple wave run-up calculation to illustrate the uncertainties in the result

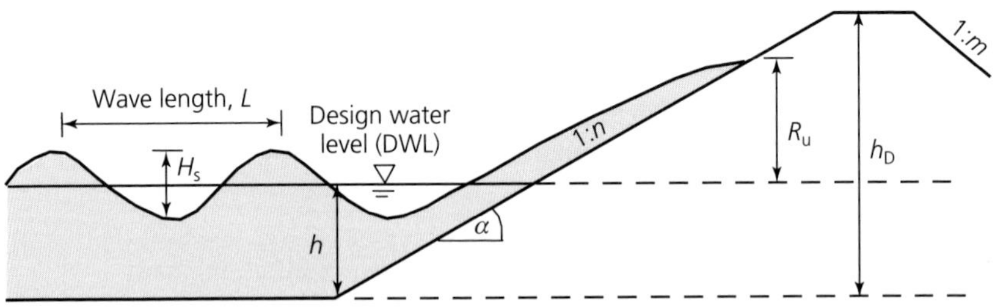

$$R_u = c_1 \sqrt{\frac{g}{2\pi}} \sqrt{H_s}\, T_p \tan \alpha$$

H_s	= significant wave height at toe (m)	= 3.0 m
T_p	= peak period of waves (s)	= 6.0 s
h_D	= height of dike (m)	= 6.0 m
R_u	= wave run-up height (m)	
h	= water level at dike toe (m)	
1:n	= seaward slope (–)	= 1:6
c_1	= empirical coefficient (–)	= 1.5

of flood defences can be a very valuable tool for design, because uncertainties are implicitly considered.

Probabilistic design tools also allow for an evaluation of the importance of some of the related input parameters for the limit state equations used. In many of the formulae used for the design of coastal flood defences, the water level and its statistical distribution is essential for the outcome of the probabilistic design. Therefore, the correct estimation of the design water level and an appropriate extreme value distribution may have considerable influence on the resulting failure probability of the flood defence, or any fragility curve derived from this. In addition, any accelerating sea level rise (e.g. resulting from climate change) will have a very strong influence on the design of the coastal defence structures for mainly two reasons: (i) the water level has a direct effect on the height of the flood defence; and (ii) higher water levels will also influence the waves in front of the defences, where usually there is less wave breaking with higher water depths in front of the defences.

The latter point also raises a more complex issue, i.e. the relationship between the water level and the sea state parameters in front of the sea defences. It is well known that these parameters are correlated, and this usually has to be considered in any model or limit state equation describing the failures of flood defences. This may be done by employing either joint probabilities of these parameters (e.g. water levels and wave heights, or wave heights and wave periods; see e.g. Hawkes *et al.* (2008)) or by using models that describe the relation of these parameters.

Therefore, as most of the processes associated with coastal flood defences are stochastic in nature, uncertainties should always be considered. Most simply, the use of a mean value and a standard deviation for each of the parameters and models involved should be used, and, ideally, a full probability distribution, or joint probability distributions in case of dependencies of the parameters, should be provided.

REFERENCES

Allsop W, Kortenhaus A, Morris M *et al.* (2007) *Failure Mechanisms for Flood Defence Structures*. Integrated Flood Risk Analysis and Management Methodologies, FLOODsite. Report No. T04-06-01.

Andrews JD and Moss TR (1993) *Reliability and Risk Assessment*. Longman Scientific and Technical, Harlow, UK.

Bakker WT and Vrijling JK (1980) Probabilistic design of sea defences. *Proceedings 17th International Conference Coastal Engineering (ICCE)*, ASCE, Sydney, Australia, **3**: 2040–2059.

Barlow RE (1998) *Engineering Reliability*. Society for Industrial and Applied Mathematics (SIAM), Philadelphia, USA.

BAW (Bundesanstalt für Wasserbau) (2005) Merkblatt Standsicherheit von Dämmen an Bundeswasserstraßen (MSD). Karlsruhe, Germany.

Buijs F (2007) Time-dependent reliability analysis of flood defences. PhD thesis, University of Newcastle upon Tyne, Newcastle, UK.

Burcharth HF (2001) Reliability based design of coastal structures. In *Coastal Engineering Manual, Part VI: Design of Coastal Project Elements* (Hughes S (ed.)). US Army Corps of Engineers, Washington, DC. Engineering Manual EM 1110-2-1100.

Burcharth HF and Liu Z (1996) Exploitation Report – Rubble Mound Breakwater Failure Modes (RMBFM). Department of Civil Engineering, Aalborg University, Denmark. Final Report MAS2-CT92-0042.

CIAD Project Group (1985) *Computer Aided Evaluation of the Reliability of a Breakwater Design*. Final Report. CIAD Association, Zoetermeer, The Netherlands.

CIRIA (Construction Industry Research & Information Association) (2010) *Beach Management Manual*. CIRIA, London, UK. CIRIA C685B, 2nd edn.

CIRIA/CUR/CETMEF (2007) *The Use of Rock in Hydraulic Engineering* (*The Rock Manual*). CIRIA, London, UK. CIRIA C683, 2nd edn.

CUR (1990) *Probabilistic Design of Flood Defences*. Centre for Civil Engineering Research, Gouda, The Netherlands. Codes and Specification No. 141.

D'Eliso C, Oumeraci H and Kortenhaus A (2007) A model system for breaching of sea dikes initiated by wave overtopping. *Proceedings 5th International Conference on Coastal Structures*, ASCE, Vol. I, Venice, Italy, pp. 514–525.

De Mello VF (1975) Some lessons from unsuspected, real and fictitious problems in earth dam engineering in Brazil. *Proceedings 6th Regional Soil Mechanics and Foundation Engineering Conference for Afrika*. Balkema, Rotterdam, The Netherlands, pp. 285–304.

DIN 4084 (1983) DIN 4084: Gelände- und Böschungsbruchberechnungen. Beuth-Bauverlag, Berlin, Germany. S. 266–301.

DIN 1054 (1996) DIN 1054: Sicherheitsnachweise im Erd- und Grundbau. Beuth-Verlag, Berlin, Germany, pp. 109–191.

DIN 4084-100 (1996) DIN 4084-100: Böschungs- und Geländebruchberechnungen. Teil 100: Berechnung nach dem Konzept mit Teilsicherheitsbeiwerten. Beuth-Verlag, Berlin, Germany. 1. Auflage, S. 215–234.

DVWK (1986) Flußdeiche. Nr. 210, DVWK-Merkblätter zur Wasserwirtschaft. Paul Parey, Berlin, Germany.

EAK (2002) Empfehlungen für Küstenschutzwerke. *Die Küste. Archiv für Forschung und Technik an der Nord- und Ostsee* **65**: 589 S.

EAU (2004) Empfehlungen des Arbeitsausschusses 'Ufereinfassungen', Häfen und Wasserstraßen. *Hafenbautechnische Gesellschaft, Deutsche Gesellschaft für Erd- und Grundbau* **10**: 664 S.

Hawkes PJ, Gonzalez-Marco D, Sánchez-Arcilla A and Prinos P (2008) Best practice for the estimation of extremes: a review. *Journal of Hydraulic Research* **46(Extra Issue 2)**: 324–332.

ISO (2007) ISO 21650: Actions from waves and currents on coastal structures. International Organization for Standardization, Geneva, Switzerland.

Johnson FA and Illes P (1976) A classification of dam failures. *Water Power and Dam Construction.* **28**: 43–45.

Johnson M and Matalo F (2004) Dam failure incidents – Tanzanian experience. *International Seminar – Stability and Breaching of Embankment Dams*, Oslo, Norway.

Kim YC (2010) *Handbook of Coastal and Ocean Engineering.* World Scientific, Singapore.

Klein Breteler M and Bezuijen A (eds) (1998) *Design Criteria for Placed Block Revetments.* Balkema, Rotterdam, the Netherlands, pp. 217–248.

Klein Breteler M, Pilarczyk KW and Stoutjesdijk T (1998) Design of alternative revetments. *Proceedings International Conference Coastal Engineering (ICCE)* **2(26)**: 1587–1600.

Kortenhaus A (2003) Probabilistische Methoden für Nordseedeiche. PhD thesis, Fachbereich Bauingenieurwesen, Leichtweiß-Institut für Wasserbau, Technische Universität Braunschweig, Braunschweig, Germany. Dissertation No. 154 S.

Kortenhaus A and Buß J (2008) Druckschlagbemessung eines Seedeiches bei Elisabethgroden. *Wasser und Abfall* **10(3)**: 39–43.

Kortenhaus A, Oumeraci H, Weißmann R and Richwien W (2002) Failure model and fault tree analysis for sea and estuary dikes. *Proceedings 28th International Conference Coastal Engineering (ICCE)*, ASCE, Cardiff, UK, **2**: 2386–2398.

Kuijper H and Vrijling JK (1998) Probabilistic approach and risk analysis. In *Dikes and Revetments* (Pilarczyk KW (ed.)). Balkema, Rotterdam, The Netherlands, pp. 443–462.

Larson M, Erikson L and Hanson H (2004) An analytical model to predict dune erosion due to wave impact. *Coastal Engineering* **51(11)**: 675–696.

Modarres M (1993) *What every Engineer should know about Reliability and Risk Analysis.* Marcel Dekker, New York, USA.

Morris M and Hassan AM (2002) *Breach Formation through Embankment Dams & Flood Defence Embankments: A State of the Art Review.* HR Wallingford, IMPACT Project Workshop, Wallingford, UK.

Morris MW, Hassan M, Kortenhaus A and Visser PJ (2009) *Breaching Processes: A State of the Art Review.* FLOODsite – Integrated Flood Risk Analysis and Management Methodologies, Report No. T06-06-03.

Nagy L (2004) Investigation some details of historical dike breaches. *International Seminar – Stability and Breaching of Embankment Dams*, Oslo, Norway.

Oumeraci H (1994) Review and analysis of vertical breakwater failures – lessons learned. *Coastal Engineering* **22(1/2)**: 3–29.

Oumeraci H, Schüttrumpf H and Bleck M (1999) *Untersuchungen zur Ermittlung der mittleren Wellenüberlaufrate ohne Freibord bei Stromdeichen*. Berichte Leichtweiß-Institut für Wasserbau, Technische Universität Braunschweig, No. 842, Braunschweig, Germany, 43 S.

Oumeraci H, Kortenhaus A, Allsop NWH *et al.* (2001a) *Probabilistic Design Tools for Vertical Breakwaters*. Balkema, Rotterdam, The Netherlands.

Oumeraci H, Schüttrumpf H, Möller J and Kudella M (2001b) *Loading of the Inner Slope of Seadikes by Wave Overtopping – Results from Large Scale Model Tests*. Berichte Leichtweiß-Institut für Wasserbau, Technische Universität Braunschweig, Braunschweig, Germany. Report No. 858.

Paté-Cornell ME (1984) Fault trees vs. event trees in reliability analysis. *Risk Analysis* **4(1)**: 177–186.

PIANC (Permanent International Association of Navigation Congresses) (1992) Analysis of rubble mound breakwaters. Supplement to Bulletin, vol. 78/79.

Pullen T, Allsop W, Bruce T *et al.* (2007) *EurOtop. Wave Overtopping of Sea Defences and Related Structures: Assessment Manual*. Environment Agency, Wallingford, UK; Expertise Netwerk Waterkeren (ENW), Lelystad, The Netherlands; Kuratorium für Forschung im Küsteningenieurwesen (KFKI), Hamburg, Germany. Available at: http://www.overtopping-manual.com (accessed 2 February 2012).

Rajabalinejad M (2009) Reliability methods for finite element models. PhD thesis, Delft University of Technology, Delft, The Netherlands.

Richwien W (1995) Versagensformen und Nachweiskonzepte – Kritische Beanspruchungen von Deichböden. HTG. *Jahrbuch der hafenbautechnischen Gesellschaft* **50**: 226–228.

Richwien W and Weißmann R (1999) *Quantifizierung der Sicherheitsreserven von Seedeichen bei Wellenüberlauf*. Abschlußbericht. Institut für Grundbau und Bodenmechanik, Universität und Gesamthochschule Essen, Essen, Germany, Teil IV, 55 S.

Rijkswaterstaat (1994) *Design Plan Oosterschelde Storm-surge Barrier – Overall Design and Design Philosophy*. Ministry of Transport, Public Works and Water Management/Balkema, Rotterdam.

Rose CW, Williams JR and Barry DA (1983) A mathematical model of soil erosion and deposition processes: I. Theory for a plane land element. *Journal Soil Science Society of America* **47**: 991–995.

Scheuermann A and Brauns J (2001) Die Durchströmung von Flussdeichen in Abhängigkeit von Vorbedingungen. In *Zweites Forum Katatrophenvorsorge: Extreme Naturereignisse – Folgen, Vorsorge, Werkzeuge*. DKKV 8 S.

Schiereck GJ (1998) Filter structures. In *Dikes and Revetments*, Balkema Rotterdam, The Netherlands, pp. 177–190.

Schneider J (1994) *Sicherheit und Zuverlässigkeit im Bauwesen*. Grundwissen für Ingenieure. Verlag der Fachvereine, Zürich, Switzerland. 182 S.

Schüttrumpf H (2001) *Wellenüberlaufströmung bei Seedeichen – experimentelle und theoretische Untersuchungen*. PhD thesis, Mitteilungen Leichtweiß-Institut für Wasserbau der Technischen Universität, Braunschweig, Germany.

Schüttrumpf H and Oumeraci H (2004) Learning from seadike failures. PIANC, *Bulletin* **117**: 47–60. Permanent International Association of Navigation Congresses (PIANC), Brussels, Belgium.

Schüttrumpf H and Oumeraci H (2005) Layer thicknesses and velocities of wave overtopping flow at seadikes. *Coastal Engineering* **52(6)**: 473–495.

Singh VP (1996) *Dam Breach Modeling Technology*. Kluwer Academic, Louisiana, USA.

Sörensen JD (1995) Reliability estimation by FORM/SORM techniques. PhD course, Institute of Building Technology and Structural Engineering, University of Aalborg, Aalborg, Denmark.

Stanczak G (2008) Breaching of sea dikes initiated from the seaside by breaking wave impacts. PhD thesis, Civil Engineering Faculty, Leichtweiß-Institut für Wasserbau, Technical University Braunschweig, University Florence, Braunschweig, Florence.

Stewart MG and Melchers RE (1997) *Probabilistic Risk Assessment of Engineering Systems*. Chapman & Hall, London, UK.

TAW (Technische Adviescommissie voor de Waterkeringen) (1995) *Under Pressure*. TAW, Delft, The Netherlands.

TAW (2000) From probability of exceedance to probability of flooding. Towards a new safety approach. TAW, The Hague, The Netherlands.

Townend IH (1994) Risk assessment of coastal defences. *Proceedings of the Institution of Civil Engineers (ICE) – Water, Maritime and Energy* **106**: 381–384.

Tuan TQ (2007) Seasonal breaching of coastal barriers. PhD thesis, Technische Universität Delft, Delft, the Netherlands.

USACE (2002) *Coastal Engineering Manual*. US Army Corps of Engineers, Washington, DC, USA. Engineer Manual 1110-2-1100, 6 vols.

Van Agthoven AM, Den Heijer F and Kraak AW (1997) The way to a flood risk based safety concept – four case studies. *Proceedings of the 1st RIBAMOD workshop: River Basin Modelling, Management and Flood Mitigation, Concerted Action*, Delft, The Netherlands, pp. 267–277.

Van der Meer JW (1998) Application and stability criteria for rock and artificial units. In *Dikes and Revetments* (Pilarczyk KW (ed.)). Balkema Rotterdam, The Netherlands, pp. 191–215.

Visser PJ (1995) A model for breach growth in sand-dikes. *Proceedings International Conference Coastal Engineering (ICCE)*, ASCE, Kobe, Japan, **3(24)**: 2755–2769.

Voortman HG (2002) Risk-based design of large-scale flood defence systems. PhD thesis, Delft University of Technology, Delft, The Netherlands.

Vrijling JK (1993) Development in probabilistic design of flood defenses in the Netherlands. In *Reliability and Uncertainty Analysis in Hydraulic Design*. ASCE, New York, USA, pp. 133–178.

Vrouwenvelder ACWM, Steenbergen HMGM and Slijkhuis KAH (1999) *Theoriehandleiding PC-Ring. Deel A: Mechanismebeschrijvingen*. TNO 98-CON-R1430, Delft, The Netherlands.

Weijers JBA and Sellmeijer JB (1993) A new model to deal with the piping mechanism. In: *Filters in Geotechnical and Hydraulic Engineering, Proceedings of the First International Conference (Geo-Filters)* (Brauns HSE (ed.)). Balkema, Rotterdam, The Netherlands, pp. 349–355.

Weißmann R (1999) Infiltration von Wellenüberlauf in Deichbinnenböschungen. Workshop 'Strömungen in Böden', Universität Essen, Essen, Germany.

Zeidler RB (1994) Example of reliability assessment for real case revetment. MAST Advanced Study Course: Probabilistic Approach to the Design of Reliable Coastal Structures, Bologna, Italy.

FURTHER READING

Bedford T and Cooke RM (2001) *Probabilistic Risk Analysis – Foundations and Methods*. Cambridge University Press, Cambridge, UK.

Guedes Soares C (ed.) (1997) *Probabilistic Methods for Structural Design*. Kluwer Academic, Dordrecht, Germany.

Joint Committee on Structural Safety, RILEM (2001) *Probabilistic Assessment of Existing Structures*. RILEM, Bagneux, France.

Möller B and Beer M (2003) Fuzzy Randomness. Uncertainty in Civil Engineering and Computational Mechanics. Springer-Verlag, Berlin, Germany.

Nikolaidis E, Ghiocel DM and Singhal S (2004) *Engineering Design Reliability Handbook*. CRC Press, Boca Raton, FL, USA.

Oumeraci H, Kortenhaus A, Allsop NWH *et al.* (2001) *Probabilistic Design Tools for Vertical Breakwaters*. Balkema, Rotterdam, The Netherlands.

Reeve D (2010) *Risk and Reliability: Coastal and Hydraulic Engineering*. Spon, London, UK.

Rychlik I and Rydén J (2006) *Probability and Risk Analysis – An Introduction for Engineers*. Springer-Verlag, Berlin, Germany.

USEFUL WEBSITES

A. James Clark School of Engineering, Department of Mechanical Engineering, University of Maryland: http://www.enre.umd.edu/tools.htm (accessed 2 February 2012).

CIRIA (Construction Industry Research & Information Association, *The Rock Manual*, available at: http://www.ciria.org/service/bookshop/core/orders/product.aspx?prodid=154 (accessed 2 February 2012).

CRESS (Coastal and River Engineering Support System): http//:www.cress.nl (accessed 2 February 2012).

FLOODsite, HR Breach model: http://www.floodsite.net/html/HR_Breach_Model.htm (accessed 2 February 2012).

FLOODsite project: http://www.floodsite.net (accessed 2 February 2012).

Fault Tree: http://www.faulttree.net (accessed 2 February 2012).

PLAXIS bv: http://www.plaxis.nl (accessed 2 February 2012).

Veri-Tech Inc.: http://www.veritechinc.com (accessed 2 February 2012).

Flood Risk
ISBN: 978-0-7277-4156-1

http://dx.doi.org/10.1680/fr.41561.125

Chapter 6
Performance of fluvial defences

Jaap Flikweert
Royal Haskoning, Peterborough, UK

This chapter describes the performance of the assets that play a role in fluvial flood risk management: flood defences and channels. It sets out the types of assets and their failure modes under extreme loading, and the deterioration mechanisms that reduce asset performance over time. The chapter provides references to publications with more specific design guidance, calculation methods and tools, and identifies areas of ongoing research.

6.1. Introduction

This section addresses the performance of fluvial defences. This includes all man-made or managed assets that have a function to manage the risk of flooding from rivers.

Performance can only be defined in relation to a chosen function. Fluvial defences often have a range of functions, and one of the key challenges of design and asset management planning is to find the right balance between these functions. This section focuses primarily on the function of the defences to reduce flood risk. However, the text refers to other functions where relevant. Chapter 4 provides further detail and references on this topic.

Section 6.2 defines and structures the asset types in fluvial flood defence. This is followed by Sections 6.3 and 6.4 on, respectively, barrier assets and channels, which describe failure modes, fault trees and process models for these two overarching fluvial asset types, and point the reader toward further aids for design and performance assessment.

6.2. Defence types

The first distinction to make is the one between raised defences and channels. Their flood defence function is fundamentally different: whereas raised defences act as a barrier against flood waters reaching receptors, the flood defence function of channels is to convey the river water from source to sea.

Within the category of raised defences, the primary distinction is whether the asset has an additional function to let water through (typically point assets such as gates and pumping stations). Defences that do not have this additional function are usually linear assets: they have an approximately uniform cross-section over a significant length. There are exceptions, where very localised solutions are needed to fit a defence into its surroundings. The next level of classification is the type of structure: for linear assets these are earth embankments, concrete and sheetpile walls; for point assets these are gates, pumping stations, barriers and locks. Below this level, further distinction is possible based on materials, size, presence of particular asset elements (e.g. revetments on embankments), and many other variables.

Within the category of channels, the main functional distinction is between open water channels and culverts (which are treated separately in Chapter 7). Within the open water channels, a more detailed classification is possible based on a range of characteristics: whether they are artificial or natural; whether or not they are pumped; their dimensions and bed materials.

Due to the fundamental functional difference between barrier assets and conveyance assets, the rest of this chapter deals separately with these two types (Table 6.1).

Table 6.1 Types of flood risk management assets

Barrier assets	No additional function to let water through (often linear assets)	 Earth embankments (photograph courtesy of Stefan Lombardo)  Concrete/masonry walls (photograph courtesy of Stefan Lombardo)  Sheet-pile walls (photograph courtesy of Scott Underwood)

Table 6.1. Continued

Barrier assets	Additional function to let water through (often point assets)	 Gates (photograph courtesy of Chris Fosbeary)  Pumping stations  Barriers (contains Environment Agency information © Environment Agency and database right)  Locks (contains Environment Agency information © Environment Agency and database right)

Table 6.1. Continued

Conveyance assets	Open water channels (further distinction is based on a range of characteristics)	
		 Gates (photograph courtesy of Paul Winfield)
	Culverts (see Chapter 7)	

6.3. Barrier assets

6.3.1 Failure modes and standard fault trees

At the most generic level, five broad failure modes can be distinguished for barrier assets, as shown in Figure 6.1. The defence in these pictures can be any type of barrier asset.

There are various connections between these high-level modes. In particular, all mechanisms 1 to 4 in Figure 6.1 are important causes of mechanism 5 (structural failure):

Figure 6.1 Generic flood defence failure modes

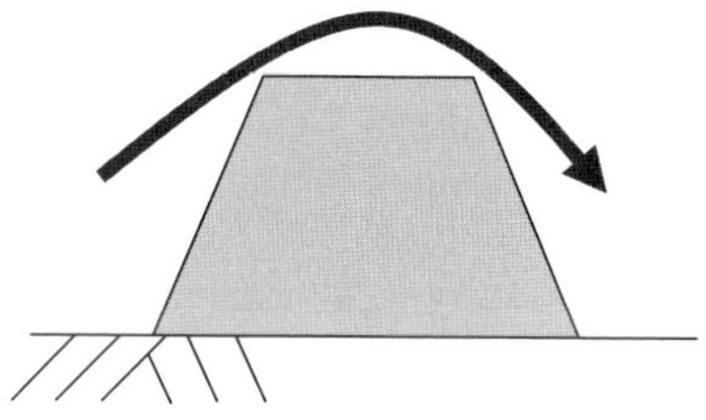

1. Water flowing over the defence, causing flooding while the defence remains intact. This is sometimes called 'functional failure'

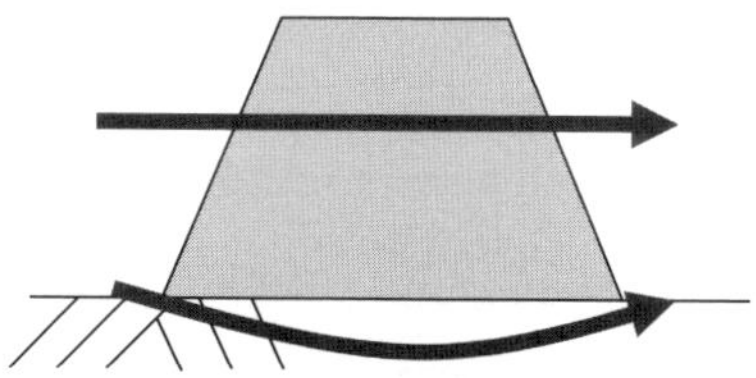

2. Excessive permeability, allowing water to flow through or under the structure

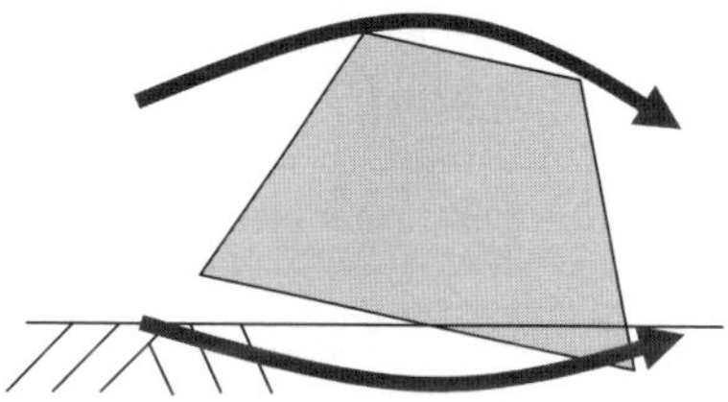

3. Displacement of the whole structure allowing water to flow over, through or under the defence

4. Operational failure, when human or automated action required for flood defence is not carried out, allowing water to flow across. This is sometimes called 'failure on demand'

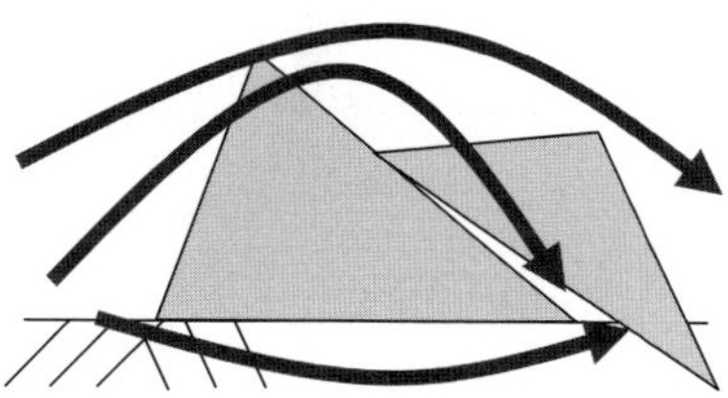

5. Structural failure, causing collapse of the defence structure and allowing water to flow across

overtopping causing erosion of the crest or landward slope; seepage causing piping, etc. Structural failure and displacement can also be caused directly by hydraulic loading (water level differential, waves and currents), or by other factors such as loading by structures, ships, ice and debris, and the actions of people or animals.

For each type of barrier asset, these five generic failure modes can have different forms. In practice, the sets of failure modes used for design and assessment reflect the most critical factors for specific asset types.

- For embankments, the emphasis is on generic modes 1, 2 and 5. Structural failure is divided into various mechanisms; for example, distinguishing piping, surface erosion, deep and shallow slips, and distinguishing failure toward the landward and the 'waterward' side. Displacement of the whole structure can also be relevant in particular cases. Examples are the horizontal sliding of dried-out peat embankments in The Netherlands in the summer of 2003, or the potential collapse or liquefaction of deep estuary channels causing the loss of tidal defences in the Scheldt Estuary.
- For concrete and masonry walls, and to some extent also for sheet-pile walls, generic mode 3 (displacement of the whole structure) can be very relevant, and could involve rotation or sliding in various directions. Deterioration of materials can lead to structural failure. Given the finite size of the concrete, steel or stone elements, walls typically contain transitions. These in themselves are important potential causes of failure, allowing flow through the structure, which can cause structural failure. A specific failure mode for sheet piles can be anchor failure.
- For defences with the additional function to let water through (often point assets), all five generic modes are relevant. Most of these defences have mechanical elements that can fail structurally. Many also have an operational element, and failure to operate them can cause flooding.

Different publications use different lists and categories of failure modes. Section 9.3 of *The Fluvial Design Guide* (Environment Agency, 2010) gives a practical overview of the most important failure modes for walls and embankments. A structured and systematic overview of all flood defences types and loading types is provided in the FLOODsite report *Failure Mechanisms for Flood Defence Structures* (Allsop *et al.*, 2007). This chapter does not include further discussion of wave-related failure modes, because they are covered in detail in Chapter 5.

Fault trees can be helpful tools when analysing failure processes of complex systems. They can help to visualise the logical relationship between various failure mechanisms, analyse the probability of failure and assess measures to prevent failure. The general concept is described in detail in Chapter 5.

6.3.2 Supporting equations and models and design aids and tools (references)

The assessment of the design and performance of defences is typically based on equations and models that represent the resistance of the asset against each relevant failure mode.

The Fluvial Design Guide (Environment Agency, 2010) gives practical guidance for the design of all types of fluvial flood defence assets. It refers to other sources for supporting equations and models; for example, for embankments, to the report

Management of Flood Embankments – A Good Practice Review (Defra/Environment Agency, 2007), and for temporary and demountable defences, to the report *Temporary and Demountable Flood Protection: Interim Guidance on Use* (Ogunyoye and Van Heereveld, 2002). The FLOODsite report *Failure Mechanisms for Flood Defence Structures* (Allsop *et al.*, 2007) gives a systematic overview of equations, focused particularly on their use for flood risk system analysis, but also applicable to the design and assessment of individual assets. Figure 6.2 shows an example page from the FLOODsite report.

The equations and models from these sources are typically included in design guidance documents, and are often incorporated in tools and software packages. In practice, there is not always a need to do a full computational analysis for all failure modes. There are different stages to a design, and simplified approaches may suffice for the early stages. In addition, it will often be possible to identify the failure modes that are most likely to determine design, which could allow the use of simpler, more conservative approaches to check the design against the other, less dominant modes.

Recent advances are particularly focused on the following.

- Fine fissuring – this is actually a deterioration mechanism where desiccation of clay causes fissuring, which then increases permeability and reduces the cohesion of surface layers. Section 6.3.3, on deterioration, provides further details and references.
- Drought-induced failure of peat embankments, as experienced in The Netherlands in the summer of 2003 – drought reduced the weight of the embankment, while shrinkage of the peat allowed water from the channel to infiltrate, causing uplift pressure.
- Erosion of vegetated slopes – the Environment Agency has carried out embankment vegetation trials to assess the impact of maintenance regimes on vegetation (Environment Agency, 2009a). This study concluded that maintenance has an overriding influence on vegetation composition (over factors such as starting condition, orientation, soil composition and presence of invasive species). Internationally, research is ongoing into the erosion resistance of vegetated slopes, particularly in The Netherlands and the USA, but this is focused on wave overtopping, and therefore is more relevant to coastal defences.

6.3.3 Deterioration

The general concept of asset deterioration is discussed in Chapter 4. As indicated there, understanding deterioration is crucial for making decisions about design and about asset management. The deterioration rate determines how the benefits of the asset decrease over time. It also determines the level of maintenance required to keep meeting the performance targets of the asset.

In an ideal world, decisions about design and maintenance would be based on a whole-life analysis of both the costs and the benefits, as shown schematically in Figure 6.3. The maintenance philosophy and performance targets are based on the asset manager's

Figure 6.2 Overview of equations and parameters for failure modes – Example for piping
From Allsop *et al.* (2007), reproduced by permission of the members of the FLOODsite Consortium

Summary: Primarily flow through and/or under embankment. May lead in turn to internal erosion

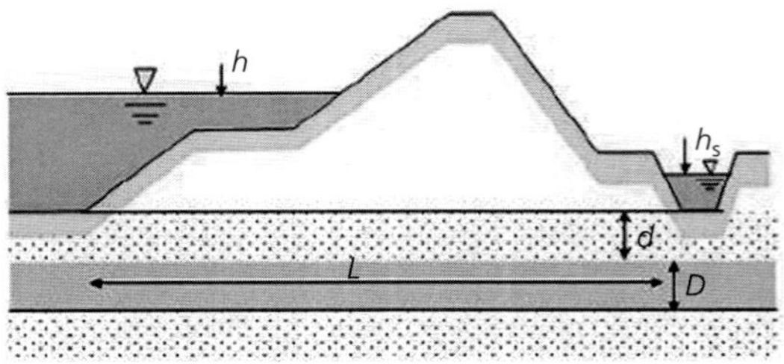

Reliability equation:
The reliability function is expressed by:

$$z = h_{crit} - \Delta h$$

where:

h_{crit} = critical pressure head gradient for piping [–]
Δh = existing pressure head gradient [–]

Loading equations:

$$\Delta h = h_w - h_h$$

Resistance (strength) equations:

$$h_{crit} = \alpha c l_D \left(\frac{\rho_s}{\rho_w}\right) \tan\theta\,(0.68 - 0.10 \ln c)$$

with

$$\alpha = \left(\frac{D}{l_D}\right)^{\left(\frac{0.25}{\left(\frac{D}{l_D}\right)^{2.3}} - 1\right)} \qquad c = \eta d_{70}\left(\frac{1}{\kappa l_D}\right)$$

Parameter definitions:

- D = thickness of sand layer [m]
- l_D = length of seepage line [m]
- h_{crit} = critical water level difference between water level in front and behind dike
- d_{70} = diameter of sand particle [mm]
- ρ_s = density of sand particle [t/m^3]
- ρ_w = density of water [t/m^3]
- κ = internal permeability [–]
- θ_h = bedding layer [*]
- η_s = drag coefficient [–]
- V = kinematic viscosity of water [Pa s]
- g = gravitational content [m/s^2]
- h_w = water depth at dike foot [m]
- H_s = significant wave height [m]
- ΔH = overall head difference [m]
- L_s = seepage length [m]
- c_k = creep factor

Sources of failure mechanism equations/methods:

Bligh, W.G. (1910): Dams, barrages and weirs on porous foundations. *Engineering News*, vol. 63/64, pp.708.

Sellmeijer, J.B. (1988): On the mechanics of piping under impervious structures. Ph.D. thesis, LGM mededelingen/Laboratorium voor Grondmechanica, Stichting Waterbouwskundig Laboratorium, LGM, Delft, The Netherlands, 43 pp.

Weijers, J.B.A.; Sellmeijer, J.B. (1993): A new model to deal with the piping mechanism. *Filters in Geotechnical and Hydraulic Engineering, Proceedings of the First International Conference (Geofilters)*, Brauns, Heibaum & Schuler (eds.), Rotterdam, Holland: Balkema, Karlsruhe 1992, S. 349–355.

TAW (1999): 'Technical Report on Sand Boils (Piping)'.

Sources of uncertainties in failure equations/input parameters:

Remarks:

Figure 6.3 Benefits and costs of maintenance
Environment Agency (2009b). Contains Environment Agency information © Environment Agency and database right

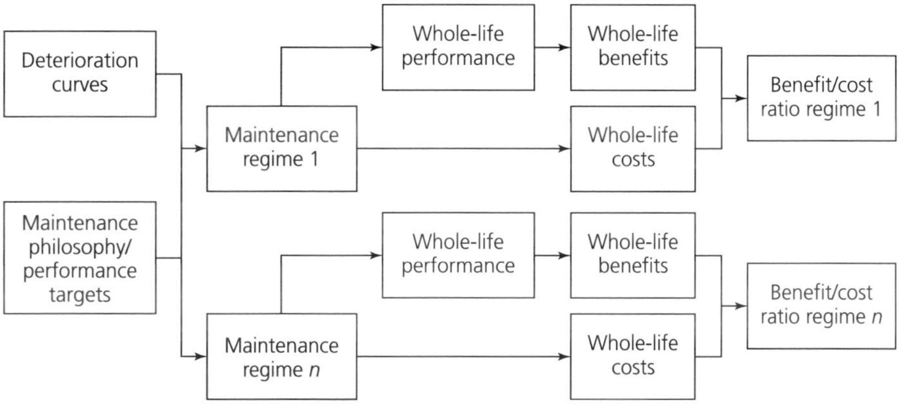

organisational strategy and objectives, and are, for example, formulated in terms of target condition. For a given asset, combining this with knowledge about deterioration produces one or more possible maintenance regimes, which are sets of activities needed to meet or achieve the performance targets. The whole-life costs of each regime can be calculated based on the activities. The impact of the maintenance regime on the deterioration rate determines how the performance of the asset varies throughout its life, and this can be translated into the benefits (risk reduction) that the asset provides. Comparing the benefit/cost ratio of different regimes would be useful information in comparing options for the maintenance regime.

In practice, this analysis of whole-life costs and benefits is complicated by limited knowledge of the processes and lack of suitable tools. In particular, understanding is required of the following four elements.

- The processes that cause deterioration.
- How maintenance influences deterioration.
- Relating deterioration of condition to defence performance and risk reduction.
- Whole-life costs of typical maintenance regimes.

Some of these elements have been developed in recent years, and the Environment Agency's project on deterioration and whole-life costing (Environment Agency, 2009b) is developing an integrated tool that includes maintenance regimes, whole-life performance and whole-life costs.

It is interesting to note that in general infrastructure asset management, combined use of deterioration modelling and whole-life costing is reasonably well developed. However, the application of these methods to flood defence asset management is complicated by, among other things, the following factors.

- It is more difficult to predict the deterioration of flood defences because they are often built of natural materials and their loading is very variable. This introduces the need to deal with uncertainty, and makes the outcomes of the analysis more difficult to understand and apply.
- There is a fundamental difference from most other infrastructure types (such as bridges and buildings), the condition of which is not directly related to the benefits they provide. As a result, the aim of the analysis is to determine the minimum whole-life cost to achieve the performance target. For flood defences, better performance directly leads to higher benefits, which leads to a more complex optimisation problem.

For fluvial defences, the most relevant causes of deterioration are as follows.

- Scour, which can damage toe structures and affect the geotechnical stability of embankments and walls.
- Settlement, which can reduce crest height, increasing the likelihood of overtopping and associated erosion of the crest and landward slope. In design it is important to differentiate between the settlement of the fill and of the subsoil, and it is essential to estimate these with care. These estimates are often based on historical values, or can be based on various levels of modelling.
- Desiccation of clay, which is a natural process that increases permeability and reduces cohesion of the surface layers. Recent research (Sentenac and Zielinski, 2009) showed that desiccation can occur within 2 years of construction and down to a depth of 1 m. Desiccation fissuring can be monitored through inspection, but this is not straightforward. The report lists a number of measures to limit, prevent and cure fissuring.
- Corrosion of sheet piles and of reinforcement steel, which can cause structural failure of walls. There are various ways of incorporating this into design (e.g. protection measures, surplus steel thickness).
- Third-party use of defences, for example, consisting of footpath use causing rutting of the crest, unauthorised excavations and interference with closure operations. Much of this can be managed through consent by the defence manager and monitored through regular inspection.
- Uncontrolled vegetation, particularly woody vegetation, which can have an effect on the structure of embankments: it can cause piping, obscure the defence for inspection, and there is a risk of damage to the slope if a large tree were to collapse.
- Animal burrowing, which can be an important source of failure due to the creation of pipes and potential instability.

6.4. Channels

6.4.1 Failure modes and standard fault trees

The flood defence function of channels is to convey the river water from source to sea. Conveyance is a quantitative measure of the discharge capacity of a watercourse. It depends on channel roughness, channel shape (section and plan form) and cross-sectional area. Conveyance describes the relationship between the gradient and the discharge: a higher conveyance leads to a higher discharge under a given hydraulic gradient. For further details and equations, see Knight *et al.* (2010).

It is not common to describe the performance of channels in terms of failure modes, but there are clear parallels with the framework for barrier assets, as described in Section 6.3. There is only one high-level failure mode, which is that the discharge exceeds the channel's capacity. Given that the channel gradient is normally a physical constraint, insufficient conveyance will be the result of an insufficient combination of channel cross-sectional area and shape and channel roughness. This can be caused by inadequate design (either for the whole channel or for localised constrictions, such as a bridge or culvert). However, in practice, the main issue is often the gradual deterioration of conveyance.

- A reduction in the cross-sectional area due to siltation, collapse of channel banks or structures, deposition of debris, vegetation growth or ice; an important related concept is afflux – the local increase in water level caused by the presence of a constriction.
- An increase in roughness, often due to vegetation growth.

6.4.2 Supporting equations, models and tools

Channel conveyance is governed by the standard equations of channel hydraulics (see Knight *et al.*, 2010). A range of methods is available for hydraulic analysis, from simple to advanced. The choice of method depends primarily on the question that needs to be answered, and is influenced by availability of data and resources. The available methods are as follows.

- Backwater calculations (without computer modelling) – this can be sufficient for the calculation of the impact of local interventions, but only for simple systems and steady-state situations.
- Steady-state modelling or dynamic modelling – steady-state modelling is easier and cheaper but cannot be used if the shape of the hydrograph is relevant, when assessing the influence of attenuation due to storage, or when modelling the performance of the system in an historic flood event.
- One-dimensional or two-dimensional modelling – two-dimensional modelling has the benefit of more detailed representation of flood-plain flow and the possibility to determine flow patterns and velocities.

Recent developments in channel performance have focused on improving modelling of the changes that channels undergo over time: geomorphology, vegetation and debris. Each of these can influence both the channel profile and the roughness. An important element of recent research has also been to find the right balance between habitats and flood risk in channel management approaches.

The Conveyance Estimation System (CES) contains a Roughness Advisor, which helps modellers to determine roughness as a function of vegetation type, combining different segments of the channel cross-section and including seasonal influences, which makes it a useful tool to support channel management.

Afflux is included in all relevant one-dimensional hydraulic models, and the Afflux Estimation System (AES) is available as an integrated package with the CES. There is also a separate spreadsheet application called the Afflux Advisor, which can provide a relatively quick calculation for simple structures in a uniform channel.

6.5. Conclusion

Flood defences continue to play an important role in the management of flood risk, as part of a wider portfolio of intervention types. This not only includes barrier assets such as embankments, walls and gates, but also the managed channels that convey flood water. The management of these assets is moving toward a risk-based approach, in which decisions about interventions are based on their impact on performance. This requires an understanding of failure modes, how these interact and how each intervention affects performance, quantitatively where needed.

REFERENCES

Allsop W, Kortenhaus A, Morris M *et al.* (2007) *Failure Mechanisms for Flood Defence Structures*. Integrated Flood Risk Analysis and Management Methodologies, FLOODsite. Report No. T04-06-01.

Defra/Environment Agency (2007) *Management of Flood Embankments – A Good Practice Review*. Environment Agency, R&D Technical Report FD2411/TR1. Available at: http://sciencesearch.defra.gov.uk/Document.aspx?Document=FD2411_6509_TRP.pdf (accessed 2 February 2012).

Environment Agency (2009a) *Flood Embankment Vegetation Management Trials – Final Report*. Environment Agency, Science Report SC030228/SR1. Available at: http://publications.environment-agency.gov.uk/PDF/SCHO0909BQYW-E-E.pdf (accessed 2 February 2012).

Environment Agency (2009b) *Assessment and Measurement of Asset Deterioration including Whole Life Costing*. Environment Agency, Science Report SC060078/SR. Available at: http://publications.environment-agency.gov.uk/PDF/SCHO0509BQAV-E-E.pdf (accessed 3 May 2012).

Environment Agency (2010) *The Fluvial Design Guide*. Available at: http://evidence.environment-agency.gov.uk/FCERM/en/FluvialDesignGuide.aspx (accessed 2 February 2012).

Knight DW, McGahey C, Lamb R and Samuels P (2010) *Practical Channel Hydraulics; Roughness, Conveyance and Afflux;* CRC Press, Boca Raton, FL, USA.

Ogunyoye F and van Heereveld M (2002) *Temporary and Demountable Flood Protection: Interim Guidance on Use*. Environment Agency, R&D Publication 130/1. Available at: http://sciencesearch.defra.gov.uk/Document.aspx?Document=TDFD_phase_1__1628.PDF (accessed 2 February 2012).

Sentenac P and Zielinski M (2009) *Influence of the Desiccation Fine Fissuring on the Stability of Flood Embankments. Final Report including Laboratory Studies*. FRMRC Research Report II UR14. Available at: http://www.floodrisk.org.uk/images/stories/Phase1/UR14_2009%20Final%20version_nso.pdf (accessed 2 February 2012).

FURTHER READING

International Levee Handbook: this is currently being developed as a partnership between flood risk management organisations from the USA, Great Britain, France and The Netherlands, among others. It will be published in 2013 and aims to provide a compendium of good practice, offering comprehensive guidance on the design, construction, maintenance and improvement of levees and describing the international state of the art on these matters. Available at: http://www.leveehandbook.net (accessed 2 February 2012).

Flood Risk
ISBN: 978-0-7277-4156-1

http://dx.doi.org/10.1680/fr.41561.139

Chapter 7
Urban drainage systems

David Stewart
Torbay Council, Torbay, UK

Historically, as urban areas developed, drainage systems were constructed to convey both foul and surface water away from properties. These systems discharged untreated effluent to receiving waters. However, in many locations the drainage systems were very poorly designed, resulting in flooding, which caused human health issues, including cholera.

In Victorian times legislation for both water supply and sewerage services was introduced, giving powers to local authorities to provide potable water and sewage services to urban areas. This legislation has been updated over the years, resulting in major improvements to urban drainage systems. In addition to facilitating these improvements, the legislation has resulted in the responsibility for sewage services transferring from local authorities to private water companies.

In this chapter, the different types of drainage system encountered are explained, together with the historical techniques used to design the urban drainage system. Over the past 20 years, the manner of and the tools used for design have evolved significantly, and some of the new methods being used are highlighted. These include hydraulic modelling techniques (which can be used to model both underground and above-ground drainage systems), integrated urban drainage and risk-based approaches to design.

7.1. Introduction

The basic function of an urban drainage system is to remove sewage together with rainwater run-off from buildings, roads and other impermeable areas, and convey it safely to either a sewage treatment works or, in the case of surface water run-off, a watercourse, river, the sea or a ground infiltration system. A number of different methods of providing storm drainage have evolved over the years, but in general urban drainage systems consist of a network of underground pipes. These networks may be classified as either combined or separate systems. Combined systems are designed to convey both foul sewage and storm-water run-off and generally discharge to a sewage treatment works. Separate systems are designed as a two-pipe network, with one pipe conveying foul sewage to a sewage treatment works and the second pipe conveying the storm-water run-off to an ordinary watercourse, river, coastal waters or a ground infiltration system. However, the mechanism by which storm water runs off in urban areas is complex, and the measures taken to control surface water run-off can affect both the natural and the built environment in many ways.

Historically, drainage systems within many urban areas were very poorly designed, resulting in disease and illnesses such as cholera. As a result of 'The Great Stink' in London in 1858, Sir Joseph Bazalegette designed and built a new sewerage system for central London (Halliday, 1999). Bazalegette designed the new sewerage system to intercept the existing drains, which discharged into the River Thames in and around the main urban areas, and transferred the discharge away from the urban areas to the south of London. The new sewerage system was opened in 1865, although the project was not completed entirely until 1875. Following the opening of the new sewerage system, the effect on the health of the residents of London was to reduce significantly the level of cholera outbreaks, not only in the locations that no longer stank, but also wherever water supplies ceased to be contaminated by sewage.

In this chapter, the different types of drainage system encountered in an urban catchment will be explained, together with some of the historical techniques used to design an urban drainage system. Over the last 20 years, the manner of and the tools used to design the urban drainage system have evolved significantly, and in this chapter some of the new methods being used are highlighted, including hydraulic modelling techniques, integrated urban drainage, risk-based approaches to design, design assessments and tools. In addition, the various types of asset used to convey, store and control flows are identified, along with a number of failure modes that may be experienced.

7.2. Legislation related to urban drainage

Legislation for water supply and sewerage services in England and Wales dates back to Victorian times, when Acts of Parliament received Royal Assent to give local authorities the powers they required to provide potable water and sewerage services to expanding centres of population. The scope of this legislation was increased and consolidated, but without fundamental changes until 1963, when the Water Resources Act was passed (Department of the Environment (DoE), 1963). This Act created 'River Authorities' which had responsibility for enforcing the law relating to water resources, river pollution and land drainage, and introduced the concept of river basin management.

The administration of water law remained with the local authorities until the Water Act 1973 was passed (DoE, 1973). This Act created ten regional water authorities in England and Wales, the areas of which were defined by river basins. These authorities took over the responsibility for water supply, sewage disposal and river basin management. Local drainage engineers continued to provide the sewerage services within the urban areas as agents for the water authorities.

Privatisation of the ten water authorities, water supply and sewerage functions came into effect with the Water Act 1989 (DoE, 1989). The river function previously undertaken by the water authorities transferred to a newly created National Rivers Authority, which was later to become the Environment Agency as a result of the Environment Act 1995 (DoE, 1995).

Since 1989, legislation for water supply and sewerage services has been consolidated and updated as a result of the following Acts of Parliament.

- Water Industry Act 1991 (DoE, 1991b)
- Water Resources Act 1991 (DoE, 1991c)
- Water Industry Act 1999 (DoE, 1999)
- Water Act 2003 (Department for Environment, Food and Rural Affairs (Defra), 2003).

As a result of this legislation, between 1989 and 2003 agency agreements that local authorities had with the water companies to provide the sewerage services within their areas were stopped, and this work was taken back in house.

7.3. Types of urban drainage systems

Within urban areas there are a number of different types of urban drainage system, which are designed to convey flows from different sources. Some typical examples of such drainage systems are as follows.

- Foul sewers – As previously indicated, these systems deal with the foul element of sewage, and are generally designed based on an average discharge rate from each property. In order to provide self-cleansing, the minimum flow velocity within a foul sewer should be 0.75 m/s at one-third design flow. As these sewers deal with foul flows only, they are not affected by rainfall events.
- Surface water sewers – These sewers deal with the surface water run-off from impermeable areas such as roofs, driveways, roads, car parks, etc. They are designed by identifying the impermeable area serving the system and the rainfall run-off generated from these impermeable areas for a specific return period storm event. The minimum velocity within these sewers for self-cleansing should be 1 m/s at pipe full flow.
- Combined sewers – As the name suggests, these sewers cater for both the foul and surface water flows, and are generally designed using rainfall events, with an additional allowance for the foul flows from each property.
- Highway drainage – In certain locations there is separate highway drainage, and these systems are designed only for the surface water run-off from the highway. These systems are designed in the same manner as surface water sewers.

Within urban areas, in addition to the sewers and drainage systems identified above, natural surface water features exist, such as ordinary watercourses and main rivers. As development has evolved, many of these natural systems have been culverted, and as a result they now form an integral part of the urban drainage system.

7.4. Historical trends

Historically, as urban areas developed, drainage systems were constructed to convey both foul and surface water away from the properties. These systems were known as ‘combined sewers’, and they discharged without treatment to coastal waters, main rivers or ordinary watercourses. Over the years, due to pollution within the coastal waters and main rivers, legislation has been introduced to clean up the rivers and coastal waters, and as a result the majority of these combined sewers now discharge to sewage treatment works prior to the treated effluent discharging to receiving waters. In England, all combined sewer overflows that discharge to a watercourse, river or coastal waters must be consented by the Environment Agency.

As a result of flooding problems being experienced in some combined sewer systems during extreme rainfall events, combined sewer overflows (CSOs) have been installed. These CSOs reduced the risk of flooding by allowing the combined sewer system to be relieved during storm events. The excess flows discharge from the CSO to a receiving watercourse, river or coastal waters. These discharges, where uncontrolled, have resulted in many pollution incidents being reported in rivers and coastal waters, especially bathing waters. Due to recent legislation, the frequency of these discharges has been reduced by the installation of additional storage within the combined sewer system. In addition, primary treatment of the overflow discharge has been provided on many of the overflows by the installation of screens.

During the past 50 years, the installation of combined sewers has decreased, and these have been replaced by separate foul and surface water systems. The surface water system can discharge directly to the receiving waters, and therefore only the foul flows have to be treated before discharging to the receiving waters, resulting in a more sustainable system.

Traditionally, urban drainage systems have been designed for a specific storm event and, depending on the type of system being designed, the storm event used can differ considerably. Without ongoing maintenance, the existing drainage system may not be able to perform to its original design standards, due to the structural deterioration of the assets and siltation within the existing network, resulting in a chance of flooding.

By controlling the run-off from urban areas it is possible to reduce the chance of flooding within the urban environment. The measures taken to control the urban run-off can be wide ranging. However, where a well-designed urban drainage system exists, many people will not be aware that control of run-off is taking place. Unfortunately, flooding problems can occur even within a well-designed drainage system. For example, if new developments are constructed and their discharge is connected to the existing drainage system without a detailed hydraulic analysis of the existing and proposed systems, there is a chance that the additional flows generated from the new development could overload the existing drainage system. Without some form of control on these additional flows this increase in run-off will, in all probability, result in a significantly increased risk of flooding downstream of the new development.

Where surface water sewers and highway drainage exist it is common for these systems to discharge to a natural water feature, such as an ordinary watercourse, river or coastal waters. As a result, when carrying out the design of these systems it is important to consider the affect of these discharges on the downstream receiving waters, as the additional flows from the surface water system could increase the chance of flooding from the natural surface water system. Similarly, in coastal areas the capacity of the surface water outfall could be restricted during different tidal conditions, resulting in decreased discharge from the surface water system, and hence an increased risk of flooding in the upstream catchment areas during severe storm events.

In order to reduce the risk of flooding within the urban catchment it is important to control the urban run-off from development areas. The control of flows can be achieved in a number of ways, including increasing the capacity of the drainage system, reducing the flows entering the drainage system, or attenuation of flows within the drainage system.

Before hydraulic modelling was introduced for the design and analysis of urban drainage systems, it was common practice, where flooding problems occurred or new development was proposed, to increase the capacity of the drainage system by re-laying or duplicating pipes. This process, however, can be very costly and disruptive, and in many cases, unless the capacity of the system is increased to the outfall, the localised increase of capacity within the system will result in the problem being transferred downstream.

By reducing or controlling the surface water run-off in urban areas entering the sewerage system by means of sustainable urban drainage system schemes (SuDS) or attenuation tanks, the risk of flooding to downstream areas of the catchment can be reduced. Typical SuDS schemes can include soakaways, swales, permeable paving and ponds. All of these systems allow the surface water run-off to discharge to land, rather than to the surface water or combined drainage systems.

Other SuDS schemes that can be used to reduce surface water run-off include green roofs, rainwater harvesting and rainwater butts. It is possible in urban areas that retro-fitting of SuDS schemes can be undertaken in order to reduce the surface water discharges to the sewer system, and hence reduce the chances of flooding.

Where flooding problems exist from sewers or watercourses, it is possible to reduce the frequency and impact of the flooding by attenuating the upstream flows using storage tanks or ponds. These storage systems can be oversized online pipework, underground storage tanks, or, on watercourses and rivers, flood storage reservoirs. All of these will have a downstream flow control such as a hydrobrake or orifice, which restricts the continuation flow to the downstream system. Some examples of these flood storage systems are shown in Figures 7.1 and 7.2.

7.5. Emerging approaches

7.5.1 Integrated urban drainage

Integrated urban drainage (IUD) is an approach to planning or managing the urban drainage system leading to an understanding of how different elements of the drainage system interact and how different organisations must work together for it to operate effectively. IUD considers all elements of the urban drainage system that contribute to flooding issues and water quality issues, and reflects the concerns of the UK government to develop a more holistic approach to flood risk management following the July 2007 floods.

Figure 7.1 Online combined sewer attenuation tank with orifice flow control

IUD has a role in a range of situations; however, it is particularly useful in supporting the development of surface water management plans (SWMPs) (Defra, 2010a). As part of its future water strategy for England and Wales (Defra, 2008a) the UK government proposed that SWMPs will be used to assess urban flooding and identify options to resolve this flooding in the future. Formal SWMPs were developed through the Defra IUD pilot projects undertaken in 2007/8 and promoted in Sir Michael Pitt's review of the July 2007 floods (Pitt, 2008). One of the IUD pilot projects was undertaken within Torquay and details of this project are included in a case study at the end of this chapter.

Defra published guidance on the production of SWMPs in January 2009, which identifies a framework for local partnerships (local government, water companies, the Environment Agency and others) to understand the surface water flood risk and then plan a practical, sustainable and cost-effective series of options to reduce this risk.

7.5.2 The Floods and Water Management Act 2010

Following the significant flooding in England during the summer of 2007, The Pitt Review identified 92 recommendations, which the UK government endorsed in their response (Defra, 2008b). The governments major policy change has resulted in local authorities in England and Wales taking on a local leadership role for flood risk

Figure 7.2 Outlet from a main river flood storage reservoir with hydrobrake flow control located behind the trash screen

management. In order for local authorities to undertake this role, new legislation was required, and in April 2010 the Floods and Water Management Act (Defra, 2010b) received Royal Assent.

The Act aims to reduce the likelihood and impacts of flooding, clarifies the roles and responsibilities around flood risk, improves the efficiency and management of the water industry, and improves water quality. The Act has a range of implications for local authorities within England and Wales, with one of the major ones being the creation of a new role for Unitary Authorities (single-tier authorities) and County Councils (two-tier authorities) as the lead local flood authority. Under the Act, the lead local flood authority will be required to undertake a wide range of new responsibilities, including managing a local flood risk strategy, maintaining an asset register of significant flood defence structures, designating flood risk structures, and forming a SuDS Approval Body (SAB) to approve the design of SuDS in development and adopt SuDS that serve more than one property.

These new requirements under the Floods and Water Management Act link to various other new and existing pieces of legislation and policy. The most significant of these is the Flood Risk Regulations (Defra, 2009), which were introduced in 2009 to transpose the EU Floods Directive (European Commission, 2007) into domestic law in England and Wales and to implement its provisions. In particular, it places duties on lead local flood authorities to prepare a preliminary flood risk assessment, flood hazard and risk maps and flood risk management plans.

7.5.3 Sustainable urban drainage systems (SuDS)

The aim of SuDS is to: manage run-off from a development in an integrated manner, in order to reduce the quantity of water entering drains, sewers, watercourses and rivers, especially at peak periods; to improve the quality of run-off; and to promote amenity and biodiversity benefits by using water in the environment. As mentioned previously, typical techniques that can be used for SuDS include filter strips and swales, filter drains and permeable paving, green roofs and bioretention, rainwater harvesting and rainwater butts, wetlands and ponds, as well as underground storage systems such as soakaways and tanks.

Under the Floods and Water Management Act 2010, lead local flood authorities will be required to establish an SAB, which will be responsible for the following.

- SuDS approval – The SAB will be required to approve surface water drainage designs for new developments and redevelopments before construction can commence on site. The proposed surface water drainage system will be required to comply with the new national standards, which will cover design, construction, operation and maintenance.
- SuDS adoption and maintenance – The Floods and Water Management Act requires the SAB to adopt and maintain approved SuDS that serve more than one property.

It should be noted that this section of the Floods and Water Management Act is likely to be commenced in 2012.

When undertaking a SuDS design various criteria should be considered, including quantity (flows, volumes and frequency of run-off), quality (preventing and treating pollution), amenity (ensuring attractive, multifunctional and easily maintained SuDS) and biodiversity (integrating opportunities for wildlife within the SuDS design).

The national SuDS standards are currently being developed; however, they are likely to be based on *The SuDS Manual* (Woods-Ballard *et al.*, 2007), including basic principles and a range of design guidance on run-off destination, peak flow rates and volume and water quality.

7.6. Risk-based approach to flooding

In order to reduce the risk of flooding from new developments in urban areas, *Planning Policy Statement 25: Development and Flood Risk* (PPS 25) was produced in 2006 and revised in 2010 (Department for Communities and Local Government (DCLG), 2010). This statement highlights the following:

> Flooding from rivers and coastal waters is a natural process that plays an important role in shaping the natural environment. However flooding threatens life and causes substantial damage to property. The effects of weather events can be increased in severity both as a consequence of previous decisions about the location, design and nature of a settlement and land use, and as a potential consequence of future climate change. Although flooding cannot wholly be prevented, its impacts can be avoided and reduced through good planning and management.

As a result of this document, all urban areas of the country require a strategic flood risk assessment (SFRA) to be undertaken, which will identify the risk of flooding from surface water and coastal waters. In general, the SFRA will be carried out in two levels. Level 1 focuses on the flood risk to the entire catchment area, whereas the level 2 document focuses specifically on the major flood risk areas within the catchment.

Within the level 1 SFRA, flood risk maps will be produced that identify land in one of four flood risk categories.

- Flood zone 1 – low risk: less than a 0.1% annual probability of river or sea flooding in any year.
- Flood zone 2 – medium risk: a 1% to 0.1% annual probability of river flooding, or a 0.5% to 0.1% annual probability of sea flooding in any one year.
- Flood zone 3a – high risk: a 1% or greater annual probability of river flooding, or a 0.5% or greater probability of coastal flooding in any one year.
- Flood zone 3b – functional flood plain: comprises land where water has to flow or be stored in times of flood (land which would flood with an annual probability of 5% or greater in any one year) or is designed to flood in an extreme 0.1% flood event.

Within PPS 25 the recommended type of development for each of the flood zones is identified together with the process required for that development to proceed.

The level 2 SFRA focuses on the major flood risk areas within the catchment, and flood depth and vulnerability maps will be produced. The vulnerability maps are based on the risk to human life during flooding events, and are dependent on the depth of flooding and the velocity of the flood waters. An example of a flood vulnerability map is shown in Figure 7.15 later in this chapter.

As part of the SFRA various strategies may be identified that will help to reduce the risk of flooding. These strategies could include

- making space for water
- managing surface water
- sustainable drainage
- climate change

- considering ingress/egress to properties
- resistance and resilience
- opening culverts.

In addition to the risk-based approach identified in PPS 25, design of the urban drainage system is changing from the traditional design for a specific storm event to a more risk-based approach that will investigate performance across a range of storm conditions. An example of recent research into risk-based approaches is that undertaken within the Dti sponsored Sam project (HR Wallingford, 2009).

The Dti Sam project – *System Based Analysis and Management of Urban Flood Risk* – is focused on consequences rather than achieving a specific level of performance for the network. A guide has been produced summarising the proposed procedure, along with an explanation of the tools that have been developed, the data requirements and how the studies should be undertaken. This guide is available on the Dti Sam website.

7.7. Asset types

Within the urban drainage system there are many different types of assets that convey, control or relieve the flows of foul, surface or combined water through the catchment. Typical examples of some of these assets are listed in Table 7.1.

All of these assets play an important role in reducing the risk of flooding within urban areas. Even though the drainage systems are designed to recommended design standards, it is likely that sometime in the future a storm event will occur that exceeds the design standard. As a result, when designing a drainage system the engineer should carry out an assessment using storm events in excess of the design standard. As part of this design, where flooding is predicted, the exceedance flows or overland flows can be modelled using two-dimensional (2D) modelling, and by using these techniques it is possible to divert the exceedance flows away from buildings.

Table 7.1 Typical asset types

Conveyance	Control	Relieve
Pipes	Attenuation tanks	Combined sewer overflows
Culverts	Flow control devices	Emergency overflows
Tunnels	Pumping stations	Bifurcations
Manholes	Flap valves	
Open channels	Soakaways	
Natural channels	Swales	
Road gullies	Ponds	
Overland flow routes	Permeable paving	
Pumping stations	Infiltration trenches	
	Rain water harvesting	
	Water butts	

An example of this design technique is given in *Sewers for Adoption*, 6th edition (Water Research Council (WRC), 2006), where the sewers are designed for the critical 1 in 30 year storm event. However, the design is then assessed to confirm that no flooding to buildings is predicted for the 1 in 100 year storm event. This exceedance flow can be conveyed on the ground by surface pathways, which could be roads, paths, parkland or depressions on the surface. It should be noted that, when designing systems that utilise surface pathways to reduce the risk of flooding, the connection between the underground and surface systems is extremely complicated, and can only properly be represented by complex computer modelling using both one-dimensional (1D) and 2D techniques. A recent UK Water Industry Research (UKWIR) study has investigated the use of 1D and 2D modelling techniques (UKWIR, 2010a).

7.8. Failure modes

Failure of either the underground drainage system or the above-ground surface pathway systems within urban areas during major storm events can result in catastrophic flooding, with a worst-case scenario of risk to human life. Failure can result from many different mechanisms; however, the major failure mode is the rainfall event exceeding the design capacity of the system. Where the capacity of the underground system is exceeded, the hydraulic head within the system builds up, and once this hydraulic head exceeds the ground level, flooding from manholes, property connections and road gullies will occur. In addition, when the hydraulic head is nearing ground level, flows will be unable to enter the underground system, resulting in ponding, flooding and overland flows.

Typical examples of this type of flooding are illustrated in Figures 7.3 to 7.5.

Figure 7.3 Combined sewer manholes surcharging where a rainfall event exceeds the design capacity of the combined sewer system

Figure 7.4 Manhole on a combined sewer system surcharging where a rainfall event exceeds the design capacity of the combined sewer system. Flood water from this location flows down the main shopping street and increases the depth of flooding identified in Figure 7.5

Figure 7.5 Surface water run-off and combined sewage that has surcharged from manholes flowing down the main shopping street during an extreme rainfall event

Figure 7.6 Floodwater from an ordinary watercourse flowing across a main road during an extreme rainfall event. Shortly after this photograph was taken the road had to be closed due to the depth and velocity of the floodwater

In addition to the failure of the underground drainage system, both open rivers and watercourses can become overloaded during extreme rainfall events. Once the water level within the rivers and watercourses overtops the bank levels, flooding to surrounding land and properties will occur. An example of watercourse flooding is shown in Figure 7.6.

In addition to flooding occurring due to extreme rainfall events exceeding the capacity of the urban drainage system, in coastal urban areas flooding can be experienced as a result of tidal and wind conditions. During high tides and certain wind directions, sea defences can be overtopped, resulting in flooding to land and property behind these defences. An example of this type of flooding is shown in Figure 7.7. Where this type of flooding occurs in urban areas, the coastal floodwater can enter the underground drainage system via road gullies and private connections. This additional water can quickly overload the underground drainage system, and hence increase the risk of flooding from the underground drainage system.

Other failure modes that may cause flooding during storm events that do not exceed the design capacity of the system include the following.

- Collapses – In many urban areas the underground system is over 60 years old and the structural condition of the system is deteriorating. Should a collapse occur

Figure 7.7 Flooding as a result of high tides and easterly winds overtopping the existing coastal defences

(Figure 7.8), the capacity of the system will be reduced, and hence flooding can occur even during dry weather flows. As a result, it is important to inspect the critical underground network on a regular basis, and where deterioration is noted rehabilitation works should be undertaken. Further information regarding inspections and asset conditions can be found in the *Sewer Rehabilitation Manual* (WRC, 1994).

- Blockages – These can occur within the underground system due to poor maintenance, poorly designed systems and siltation, and as a result of debris entering the system. Should a blockage occur, flows will back up the system, eventually resulting in flooding (Figure 7.9).
- Poor maintenance – Many of the control assets identified previously require regular maintenance in order to function correctly. All assets should have an agreed maintenance regime and programme. Failure to undertake the maintenance could result in failure of the asset (Figure 7.10), and hence flooding to properties and open space.
- Loss of power – Where pumping stations, actuated flow-control devices or raked bar screens are installed there is a risk of failure due to loss of power. In order to minimise this risk, it is good practice that all these sites should be linked to a telemetry system in order that an alarm is received should a power failure occur. Every asset that requires power should have a risk assessment undertaken to assess how critical the asset is and how quickly power has to be restored prior to flooding or pollution incidents occurring. At the most critical assets it may be

Figure 7.8 Sewer collapsing increasing the likelihood of flooding upstream of the defect

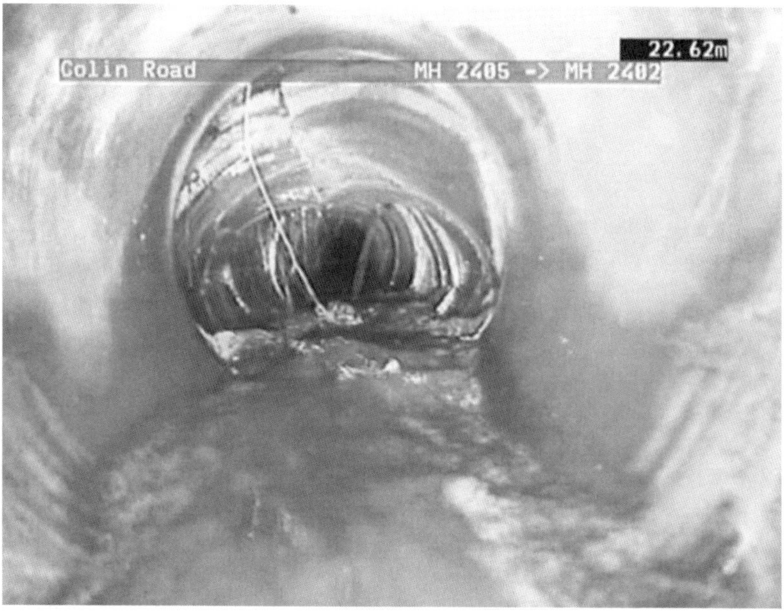

Figure 7.9 Sewer blocked by debris and fat deposits, increasing the likelihood of flooding upstream of the blockage

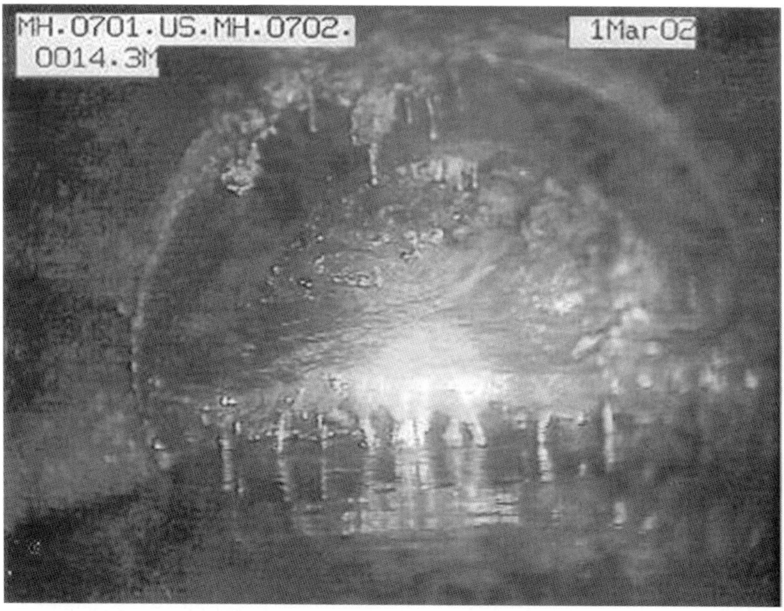

Figure 7.10 Poor maintenance of a trash screen on an ordinary watercourse, resulting in the trash screen being blinded and an increased risk of flooding

necessary to install an alternative energy source. At the less critical assets it may be sufficient to design the asset with a system for connecting a mobile generator should a power failure occur.

7.9. Design standards and assessment aids

The responsibility for drainage throughout England and Wales is divided between various organisations, each either having duties or powers for different elements. The principal organisations include local authorities, highway authorities, the Environment Agency, sewerage undertakers and riparian owners. The responsibility of each organisation is outlined below.

Under the Land Drainage Act 1991 (DoE, 1991a), local authorities are classified as the ‘Land Drainage Authority’ and they are responsible for managing the flood risk from ordinary watercourses (open and culverted). In addition, if the local authority is a coastal authority, it is responsible for the coastal flood risk. Local authorities have permissive powers to carry out improvement works to flood defences, and they provide an emergency response during major incidents. Under the Floods and Water Management Act 2010, county councils and unitary local authorities are designated as the ‘Lead Local Flood Authority’, and have statutory powers for dealing with flooding issues from all sources.

Highway authorities (usually county councils or unitary authorities) are responsible for ensuring that the highway is free from surface water flooding. Also, where ordinary watercourses pass under the public highway or within the boundary of the public highway, the highway authority is responsible for the maintenance of the watercourse, as the culverted watercourse is classified as being part of the highway structure.

The Environment Agency is responsible for the flood risk arising from main rivers and for flooding from the sea. It maintains, operates and improves flood defences, and provides a 24-hour flood warning service and emergency response.

Riparian landowners are identified as the landowners adjacent to a main river or ordinary watercourse, whether it is open or culverted. Riparian landowners have certain rights and responsibilities in relation to the main rivers or ordinary watercourses flowing through or adjacent to their property. These 'riparian rights' are based on common law, and have been defined through the results of legal cases over many years. A riparian landowner is responsible for the maintenance of the main river or ordinary watercourse through his property, and is not allowed to impede the flow of water along the length of main river or ordinary watercourse.

Sewerage undertakers are responsible for the public sewerage system that serves the urban areas and some rural areas. Normally, within England and Wales the sewerage undertaker is the local water company. The responsibilities of sewerage undertakers extend to the effectual drainage of foul and surface water flows from the land within the curtilage of a property. It should be noted that sewerage undertakers are not responsible for the effective drainage of surface water from the highway or land drainage. All sewerage undertakers within England and Wales identify a target level of service for protection against flooding from storm flows, and this is typically a 30-year return period storm event. The sewerage undertaker is not responsible for the drainage system within the curtilage of properties, as this will normally remain private and is the responsibility of the property owner.

As a result of the different responsibilities for drainage systems, it is extremely difficult to manage the effects of extreme storm events on the urban catchment area. Where flooding to a property occurs, it is sometimes difficult to assess whether or not the flood water has originated from land outside of a curtilage, a highway, the public sewer system or ordinary watercourse, and therefore it is uncertain which authority is responsible. As a result of the Floods and Water Management Act 2010, lead local flood authorities (county councils or unitary authorities) now have a statutory duty to investigate significant flooding events and identify which flood risk authority (highway authority, sewerage undertaker, etc.) is responsible.

It is evident from the above that different stakeholders are responsible for different aspects of the urban drainage system, but also for different levels of performance. There are various design guides to assist with the design and assessment of the urban catchment, and these recommend appropriate standards for the design. Table 7.2 lists some of the design guides, the performance standards and the stakeholder responsibility. It should be noted that, at present, there are no specific guidelines on the return period of events that should be used for exceedance; however, *Sewers for Adoption*, 6th edn (WRC,

Table 7.2 Drainage design guides

Design guide	Description	Return period event
Sewers for Adoption, 6th edn (WRC, 2006)	Pipes designed to run full – average ground slope greater than 1%	1 in 1 year
Sewers for Adoption, 6th edn (WRC, 2006)	Pipes designed to run full – average ground slope less than 1%	1 in 2 years
Sewers for Adoption, 6th edn (WRC, 2006)	Pipes designed to run full – sites with a severe consequence of flooding	1 in 5 years
Sewers for Adoption, 6th edn (WRC, 2006)	System should be designed not to flood any part of the site	1 in 30 years
Sewers for Adoption, 6th edn (WRC, 2006)	Adequate protection to property against internal flooding during extreme rainfall	1 in 100 years
Design Manual for Roads and Bridges (Highways Agency, 1996)	Surface water highway design	1 in 1 year
Design Manual for Roads and Bridges (Highways Agency, 1996)	No flooding of highway drains	1 in 5 years
Design Manual for Roads and Bridges (Highways Agency, 1996)	Design of channel drainage in highway	1 in 5 years
Sustainable Drainage Systems (SUDS). An Introduction (Environment Agency, n.d.)	No flooding to properties (as per PPS 25)	1 in 100 years plus an allowance for climate change
Construction Industry Research and Information Association (CIRIA): *Culvert Design and Operation Guide* (Balkham *et al.*, 2010)	Hydraulic design of culverts – no flooding to properties	1 in 100 years plus allowance for climate change
CIRIA: *The SuDS Manual* (Woods-Ballard *et al.*, 2007)	No flooding to properties Design of different types of SUDS	1 in 100 years Varies between 1 in 10 years and 1 in 100 years

2006) identifies that, for public sewers, the sewers should not flood for a 1 in 30-year event, and a check should be made that property flooding should not occur during the 1 in 100-year event. When considering a watercourse or main rivers, the design should be that no flooding should be predicted for the 1 in 100-year event.

In addition to the design guides identified above, there are numerous other design guides that can be used to assist with the design of the urban drainage system. Some of these include

- *Wallingford Procedure for Design and Analysis of Urban Storm Drainage* (HR Wallingford, 1981)
- *Flood Studies Report* (Institute of Hydrology, 1975)
- *BRE Digest 365: Soakaway Design* (Building Research Establishment (BRE), 1991)
- *Designing for Exceedance in Urban Drainage Systems – Good Practice* (Balmforth *et al.*, 2006)
- *The Fluvial Design Guide* (Environment Agency, 2010).

All design should be in accordance with BS EN 752: 2008 – Drain and sewer systems outside buildings.

7.10. Modelling equations and models

Until the beginning of the twentieth century, drainage systems were designed on the basis of an average rainfall intensity, which was assumed to be independent of duration. Following the collection of information on heavy rainfalls over short periods, the inverse relationship between average rainfall intensity and duration became well established. Lloyd-Davies was one of the first people to try and quantify this relationship for drainage-design purposes. In 1906, Lloyd-Davies introduced an approach to the design of drainage systems known as the 'rational method'. This method is based on the premise that every drainage area has a time of concentration (the time taken for flow from the furthest point on the catchment to reach the point being designed) and the peak discharge is assumed to occur when the entire catchment area contributes to the flow. The discharge is taken to be proportional to the volume of rainfall run-off during the time of concentration, and is expressed as

$$Q = CiA$$

where Q is the discharge rate, A is the upstream impermeable catchment area, i is the average rate of rainfall for time of concentration, and C is a coefficient.

This approach is known to give erroneous results under certain design conditions, and therefore further design techniques were developed, in particular for estimating the run-off hydrograph from the entire catchment area. This work resulted in the development of what is now known as the 'Transport and Road Research Laboratory (TRRL) hydrograph method', which was published in 1963. The choice of design procedure for storm drainage systems was provided in Road Note 35, which recommended that the rational method should be used for smaller schemes (outfall not exceeding 600 mm in diameter) and the TRRL method should be used for larger schemes.

The TRRL method was limited in the simulation of surcharged flows, surface flooding and antecedent catchment conditions, which caused problems in checking the performance of existing systems for deficiencies and in investigations into system performance for more severe storm events. As a result, in 1974 a working party was set up to investigate a new procedure for undertaking the design of drainage systems. This resulted in the publication of *Wallingford Procedure – Design and Analysis of Urban Storm Drainage* (HR Wallingford, 1981).

Since 1981, further works have been undertaken to improve the design procedures, and there are now a number of different computer programs that can be used to assist with the hydraulic modelling of both underground and above-ground drainage systems within urban catchments. Using these programs, a computer model of the underground system can be constructed and its performance can be assessed using either actual rainfall events or design rainfall events. In addition, some of these programs can be used to carry out 2D surface modelling, by incorporating a digital terrain model of the area.

Risk assessments, as required within a SFRA and identified in PPS 25, can be undertaken using these programs. These risk assessments can take the form of flood vulnerability maps, which are based on the risk to human life during flooding and are dependent on the predicted depth and velocity of the flood water. Further details relating to this can be found in *Flood Risk Assessment Guidance for New Development. Phase 2 – Framework for Assessing and Managing Flood Risk for New Development* (HR Wallingford, 2005).

In order to assist hydraulic modellers, within the Chartered Institution of Water and Environmental Management (CIWEM) there is a specialist urban drainage group called WaPUG, which has been responsible for the publication of a number of practice guides, design standards and technical notes. Samples of some of these publications, which are all available from the CIWEM website, are

- *Code of Practice for the Hydraulic Modelling of Sewers*, 3rd edition (CIWEM, 2002)
- *The Design of CSO Chambers to Incorporate Screens* – Version 3 (CIWEM, 2006a)
- *Guide to Quality Modelling of Sewer Systems* – Version 1 (CIWEM, 2006b)
- *River Data Collection Guide* – Version WO1 (CIWEM, 1999a)
- *River Modelling Guide* – Version W01 (CIWEM, 1999b)
- *Integrated Urban Drainage Modelling Guide* – Version 1 (CIWEM, 2009).

As mentioned earlier in this chapter, as part of the UK government's future water strategy for England and Wales, surface water management plans were proposed as a means of assessing the risk of surface water flooding. Within the Defra guidance for producing surface water management plans, integrated urban drainage modelling will be critical for the risk assessment and option components of the plan framework. In addition, integrated urban drainage modelling approaches can be used for a number of applications, including

- surface water management plans
- detailed analysis of the cause, effect and remedy of sewer flooding

- understanding of the impact of watercourse interactions on sewer system performance and the operation of CSO pumping stations and other sewer assets
- development of integrated flood risk plans for essential infrastructure and utility assets
- strategic flood risk assessments
- detailed flood risk assessments
- development of emergency response plans
- understanding pollution impacts on receiving waters
- climate change adaptation and carbon reduction strategies.

7.11. Case studies

Case study 1: Torquay integrated urban drainage pilot study

Historically, within the Torquay catchment area flooding of the main shopping streets was experienced on an annual basis. Due to the frequency of flooding being experienced, this catchment was chosen as a case study within the original *Sewer Rehabilitation Manual*. However, only the combined sewer system was studied and modelled. The study identified that the risk of flooding within Torquay town centre could be reduced by the installation of three attenuation tanks upstream of the town centre, having a total storage volume of approximately 13 000 m^3. These attenuation tanks were constructed in the early 1990s, and were designed for the 1 in 30-year critical storm event.

Since the construction of these attenuation tanks there have been four major flooding events within Torquay town centre, with the worst being recorded on 24 October 1999, when over 50 properties were flooded as a result of a 1 in 50-year storm event when over 50 mm of rainfall fell in 2 hours (Figures 7.11 to 7.13). As the drainage system for this catchment contains an ordinary watercourse, surface water drainage, combined sewers and highway drainage, an integrated flood study was undertaken as one of the 15 Making Space for Water pilot studies. This study was undertaken in a partnership between Torbay Council, the Environment Agency and South West Water.

For the first time an integrated flood study of the catchment was undertaken. The study included hydraulic modelling of the combined sewer system, surface water systems, the ordinary watercourse and the highway drainage systems. In addition to the underground system, 2D modelling was undertaken in order to model the overland flows.

The pilot study focused on the historical flooding that affects the Union Street and Fleet Street areas of Torquay. The study looked in detail at

- identifying key flooding problems using 2D hydraulic computer modelling techniques
- recommending the most cost-effective and sustainable solutions, taking into account the effects of climate change
- gauging public awareness and attitudes to urban flooding.

The existing drainage system in Torquay comprises combined and surface water sewers, an ordinary watercourse, known as the River Fleet, and highway drainage, with all parts

Figure 7.11 The Torquay catchment area, with the main study area outlined

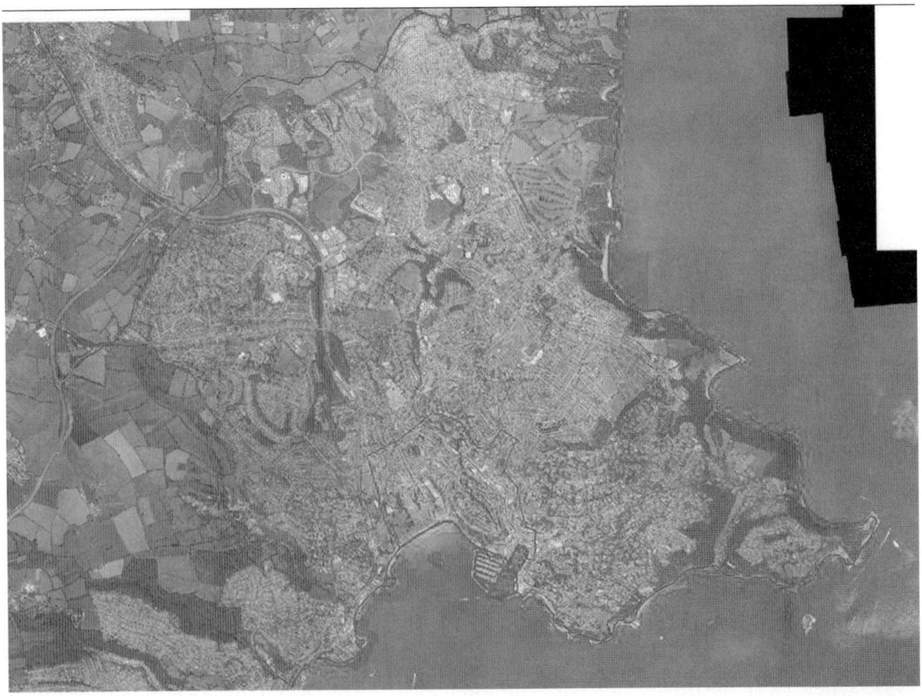

Figure 7.12 The main shopping street in Torquay in dry conditions. It should be noted that, when this street was pedestrianised, the kerbs were maintained to allow a surface flow channel in order to reduce the risk of flooding to properties

Figure 7.13 Flooding of the main shopping street in Torquay during an extreme rainfall event

of the system eventually discharging into the public combined sewer system. During heavy and prolonged rainfall the system becomes hydraulically overloaded, resulting in flooding to many commercial properties. The storms causing this flooding have been assessed as having a high return period, and therefore they are not targeted for improvement by Ofwat. In line with current design standards, without a major change in policy there is no existing driver that could be used to promote a major flood alleviation scheme. However, all the partners in this pilot study agreed that a target design standard for the town centre area of 1 in 100-year protection of property from internal flooding should be investigated.

The hydraulic model used for the study to represent the sewer network was verified in 1994, and so, in order to provide an acceptable representation of the major flooding events that had occurred since that date, the model was updated. As part of this update 'urban creep' (new impermeable areas at residential properties from paving over of gardens, building of conservatories, etc.) was investigated. This identified that, in parts of the catchment area, the effects of urban creep had increased the impermeable area by 8%. Further investigation work has been undertaken by UKWIR (2010b) on how the increase in impermeable areas will impact on the design of future sewerage systems. The results of these investigations are contained in the UKWIR publication *Impact of Urban Creep on Sewerage Systems* (UKWIR, 2010b).

A ground model using light detection and ranging (LIDAR) data was used to form the 2D digital terrain model. The LIDAR data were based on a 2 m grid; however, due to the location of buildings, canopies and narrow passageways that form above-ground flow channels, the level of accuracy of the LIDAR data was poor. As a result, further LIDAR data were acquired using the Environment Agency's Smart Rover equipment, which is a GPS-based system that provides levels of accuracy of approximately 30 mm.

By using the new LIDAR data and the updated sewer model, together with actual rainfall data for the recent flooding events, these events were modelled, and the results of this modelling accurately predicted the depth of flooding and number of properties flooded. This model was then used to identify the risk of flooding for a series of design storm events.

The flood risk problems were investigated using a range of hydraulic models and historic data. A combination of a 1D sewer network model and a 2D overland flow model was found to best represent the interaction of above- and below-ground systems and the flooding that Torquay experiences. The combined model has been used to assess the impact of design rainfall events at return periods of up to 1 in 100 years. This highlighted that flooding to properties in the town centre ranged from 50 properties during a 1 in 30-year event up to 97 properties during a 1 in 100-year event.

The study has shown that a range of integrated solutions will be required to improve the management of urban drainage. An action plan has been produced that includes the possibility of constructing a new large-diameter drainage pipe in Union Street and Fleet Street. Other measures included in the action plan include the reconstruction of Fleet Street with kerbs in order to transfer overland flows away from properties, separation of surface water flows from the combined sewer system, and the implementation of SuDS systems that return surface water run-off to the ground.

Assessment of the preferred option using hydraulic modelling predicted that the number of properties at risk from flooding in the 1 in 100-year storm event fell from 97 to one.

Key findings from the study include the following.

- To deliver an action plan, the three partners have worked together, sharing data and knowledge, and developing an agreed approach to promote integrated urban drainage management within the catchment. The partners have recognised the benefits of this joint working, and this has led to an agreement to extend the joint cooperative working on urban drainage management in Torbay beyond the completion of this pilot study project.
- Clarity of roles and responsibilities is required if multi-agency integrated urban drainage management schemes are to be delivered. This may need to include coordination of investment planning.
- The use of above-ground 2D flood flow modelling provided the final element of the catchment modelling needed to verify a suite of modelling tools that accurately replicates urban surface water flooding. The modelling provided a high

level of detail on the impacts of flooding, and allowed detailed analysis of proposed integrated urban drainage management options, including designed overland flow channels.

A copy of the final report on this project is available on the Defra website (Defra, 2008c).

Case Study 2: Torbay strategic flood risk assessment

The majority of ordinary watercourses and main rivers within Torbay flow through the urban catchment areas. Both highway drainage and surface water sewers have been connected directly to the watercourses and rivers, and as a result these systems form an integral part of the Torbay urban drainage system. Hence the identification of the flood risk areas associated with the watercourses and rivers is crucial for assessing the potential viability of land for future development and for the emergency planning process.

The SFRA provides a reference and policy document identifying the flood risk areas within Torbay, and it will be used to inform the local development framework. The SFRA provides clear guidance on the appropriate risk management measures for adoption on potential sites within flood zones 2 and 3, which are protected from flooding by existing defences. By using both the level 1 and level 2 SFRA it is possible for Torbay Council to meet its obligations under PPS 25, by applying the risk-based sequential approach to all stages of the planning process, and steering new development to areas at lowest risk of flooding. This means that land for development can be allocated in a sustainable manner.

All parts of Torbay are potentially at risk from flooding, or have the potential to make flood risk worse elsewhere. Some areas are at a higher risk of flooding than others; and many areas are at little or no risk. It is the combination of a number of factors that contribute to making an area at risk from flooding. These are: settlement location, including proximity to a main river, ordinary watercourse or the coast; climate; geology; and topography. The risk of flooding can become greater when there are extreme storm events or when taking account of the predicted effects of climate change.

It is essential that local development framework planning and development control policies address the risk of flooding. The level 1 and level 2 SFRA are the start of this process. They provide guidance on how to identify which areas lie within a flood risk area, and the implications for plan making and planning applications in taking this flood risk into account.

Within the level 1 SFRA, flood risk maps have been produced based on hydraulic modelling to identify the potential flood risk areas within Torbay. The flood maps identify the different flood zones as identified below.

- Flood zone 1 – land that has a low probability of flooding. This zone comprises land assessed as having a less than 1 in 1000 (0.1%) annual probability of river or sea flooding in any one year.

Figure 7.14 Flood Zones within Torquay. All areas not identified as Flood Zone 2 or 3 on the plan are designated as Flood Zone 1
Courtesy of Torbay Council

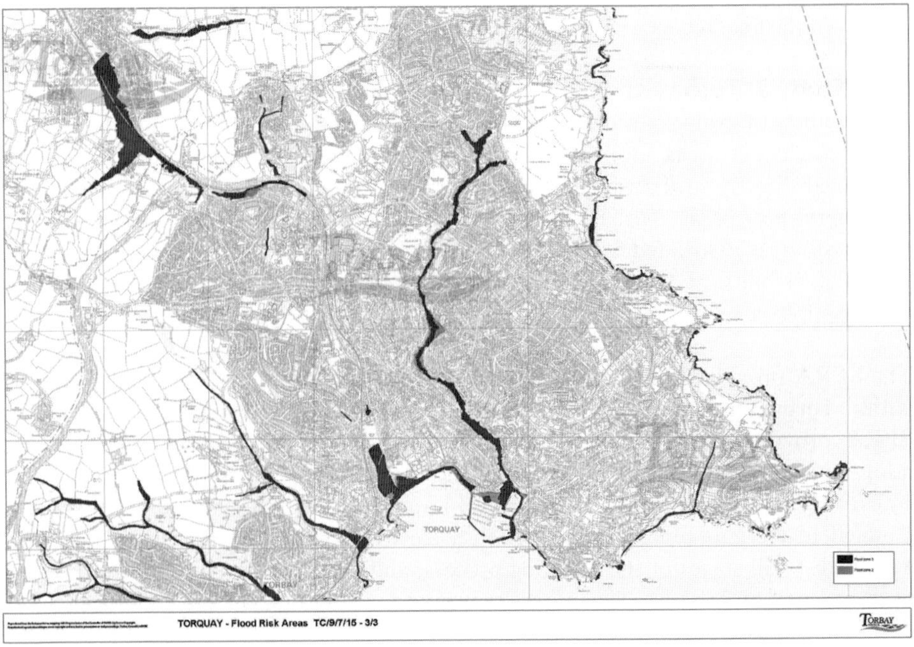

- Flood zone 2 – land that has a medium probability of flooding. This zone comprises land assessed as having between a 1 in 100 and 1 in 1000 (1–0.1%) annual probability of river flooding, or between 1 in 200 and 1 in 1000 (0.5–0.1%) annual probability of sea flooding, in any one year.
- Flood zone 3 – land that has a high probability of flooding. This zone comprises land as having a 1 in 100 (1%) or greater probability of river flooding, or a 1 in 200 (0.5%) or greater probability of flooding from the sea, in any one year.

Within the level 1 SFRA, flood zone 3 has not been divided into zones 3a (high probability) and 3b (the functional flood plain). This work will be undertaken as part of the hydraulic modelling works associated with the level 2 SFRA. A sample of the plans identifying the flood zones is identified in Figure 7.14.

The level 2 SFRA investigates the major flood risk areas further, and flood modelling work has been undertaken for a number of different scenarios to simulate flood events in order to determine and illustrate the potential flood events, depths and the areas at moderate, significant and extreme flood hazard. The scenarios that have been considered include heavy rainfall that may cause fluvial or pluvial flooding, together with breach and overtopping of flood defences. All of these forms of flooding have the potential to

Figure 7.15 Fluvial and pluvial flooding hazard map for Torquay town centre
Courtesy of Torbay Council

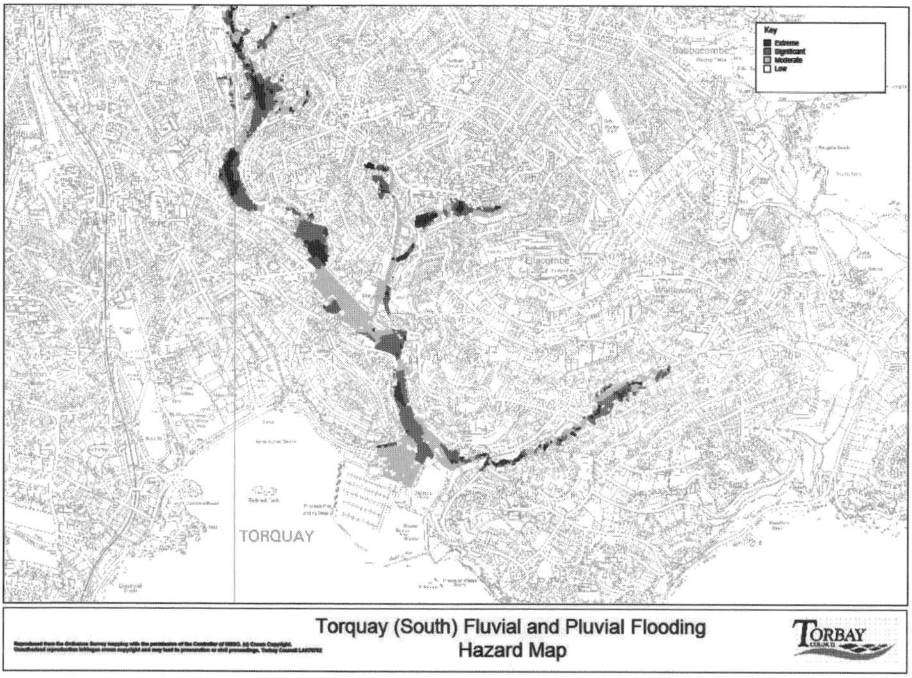

generate significant flood hazards and to cause damage to houses and infrastructure. The results of this modelling work have been identified on a series of maps contained within the final level 2 SFRA report, and a sample of these maps is shown in Figures 7.15 and 7.16.

Flood hazard is a function of the instantaneous flood depth and the velocity of the flood water, and as a result the maximum flood hazard for a given location can be expressed at any stage of a flood event. The flood hazard maps identify the maximum flood hazard predicted for each location during the specific flooding event being modelled. The flood hazard can be categorised as either low, moderate, significant or extreme, and a definition of each of these hazards is as follows.

- Low hazard – caution, flooding can occur, but the risk is low.
- Moderate hazard – danger for some, including children, the elderly and the infirm.
- Significant hazard – danger for most, including the general public.
- Extreme hazard – danger for all, including the emergency services.

In addition to the modelling works to identify the flood hazard and flood depths, hydraulic modelling work has been undertaken to identify the functional flood plain associated with each main river and ordinary watercourse in Torbay. In accordance

Figure 7.16 Predicted depth of fluvial and pluvial flooding in Torquay town centre
Courtesy of Torbay Council

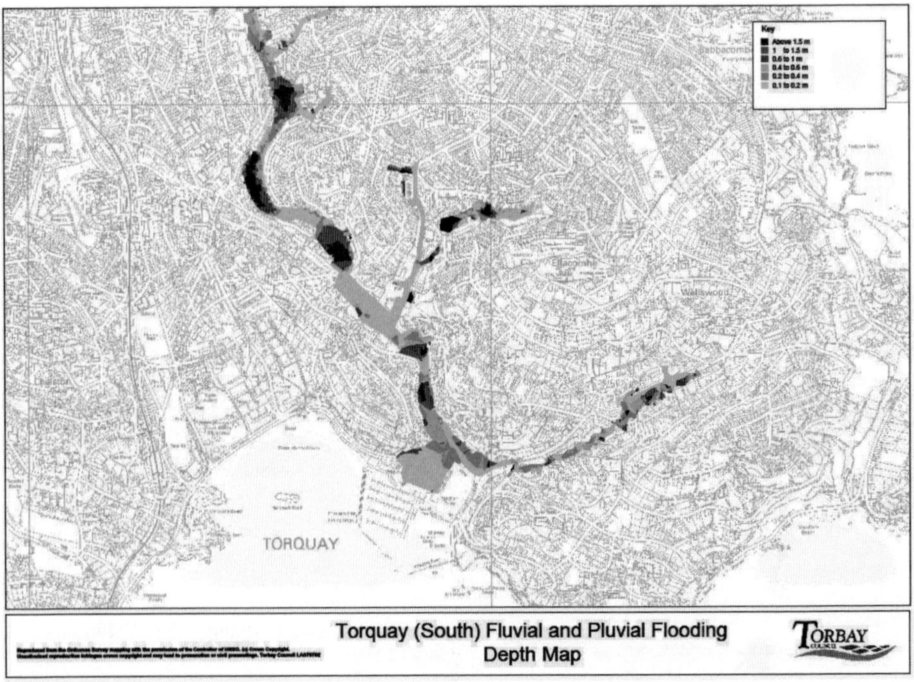

with PPS 25, these areas are identified as flood zone 3b. A sample of one of these plans is shown in Figure 7.17.

In order to mitigate the risk of flooding, the level 1 and level 2 SFRA documents make the following recommendations.

- Every planning application for development or change of land use must be considered by the planning department in terms of its potential flood risk.
- It is the developer's responsibility to provide a site-specific flood risk assessment with any planning application, as required in PPS 25.
- All site-specific flood risk assessments must be considered by Torbay Council and the Environment Agency as part of the planning consultation process, and their comments applied where possible.
- Land that is found to be unsuitable for one type of development due to its flood risk may still be suitable for other types of use. The PPS 25 guidance can be used to suggest suitable alternative land uses.
- The data contained within the SFRA constitute the best available data at the time of writing.
- The SFRA should be used to steer strategic growth and site-specific allocations in the local development framework to reasonably available sites at lowest flood risk.

Figure 7.17 Typical plan showing functional flood plain for a main river within Torbay. In accordance with PPS 25 the areas highlighted are designated as Flood Zone 3b
Courtesy of Torbay Council

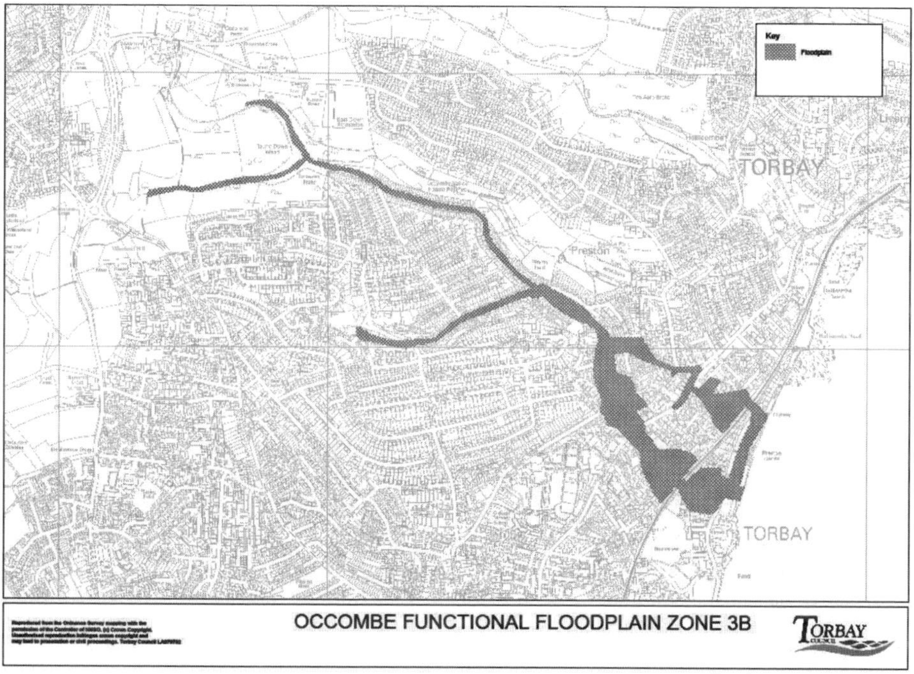

This includes investigating the impact of proposals for new development in the vicinity of, and particularly upstream of, areas sensitive to flooding.

- The Local development framework policies controlling development should reflect guidance in the SFRA documents. They may in future require amendment if the SFRA documents are updated.

7.12. Conclusions

In urban areas, all parts of the catchment area are potentially at risk from flooding or have the potential to make the flood risk worse elsewhere, even if they have been developed with a well-designed drainage system. Some areas are more at risk than others, and many areas will have little flood risk. The risk of flooding increases significantly during extreme storm events and when taking account of the predicted effects of climate change.

In order to assess the catchment area and identify the flood risks, an integrated urban drainage approach should be undertaken, where all aspects of the drainage within the catchment area are investigated and their flood risk identified. By using a risk-based approach when designing flood alleviation works on new developments, the risk of flooding can be significantly reduced. Wherever possible, it is recommended that surface water be discharged to a sustainable drainage system that mimics natural land drainage, rather than being connected to a sewer system. Where the use of a sustainable drainage system is not feasible, the discharge of surface water to the sewer system should be reduced by means of source control.

Failure to maintain assets, both within the sewer system and where sustainable urban drainage systems have been utilised, may lead to the deterioration of the asset and increase the chance of asset failure or flooding. Therefore, for all assets a risk assessment should be carried out that identifies the potential risks should failure of the asset occur. In addition, as part of this risk assessment, the recommended frequency of inspections should be identified, together with details of the routine maintenance works that should be undertaken on a regular basis. By undertaking the recommended inspections and routine maintenance works, the risk of asset failure will be significantly reduced.

REFERENCES

Balkham M, Fosbeary C, Kitchen A and Rickard C (2010) *Culvert Design and Operation Guide*. Construction Industry Research and Information Association (CIRIA), London, UK. Publication No. C689.

Balmforth D, Digman C, Kellagher R and Butler D (2006) *Designing for Exceedance in Urban Drainage Systems – Good Practice*. CIRIA, London, UK. Publication No. C635.

BRE (Building Research Establishment) (1991) *Digest 365: Soakaway Design*. BRE, Watford, UK.

BSI (2008) BS EN 752: Drain and sewer systems outside buildings. BSI, London.

CIWEM (Chartered Institution of Water and Environmental Management) (1999a) *River Data Collection Guide*. CIWEM Urban Drainage Group (WaPUG). Available at: http://www.CIWEM.org/groups/wapug (accessed 2 February 2012).

CIWEM (1999b) *River Modelling Guide*. CIWEM Urban Drainage Group (WaPUG). Available at: http://www.CIWEM.org/groups/wapug (accessed 2 February 2012).

CIWEM (2002) *Code of Practice for the Hydraulic Modelling of Sewers*, 3rd edn. CIWEM Urban Drainage Group (WaPUG). Available at: http://www.CIWEM.org/groups/wapug (accessed 2 February 2012).

CIWEM (2006a) *The Design of CSO Chambers to Incorporate Screens* – Version 3. CIWEM Urban Drainage Group (WaPUG). Available at: http://www.CIWEM.org/groups/wapug (accessed 2 February 2012).

CIWEM (2006b) *Guide to the Quality Modelling of Sewer Systems*. CIWEM Urban Drainage Group (WaPUG). Available at: http://www.CIWEM.org/groups/wapug (accessed 2 February 2012).

CIWEM (2009) *Integrated Urban Drainage Modelling Guide*. CIWEM Urban Drainage Group (WaPUG). Available at: http://www.CIWEM.org/groups/wapug (accessed 2 February 2012).

DCLG (Department for Communities and Local Government) (2010) *Planning Policy Statement (PPS) 25. Development and Flood Risk*. HMSO, London, UK.

Defra (Department for Environment, Food and Rural Affairs) (2003) Water Act. HMSO, London, UK.

Defra (2004) Making space for water: Developing a new Government strategy for flood and coastal erosion risk management in England. Department for Environment, Food and Rural Affairs, HMSO, London.

Defra (2008a) *Future Water: The Government's Water Strategy for England.* HMSO, London, UK.

Defra (2008b) *The Government's Response to Sir Michael Pitt's Review of the Summer 2007 Floods*. HMSO, London, UK.

Defra (2008c) *Making Space for Water. Integrated Urban Drainage Pilots. Torbay IUDM Pilot Study. Final Report*. Environment Agency/Torbay Council/South West Water. Project Code TRE 344. Available at: http://archive.defra.gov.uk/environment/flooding/documents/manage/surfacewater/torbayreport.pdf (accessed 2 February 2012).

Defra (2009) Flood Risk Regulations. HMSO, London, UK.

Defra (2010a) *Surface Water Management Plan – Technical Guidance Document*. HMSO, London, UK.

Defra (2010b) Floods and Water Management Act. HMSO, London, UK.

DoE (Department of the Environment) (1963) Water Resources Act. HMSO, London, UK.

DoE (1973) Water Act. HMSO, London, UK.

DoE (1989) Water Act. HMSO, London, UK.

DoE (1991a) Land Drainage Act. HMSO, London, UK.

DoE (1991b) Water Industry Act. HMSO, London, UK.

DoE (1991c) Water Resources Act. HMSO, London, UK.

DoE (1995) Environment Act. HMSO, London, UK.

DoE (1999) Water Industry Act. HMSO, London, UK.

Environment Agency (n.d.) *Sustainable Drainage Systems (SUDS). An Introduction.* Environment Agency, Bristol, UK. Available at: http://publications.environment-agency.gov.uk/PDF/GEHO0308BNSS-E-E.pdf (accessed 2 February 2012).

Environment Agency (2010) *The Fluvial Design Guide*. Environment Agency, Bristol, UK. Available at: http://evidence.environment-agency.gov.uk/FCERM/en/FluvialDesignGuide.aspx (accessed 2 February 2012).

European Commission (2007) EU Floods Directive 2007/60/EC. European Commission, Brussels, Belgium.

Halliday S (1999) The Great Stink of London: Sir Joseph Bazalegette and the Cleansing of the Victorian Capital. Stroud, Gloucestershire, Sutton Pub.

Highways Agency (1996) *Design Manual for Roads and Bridges*, Vol. 4, *Geotechnics and Drainage*, Section 2. HMSO, London, UK.

HR Wallingford (1981) *Wallingford Procedure for Design and Analysis of Urban Storm Drainage*. HR Wallingford, Wallingford, UK.

HR Wallingford (2005) *Flood Risk Assessment Guidance for New Development. Phase 2 – Framework for Assessing and Managing Flood Risk for New Development*. Environment Agency, Bristol, UK. R&D Technical Report FD2320, Technical Report 2.

HR Wallingford (2009) *System Based Analysis and Management of Urban Flood Risk – A New Procedure for Performance Assessment of Sewerage Systems*. Dti Sam, Report SR700. Available at: http://www.dti-sam.co.uk (accessed 2 February 2012).

Institute of Hydrology (1975) *Flood Studies Report*. Institute of Hydrology, Wallingford, UK.

Lloyd-Davies DE (1906) The elimination of storm water from sewerage systems. *Proceedings ICE* **164(2)**: 41–67.

Pitt M Sir (2008) *The Pitt Review: Learning Lessons from the 2007 Floods*. Cabinet Office, London, UK.

TRRL (Transport and Road Research Laboratory) (1963) *A Guide for Engineers to the Design of Storm Sewer Systems*. Transport Research Laboratory, Berkshire, UK. Road Note 35: 1st edn, 1963, 2nd edn, 1976.

UKWIR (UK Water Industry Research) (2010a) *IUD Demonstration Projects*. UKWIR, London, UK. Report No. 10/WM/17/1.

UKWIR (2010b) *Impact of Urban Creep on Sewerage Systems*. UKWIR, London, UK. Report No. 10/WM/07/14.

Woods-Ballard B, Kellagher R, Martin P, Jefferies C, Bray R and Shaffer P (2007) *The SUDS Manual*. CIRIA, London, UK. Publication No. C697.

WRC (Water Research Council) (1994) *Sewer Rehabilitation Manual*, 3rd edn. WRC, Swindon, UK.

WRC (2006) *Sewers for Adoption – A Design and Construction Guide for Developers*, 6th edn. WRC, Swindon, UK.

USEFUL WEBSITES

Chartered Institution of Water and Environmental Management (CIWEM): http://www.ciwem.org (accessed 2 February 2012).

Dti Sam: http://www.dti-sam.co.uk (accessed 2 February 2012).

Flood Risk
ISBN: 978-0-7277-4156-1

http://dx.doi.org/10.1680/fr.41561.171

Chapter 8
Dams and reservoirs – regulations, performance and past failures

Andy Hughes
Atkins, Epsom, UK

Dams are an important element of a nation's infrastructure in providing facilities for water supply, irrigation, hydropower production, navigation and flood alleviation. This chapter introduces the legislative and regulatory framework that exists in the UK, the types of dams and the inspection processes, past failures and modes of failure. In addition, inspection and design processes are discussed, as well as the topic of emergency planning and preparedness.

8.1. Introduction

This chapter introduces the subject of dams and reservoirs. Dams are an important part of every country's infrastructure, and can store water for the benefit of mankind for a number of reasons, including water supply, irrigation, hydropower, flood control, navigation and recreation. Dams, however, while being a useful asset, can become liabilities if not properly designed, monitored and maintained. Regulations and guidelines are then put in place as a means of trying to ensure reservoir safety. Unfortunately, there have been a few failures around the world, some notable ones causing significant numbers of people to lose their lives. However, there has been no loss of life due to a dam failing in the UK since 1925. This chapter seeks to describe some of the regulations, the types of dams and some of the past failures.

8.2. Legal and regulatory framework

In the Waterworks Clauses Act 1863 there was a clause whereby if a person was concerned about the condition of a dam they could complain to the justices of the peace who would then have the situation investigated.

Failures at Bilbury in 1852, and later in 1864 at Dale Dyke, led the coroner at the Dale Dyke inquest to recommend that dams should be subject to regular, sufficient and adequate inspection. Failures in 1925 at Skelmorlie and Eigeau raised public and professional concern about the need for legislation for dams, covering construction and subsequent operation, but it was not until 1930 that a specific Act relating to reservoir safety was written – the Reservoirs (Safety Provisions) Act 1930.

Due to this Act, for the first time owners were required to have their dams inspected by an independent engineer (a Panel Engineer), who would 'inspect' the reservoir and make recommendations relating to its safety. The Inspecting Engineer thus came into being – an engineer on a 'Panel' – or list – formed by experienced engineers who applied to the Secretary of State to be on the Panel.

In the UK, reservoirs are subject to legislation if they retain more than 25 000 m^3 of water above the level of the natural ground adjacent to the dam. The current legislation is the Reservoirs Act 1975 (Department for Environment, Food and Rural Affairs (Defra), 1975), which applies to about 2100 'large raised reservoirs'. The Act requires the owner or operator of the reservoir, called the 'undertaker', to register the reservoir with the Enforcement Authority. The Enforcement Authority in England and Wales is the Environment Agency, while in Scotland the Regional and Island Councils carry out the role.

An owner has the reservoir examined and reported on by an independent engineer, the Inspecting Engineer, at least every 10 years. The Inspecting Engineer inspects the reservoir and writes a report, which may include 'recommendations in the interests of safety'. These recommendations have the 'force of law' and must be carried out within a reasonable amount of time. The Inspecting Engineer also makes recommendations relating to issues of maintenance and surveillance, and the time to the next inspection – which can be no more than 10 years.

The owner is also required to employ a Supervising Engineer. This trained engineer is employed to examine the reservoir and keep the owner advised of its behaviour, and also to ensure that the owner is not in breach of the Act. This engineer can be a member of the owner's staff or may be a consultant. The Supervising Engineer would normally visit a dam, where the consequence of failure is large, at least two times a year, and will write a statement on the condition and behaviour of the dam. For any other dam the SE would visit at least once a year. If the Supervising Engineer is worried at any time, then he or she can call for an inspection from the Inspecting Engineer.

The Enforcement Authority 'polices' the Act and ensures that Inspecting Engineers are appointed at the correct time and that their reports and certificates are received, that the Supervising Engineer is appointed, and that works recommended in the interests of safety are completed within a reasonable timescale.

If an owner does not carry out works recommended in the interests of safety, the Enforcement Authority has the power to enter the owner's land and carry out the works.

At the time of writing, the Flood and Water Management Act 2010 (Defra, 2010) has been enacted but not commenced. This Act introduces the concept of risk or, more importantly, the concept of consequence of failure, and changes the capacity of the reservoir to which the legislation applies to 10 000 m^3. Those reservoirs in the range 10 000–25 000 m^3 are now called 'small raised reservoirs'.

The 2010 Act also introduces other changes, including the requirement to consider cascades of reservoirs where their combined capacity is in excess of 10 000 m^3. Incident reporting is also likely to become mandatory and recommendations are likely to become 'enforceable' (i.e. if the owner does not maintain the structure, the Enforcement Authority can arrange for the work to be done and then charge the owner for the works).

In the USA, a large numbers of the dams are owned by the United States Bureau of Reclamation or the United States Army Corps of Engineers (USACE). However, there are many tens of thousands of other dams spread across the 50 states, and regulations vary from state to state.

Regulations and legislation across Europe varies dramatically, from countries with comprehensive legislation to those with none.

8.3. Europe

This section reviews the findings of research undertaken into dam safety legislation and emergency planning practice across Europe, focusing on the procedures in Norway, Portugal, Sweden and Switzerland.

8.3.1 Legislation

Dam safety legislation and regulation in Europe has two key goals, namely: to minimise the initial risk to population, property and the environment; and to seek to control the remaining residual risk in an emergency situation. There are three aspects which in combination achieve these two goals: structural safety through design and construction; dam monitoring and maintenance; and emergency planning.

Many European countries have either specific national dam safety legislation, or more general national legislation that includes dam safety as one aspect. In general, the regulations are applied to a proportion of a country's dams, based on geometric criteria (dam height and/or capacity) and downstream hazard. The exact categorisation system varies across Europe, but is mostly used to inform design, monitoring and emergency planning requirements, and in some cases the authority responsible for ensuring the compliance of the relevant legislation and guidelines. Some countries have a national regulatory authority exclusively dedicated to dam safety, while others have authorities where dam safety is part of their broader regulatory responsibilities. In several European countries, regional (rather than national) administrations act as the primary supervisory authority on dam safety.

8.3.2 Norway

The Water Resources Act (Norwegian Government, 2000a) as approved by the Norwegian parliament is overseen by the Ministry of Petroleum and Energy, but empowers the Norwegian Water Resources and Energy Directorate (NVE) to oversee all issues relating to dam safety. The NVE issued the Regulations Governing the Safety and Supervision of Watercourse Structures (Norwegian Government, 2000b), outlining safety considerations for all phases of a dam's life cycle and containing requirements that have to be fulfilled, with enforcement mechanisms in the case of violation. NVE has also developed and issued subsidiary regulations and guidelines, setting out specifications and technical requirements in specific areas, for example,

Table 8.1 Norwegian dam classification based on consequence

Dam category	Category definition	Number of affected housing units	Approximate number of dams
Category 1	Low hazard	0	1700
Category 2	Significant hazard	1–20	540
Category 3	High hazard	More than 20	260

dam classification and flood calculation methodology. The Dam Safety Regulations are laid down by Royal Decree, but the subsidiary regulations are easily revised by NVE when necessary.

The Norwegian Regulations classify dams into three classes, based on the consequences of dam failure. Each consequence class is defined on the basis of the number of housing units affected by a dam break, with other buildings (e.g. schools and offices) converted into housing units if required. The three dam categories are presented in Table 8.1.

Potential economic and environmental downstream consequences are taken into account when classifying low-hazard dams, and when there are doubts about the final consequence category. If there are no consequences, the dam is unclassified and is not subject to public supervision.

The NVE Regulations are applicable to the following dams.

- All dams with a storage capacity exceeding 500 000 m^3.
- Category 2 and 3 dams exceeding 4 m in height.
- Category 1 dams exceeding 6 m in height.

8.3.3 Portugal

Portugal has specific national dam safety legislation, namely the Portuguese Regulations for the Safety of Dams (RSB) issued as a Decree Law in 1990 and revised and re-issued in 2007. This was the first piece of dam safety legislation in the country. The Regulations are complemented by four Codes of Practice covering the design, construction, observation/inspection and operation of dams. The Codes of Practice define the requirements and standards for the execution of the RSB. The Portuguese Code of Practice for Observation and Inspection of Dams (Portuguese Government, 1990) specifically provides methodologies for evaluating the safety of existing dams. The Codes of Practice are mandatory and constitute legal obligations for all owners with dams that fall within the scope of the regulations (International Commission on Large Dams (ICOLD), 2007).

The RSB applies to the following 'High Dams'.

- Dams exceeding 15 m in height.
- Dams with a reservoir capacity exceeding 100 000 m^3.
- Class 1 dams with a high damage potential in the downstream inundation zone.

The remainder are referred to as 'Small Dams' and are covered by The Regulations for Small Dams (rather than the RSB). Compliance with these Regulations is not strictly mandatory for dams less than 8 m high.

The four codes of practice relevant to High Dams use a further classification of dams into three classes, based on a Global Risk Index. This takes into account three factors: external and environmental conditions; dam condition and reliability; and human–economic hazards. Each of these factors is composed of a number of components that are evaluated by dam inspection. The index values are used to determine dam safety priorities and procedures. Class 1 dams have the highest damage potential in the downstream inundation zone.

There are three government departments involved in dam safety legislation.

- National Institute of Water (NIW) (part of the Ministry of Environment and Natural Resources) – the main dam safety authority, with responsibility for approving and supervising dam construction and operation, including compliance with the regulations.
- National Service for Fire and Civil Protection (SNBPC) – responsible for emergency planning and the coordination of rescue and relief.
- Commission for Safety of Dams (CSD) – prepares standards and provides opinions on dam safety processes across the country.

In addition, the National Laboratory of Civil Engineering (NLCE) provides technical advice and support to NIW and carries out studies for dam owners.

8.3.4 Sweden

Sweden has no specific dam safety regulations, but an Environmental Code includes generally formulated rules, including the stipulation that a dam owner has the main responsibility for measures to ensure dam safety and is strictly liable for the consequences in the case of dam failure. The county administrations are the supervisory authorities on dam safety and oversee the dam operator's normal dam management performance. The Water Rights Court gives permission for dam building and determines the compensation for a third person's property (Finnish Environment Institute, 2001).

Dams included on the national inventory are those that have a height exceeding 5 m or a storage capacity exceeding 50 000 m^3. Dams are also classified according to the potential downstream hazard, taking into account loss of human life, economic loss, and infrastructural and environmental damage.

In 1998, the Swedish Power Association established a voluntary risk classification scheme that was soon adopted by many hydroelectric dam facilities (Finnish Environment Institute, 2001). The classification is based on downstream consequences, and divides dams into one of four classes as presented in Table 8.2.

Guidelines were developed for each dam class, with different recommendations for operation and supervision for each. However, this classification is not contained in

Table 8.2 Swedish dam classification based on downstream consequences of dam failure

	Dam class	Consequence of dam failure
Loss of life or serious injury	1A	Obvious risk to life
	1B	Non-negligible risk to life
Damage to infrastructure, property or the environment	1A	Obvious risk of serious damage to important assets
	1B	Considerable risk of serious damage to important assets
	2	Non-negligible risk of considerable damage
	3	Negligible risk of considerable damage

Swedish legislation, and different dam owners can have different agreed standards, specifications and methods.

8.3.5 Switzerland

The federal law regarding Supervision of Hydraulic Structures was first passed in 1877 and was amended in 1953 to stipulate that the Federal Council was the highest supervisory authority responsible for dam safety for all the dams in Switzerland (Wieland and Mueller, 2009). An executive decree in 1998 enabled the delegation of this responsibility to the relevant cantonal authorities for the smaller Swiss dams. The regulations it contains are applicable to the following.

- Dams with a height greater than 10 m.
- Dams with a height greater than 5 m and a capacity exceeding 50 000 m^3.
- Dams that present specific safety concerns.

Smaller dams and/or those that do not represent a danger to people or property are not subject to this legislation.

The Swiss Federal Office of Water Management (Dams section) is responsible for the enforcement of the dam safety legislation, and is directly responsible for the supervision of the following 'large' dams.

- Dams with a height greater than 25 m.
- Dams with a height greater than 15 m and a capacity exceeding 50 000 m^3.
- Dams with a height greater than 10 m and a capacity exceeding 100 000 m^3.
- Dams with a capacity exceeding 500 000 m^3.

This amounts to 195 reservoirs and 217 dams. The several hundred dams that are smaller than these specifications are under the supervision of the cantons (states).

8.4. Australia

In Australia, dam safety regulation is undertaken on a state level, and legislation and regulations pertaining specifically to dam safety exist in most states. In general, the regulations are applied to dams greater than a certain capacity or height, or to dams that pose a threat to the community or the environment. The categorisation of the dams covered by the regulations is generally based on consequence and, in particular, the population at risk (PAR) downstream of the dam. The PAR is estimated based on the results of dam-break analysis.

The Australian National Committee on Large Dams (ANCOLD) is responsible for preparing dam safety guidelines, on which the Australian states base their regulations. ANCOLD issued its Guidelines on Dam Safety Management in 1994 (ANCOLD, 1994). These guidelines apply to 'referable' dams, which are defined as dams that are either 10 m or greater in height and have a storage capacity of 20 000 m^3, or are 5 m in height and have a storage capacity of at least 50 000 m^3.

ANCOLD suggests that the government is to enact legislation that includes who has the regulatory authority and responsibility to ensure dam safety. The legislation is to include criteria for classifying dams. ANCOLD also suggests that the regulatory authority has the power to ensure a dam is designed and operated in accordance with current standards in relation to dam operation, maintenance and surveillance. The authority is also to maintain a register of dams that includes information on the dam owner, type, construction and hazard category.

In the state of New South Wales (NSW), a separate body has the sole responsibility for the regulation of dams. This body was constituted by the Australian government under the dams safety act to ensure the safety of dams in NSW. The NSW Dam Safety Committee (NSW DSC) consists of part-time members from major dam authorities and the Institution of Engineers Australia. The responsibility of the committee is to ensure that all dams are in such a condition as to not pose an unacceptable danger to downstream communities. To do this, the NSW DSC requires dam owners to undertake regular monitoring and surveillance, ongoing assessments and regular reviews of compliance with current standards, and for the information to be reviewed by experienced personnel. If the committee feels a dam may be unsafe, or potentially unsafe, it will give notice requiring the dam owner to take relevant action. The NSW DCS regulates all 'prescribed' dams. Prescribed dams include those with high potential consequences (loss of life and damage) or, generally, those greater than 15 m high.

In the state of Queensland, the Water Resources Act of 2000 (Queensland Government, 2000) includes provisions for dam safety that are subject to the jurisdiction of the Chief Executive of the state Department of Natural Resources and Mines. These provisions state that a dam is referable if an accepted failure impact assessment demonstrates that there will be a PAR in the event of dam failure. An impact assessment is generally required if a dam is either more than 8 m high and has a storage capacity of more than 500 000 m^3, or is more than 8 m high and has a storage capacity of at least 250 000 m^3 and a catchment area no more than three times its maximum surface area at full supply. The Chief Executive also has the power to order an impact assessment for dams below these height and volume limits.

8.5. USA

In the USA, the legislation for emergency planning for dams varies depending on the owner and regulator responsible for the dam. In terms of the owner/regulator, the dams can be subdivided as

- federal dams
- non-federal, state-regulated dams
- non-federal, non-state-regulated dams (non-power, private small dams).

The main federal owners and regulators for dams are:

- the United States Bureau of Reclamation (USBR), which regulates many power and irrigation dams
- the United States Army Corps of Engineers (USACE), which owns and regulates mainly power dams
- the Federal Energy Regulation Commission (FERC), which regulates non-federal power dams.

Other dams are either owned privately and regulated by a state, or are both owned and regulated by a state. Depending on the regulation of the specific state, the state might also require to regulate the federal dams in its jurisdiction (Senate of the US, 2002).

The federal agency that supervises planning for all types of emergencies, including producing guidance for emergency planning for dams, is the Federal Emergency Management Agency (FEMA). FEMA is responsible for coordinating the federal emergency response (FEMA, 2007).

Different dam classifications are also adopted depending on the state and the federation. In general, dams are classified according to the hazard that they might pose, and categories include High, Significant and Low Hazard. High and Significant Hazard potential dams are defined as those 'dams which, if they were to fail, would be likely to cause loss of life or significant property damage' (FEMA, 2004).

In the state of Colorado, High Hazard dams are those the sudden failure of which would cause loss of life, and the State Engineer would request plans for these dams. Significant Hazard dams are those the sudden failure of which would not cause loss of life but would cause extensive property damage. The owner of the dam is responsible for damages caused by the uncontrolled release of water from a dam failure, and a written plan is recommended. Low Hazard dams are those the sudden failure of which would cause little loss other than to the structure itself. For these dams a written plan is not compulsory, but still it is recommended.

For the state of Washington, the classification applies to dams that can impound a 'volume of ten acre-feet or more and more than six feet of height' (State of Washington,

1991). Different responsibilities correspond to each class of dam. In those cases where a failure of the dam could pose a threat to life, the owner shall develop and maintain an emergency action plan (EAP) acceptable to the department. The owner is also responsible for coordinating the development of the EAP together with representatives from the local emergency services' staff, state department of community development, emergency management division, and appropriate local authorities. The owner is required to implement the EAP and to exercise each component of the plan.

The government department will coordinate and solicit review comments from the local emergency services' office and the state emergency management division on the acceptability of proposed EAPs. These comments will constitute the primary basis for accepting or requesting modifications to a proposed EAP. Owners are also required to notify the department when incidents occur, or when problems or conditions arise that may pose a threat to life or property or a threat to the integrity of the dam. If the government department envisages that an emergency condition exists that poses an imminent threat to life, the government may take actions to eliminate or mitigate the hazard and potential consequences. The dam owner can be directed to take actions, and if he fails to do so, the department may take control of the project and take actions, such as: '(a) Altering the operation of the project; (b) Lowering the reservoir water level; (c) Draining the reservoir; (d) Making emergency repairs or modifications to the project; (e) Enlisting the services of federal, state, or local authorities to make emergency repairs or modifications to the project; (f) Breaching the dam. All costs incurred by the department as a result of taking control of the project will be charged to the owner' (State of Washington, 1991).

For FERC-owned dams, the licensees need to adhere to specific regulations. The licensees of the dams have the responsibility to undertake regular planning inspections and to commission periodic inspections from independent consultants. The licensees must notify any issue or condition that compromises the safety of the project, and install monitoring and safety systems, such as warning systems, which they are responsible for maintaining. The licensees should provide maintenance and on-site public safety plans and also off-site EAPs, and these must be resubmitted each year (FERC, 2004).

8.6. Canada

Approximately 64% of the dams in Canada are currently covered by dam safety regulations. When regulations are enacted in Ontario, this number is likely to increase to approximately 80%. If legislation were implemented in Manitoba and Newfoundland, 90% of the dams would be covered by dam safety regulations (Canadian Dam Association (CDA), 2003). The CDA published the CDA Dam Safety Guidelines in 1995, revised in 2007 (CDA, 2007), the aim of which was to provide a basis for the development of provincial dam safety legislation and regulations. However, none of the provinces has entirely adopted the CDA guidelines in their dam safety regulations, although there are references to some aspects of the guidelines (CDA, 2003).

The CDA Dam Safety Guidelines propose a classification of dams according to dam dimensions and potential consequences of failure.

The classification based on dimensions differentiates between High Capacity and Low Capacity dams.

- High Capacity dams: 2.5 m or more high, with at least 30 000 m^3 of reservoir capacity, or 7.5 m or more high irrespective of the reservoir capacity, or 1 m or more high with a reservoir capacity greater than 1 000 000 m^3.
- Low Capacity dams: 2 m or more high, but not a High Capacity dam.

The second classification is based on potential consequences of a failure to people (in terms of fatalities) and socio-economic, financial and environmental consequences (in terms of damages). The dams are classified as Very High, High, Low and Very Low.

The CDA classification in terms of life safety is

- Very High (large number of fatalities)
- High (some fatalities)
- Low (no fatalities anticipated)
- Very Low (no fatalities).

In terms of socio-economic, financial and environmental consequences, the CDA (2007) classifies dams as

- Very High (extreme damages)
- High (large damages)
- Low (moderate damages)
- Very Low (minor damages beyond owner's property).

Not every Canadian region has fully adopted the CDA classification, although some regions have adopted an adapted version of it.

For example, dam regulation in British Columbia uses the same CDA classification in terms of dimensions to distinguish those dams to which the safety regulations apply. However, regardless of the dimensions, dams with a Low, High or Very High downstream consequence classification are included in dam safety regulation. The British Columbia classification based on potential consequences uses the same four categories of the CDA classification, but also contains detailed descriptions of the types of loss in each category in order to assist further dam classification. This classification is based on loss of life, social loss and cultural losses, but how these consequences are assessed in practice it is not specified (British Columbia Government, 2000).

The classification is self-assigned by the owner and checked by the official auditor appointed by the British Columbia government (British Columbia Ministry of Environment, 2008). The dam owner declares a particular level of 'consequence rating' and 'failure probability rating', and the auditor assesses whether these levels are appropriate. Both the 'likelihood of failing' rate and the 'consequence' rate are audited, and this implies that the government has adopted an approach that is risk based and involves both updatable monitoring of the state of the dam and a full understanding of the possible consequences of a failure.

Alberta has adopted the CDA dam classification in terms of incremental consequences of failure (life safety, and economic, social and environmental losses) and the four consequence categories from Very Low to Very High (Alberta Environment Regional Services, 2003).

In Quebec (Quebec Government, 2002), dams are classified in terms of dimensions and potential consequences in the case of failure. The classification in terms of dimensions is the same as the CDA classification. The classification on the basis of the degree of risk posed to persons and property is measured through the value P, which is obtained by multiplying the dam vulnerability (V) by the potential consequences of a dam failure (C). The regulations specify how these values are to be calculated, the vulnerability V depending on the physical characteristics of the dam (such as dimensions, age, dam condition) and the consequence C being determined through a detailed analysis that includes

- dam failure analysis
- mapping of the inundated area using a recognised method, and to a level of detail depending on the potential consequences
- characterisation of the area affected by the failure.

The Quebec legislation (Quebec Government, 2002) also prescribes that the consequence category must be determined or reviewed prior to authorisation for construction, structural alteration, change in use or stopping of the operation of a dam, and following a dam safety review.

Regarding responsibilities, British Columbia legislation (British Columbia Government, 2000) prescribes that dam owners are responsible for inspection and reporting 'to ensure that their dams are being maintained to a standard of care and due diligence that minimises the risk associated with their dams'. This legislation applies to both new and existing dams. The dam owner must prepare and submit an Emergency Preparedness Plan if the consequence classification is High or Very High, and an Operation Maintenance and Surveillance Plan for Low, High and Very High consequence dams. These plans are to be approved by the dam safety officer.

Within the legislation for Alberta (Alberta Government, 2000), an emergency plan must be prepared for all dams, except those in the Very Low and Low consequence categories, to ensure that appropriate action will be taken to manage emergencies and mitigate the effects of a dam failure.

8.7. Dam types

There are numerous types of dams, and their classification is often not very well defined. A broad classification can be made in terms of the construction materials used.

- *Embankment dams* are constructed of earthfill, rockfill or both. The structure is essentially a triangular-shaped body of fill, with upstream and downstream slopes of

Table 8.3 Principal types of dams

Group	Type
Embankment dams	Earthfill Rockfill
Concrete dams (including masonry dams)	Gravity Arch Buttress Multiple arch

a similar and moderate angle. As a result, these dams have a wide base, imposing low stresses on the foundation, and involve large volumes of fill material.

- *Concrete dams* are constructed of mass concrete. The downstream face is usually very steep, and the upstream face is usually vertical or near vertical.

The principal dam types within these two generic groups are listed in Table 8.3 and discussed in turn below.

8.7.1 Embankment dams

Embankment dams are numerically dominant for both technical and economic reasons and account for over 85% of all dams built in the UK and worldwide.

The embankment dam can be defined as a dam constructed from natural materials excavated or obtained nearby. Often a quarry is opened up within the reservoir basin to win material, as for every cubic metre of fill excavated one cubic metre of storage is required, and this approach reduces the environmental impact of the dam construction. The materials available are used to the best advantage in relation to their characteristics as bulk fill in the zones within the dam section. A typical cross-section for an embankment dam is shown in Figure 8.1.

Figure 8.1 Typical cross-section for an embankment dam
Courtesy of the British Dam Society

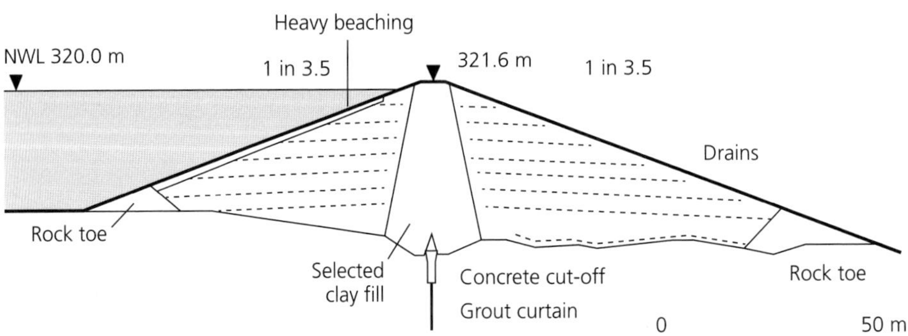

Figure 8.2 An embankment dam
Courtesy of ICOLD

The fill materials are placed and compacted, using high-capacity mechanised plant. Embankment construction is consequently an almost continuous and highly mechanised process, and plant rather than labour intensive. As a result, the unit cost of embankment dam construction has remained essentially the same over the years. A typical embankment dam is shown in Figure 8.2.

As stated above, embankment dams fall into two classes – earthfill and rockfill. However, the division between the types is often not distinct, with many modern dams utilising all the materials available on site in designated zones within the embankment.

An embankment dam is usually defined as one in which compacted soils account for over 50% of the placed material. The materials used are usually engineering soils compacted uniformly in relatively thin layers (150–300 mm) at a controlled moisture content.

A rockfill embankment (Figure 8.3) usually has a discrete impervious element to make the structure capable of retaining water. This membrane can be clay, bituminous materials or, in some cases, concrete, either as a core or an upstream facing. A rockfill dam is defined as a dam in which 50% of the fill material may be classified as rockfill, i.e. coarse-grained frictional material. Modern practice is to specify a graded rockfill and compact it in relatively thick layers (500–1000 mm) by heavy plant.

Figure 8.3 A rockfill dam
Courtesy of ICOLD

The construction method of a rockfill embankment is therefore very similar to that of an earthfill embankment but, because the slopes of the shoulders are steeper, there is less material in a rockfill embankment compared with an earthfill embankment.

The terms 'zoned rockfill' and 'zoned earthfill' are used to describe embankments incorporating relatively wide impervious or semi-impervious zones. Rockfill dams can employ a thin membrane of clay, asphalt, or reinforced concrete to create a watertight element.

The embankment dam has many advantages, which ensure its dominance.

- It uses natural materials local to the site.
- It uses a highly mechanised almost continuous construction process.
- Its costs have remained essentially the same over many decades.
- It is suitable to most dam sites – wide or narrow.
- It can adapt to a wide range of foundation conditions.
- The design is extremely flexible.

The only disadvantages of an embankment dam are that it is susceptible to damage or destruction by overtopping flows and. perhaps, in some circumstances is susceptible to piping failure. Piping failure is a situation where leakage increases through the fill such that a natural pipe is formed, releasing the water from the reservoir. Careful design and construction can reduce the risk due to piping, and filters can be retrofitted to dams to try to stop piping progressing to failure. Sufficient spillway capacity and or freeboard can protect against overtopping.

A recent development, particularly favoured in some parts of the world, is the concrete-faced rockfill dam where a reinforced-concrete face is formed on the upstream face to form the waterproof membrane.

8.7.2 Concrete dams

Concrete gravity dams rely entirely on their own mass for stability. One of the earliest forms of concrete gravity dams involved masonry, either as rubble masonry or random masonry dams. Later, masonry-faced cyclopean dams appeared, where large individual stones or displacers were used to displace concrete in the body of the dam and thus reduce cost. One of the last dams of this type to be built in the UK was the Venford Dam, which is owned by South West Water and situated near Holne on the edge of Dartmoor (Figure 8.4).

The next phase in the development of gravity dams was mass concrete, sometimes with displacers (large lumps of rock dropped in the concrete to displace the concrete) in order to use less concrete, for the ease of construction of complex shapes and also because of the costs of masonry and the time involved in construction.

In the UK, mass concrete and concrete dams of all types were favoured during the hydro-power development era of the 1950s in Scotland. Gravity dams, buttress dams and arch dams are all present. Additives, such as pulverised fuel ash (PFA) and ground blast furnace slag (GBFS) started to be introduced from this time, to reduce costs and to control potential problems with heat generation.

Figure 8.4 Venford Dam
Courtesy of South West Water

Figure 8.5 Cross-section of a gravity dam
Courtesy of the British Dam Society

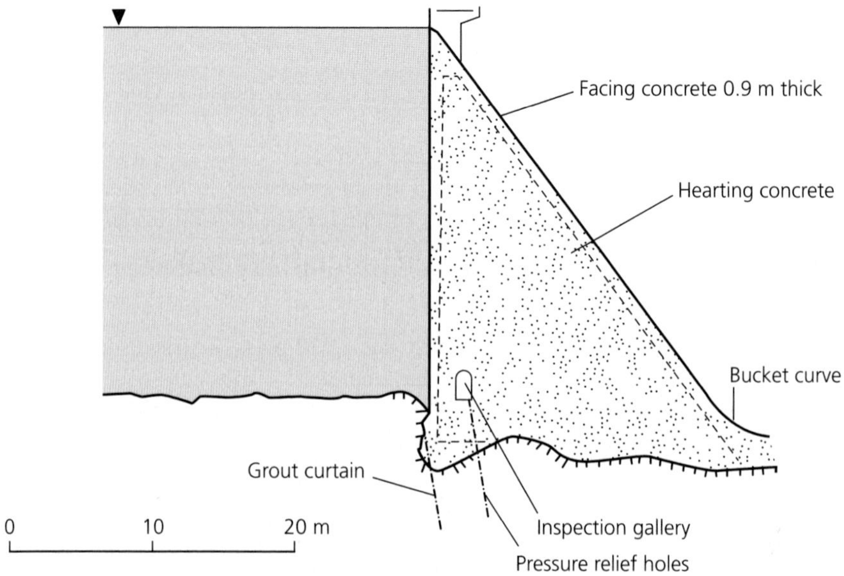

The main forms of concrete dam construction are

- gravity dams
- roller compacted concrete (RCC) dams
- arch dams
- double-curvature arch dam (or cupola)
- buttress dams.

The structure of a *gravity dam* is entirely dependent on its own mass for stability. A typical concrete gravity dam section is shown in Figure 8.5. The shape of the dam is essentially, and in its simplest form, triangular, with a vertical upstream face and a downstream face with a slope of 1 vertical to 0.8 horizontal. Some gravity dams are built with a 'heel' (an extension via a concrete-filled trench) and a foundation cut-off (usually formed by a grout curtain). A typical concrete gravity dam is shown in Figure 8.6.

A modern variant of the mass concrete gravity dam is the *roller compacted concrete (RCC) dam* (Figure 8.7), which is built from a low-strength concrete spread in layers, usually by bulldozers, and then compacted with rollers. The upstream and downstream faces, particularly where overtopping can occur (where a spillway is formed), are sometimes formed from pre-cast units or stronger/grout-enriched concrete, to provide increased durability.

If the dam is built to a slight curve in order to take advantage of the arching action of the curve, it is then known as an *arch gravity dam*. An *arch dam* (Figures 8.8 and 8.9) usually

Figure 8.6 Gravity dam
Courtesy of ICOLD

Figure 8.7 Roller compacted concrete
Courtesy of ICOLD

Figure 8.8 Concrete arch dam
Courtesy of ICOLD

has a significant curvature towards the reservoir. Structurally it functions as an arch, with the reservoir loading providing a thrust into the abutments. It is therefore a prerequisite that the abutments are of a good quality in order that they are able to take the loading, and abutment stability is critical to the safety of the structure.

The arch is a very efficient structure, allowing less concrete to be used and enabling a thin dam to be formed. A more complex structure than the basic arch dam is the *double-curvature arch dam*, or *cupola* (Figures 8.10 and 8.11), which has arches from top to bottom as well as across the valley from side to side.

A *buttress dam* has an upstream concrete 'wall' which depends on the support of a series of vertical supports called 'buttresses'. The buttresses run along the downstream side of the wall and usually have a slope of about 45°. The buttresses serve to transfer the load of the reservoir on the wall down into the foundation (Figure 8.12). The reduced downstream contact area of the buttress dam imposes significantly higher foundation stresses than occur in the equivalent gravity structure. Again the foundation can be sealed via a grout curtain.

The advantages of concrete dams include the following.

Figure 8.9 Cross-section of an arch dam
Reproduced courtesy of the British Dam Society

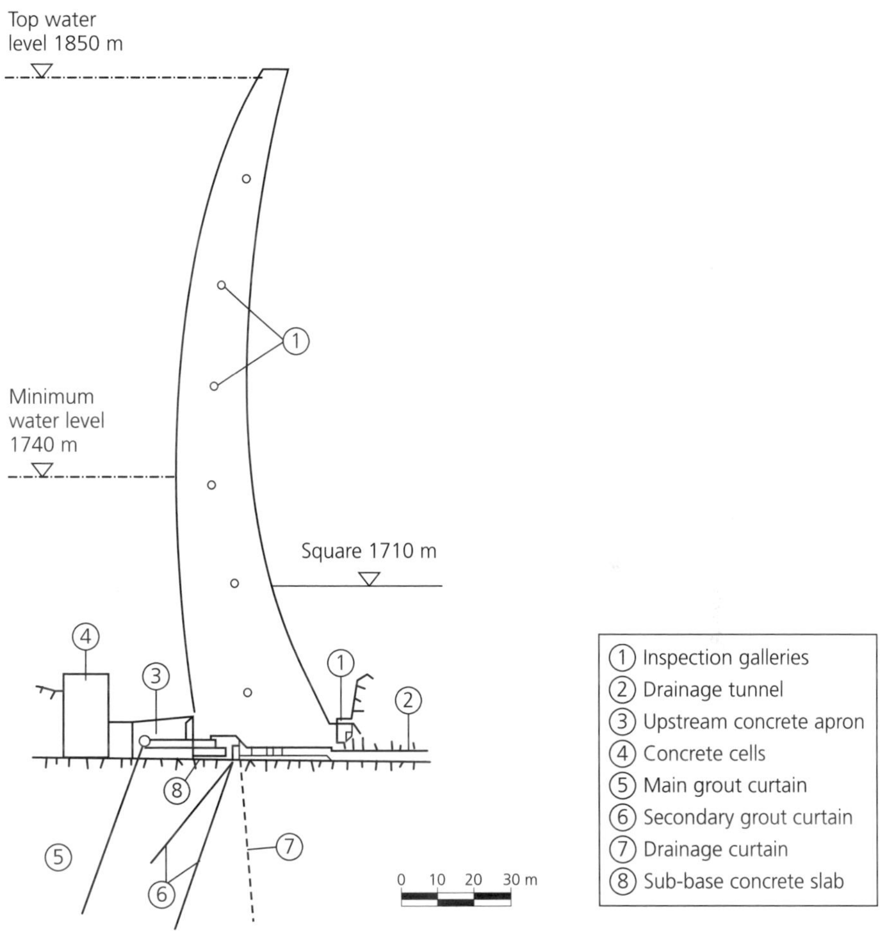

- They are not susceptible to damage by overtopping.
- Pipework, valves, conduits, chambers and galleries can all be accommodated within the body of the dam.
- The spillway can be created as part of the dam, avoiding the cost of a separate spillway.
- They can be built within wide or narrow valleys, as long as a competent rock foundation exists.

The disadvantages of concrete dams include the following.

- They are demanding with respect to their foundations – requiring a rock foundation.

Figure 8.10 A double-curvature arch dam
Courtesy of ICOLD

Figure 8.11 A double-curvature arch dam
Courtesy of ICOLD

Figure 8.12 A buttress dam
Courtesy of ICOLD

- They require processed natural materials as aggregates in large quantities and the importation of cement and PFA, etc.
- They are relatively discontinuous and labour intensive, requiring skilled labourers to make the formwork and place the concrete.
- The unit cost of concrete is significantly higher than the unit cost of embankment fills.

8.8. Inspection approaches

Frequent monitoring and surveillance of dams is vital to ensuring the continuing adequate performance of dams and to avoid failure. A surveillance programme is designed to avoid failure by identifying a problem sufficiently early to enable action to be taken. This allows actions to be taken to try to stop the failure process, to take remedial action, and to give early warning to people downstream and to allow evacuation.

Regular monitoring during first filling is particularly important, as statistics show that the 'early life' of a dam is often the most critical period in its lifetime, and many problems become apparent at this stage. At this stage, one might consider daily visits to the dam, whereas as the dam becomes more mature one might reduce the frequency of monitoring and surveillance. However, the frequency of monitoring must be chosen to reflect the type of dam and the possible failure modes that might exist which would influence the probability of failure. Other important criteria would be the consequence of failure, with a more frequent programme for high-consequence dams, the size and capacity of the dam, and its current condition. In the UK, it is normal practice for high-consequence and high-risk earthfill dams to be examined at least every 48 hours, with some very high consequence dams visited daily. However, it is the owner's responsibility to adopt an appropriate frequency for their dam, although they might get expert advice from an engineer. It is important that those carrying out the monitoring and surveillance programme are trained and aware of the early signs of a potential problem.

One might devise a two-level surveillance programme comprising a detailed, high-level examination using relevant equipment and techniques, and a second, more frequent, visual examination aimed at detection of unusual changes during the period between more thorough examinations.

The primary purpose of the programme is to monitor the condition and performance of the dam and its surroundings and identify any changes that may be occurring.

The procedure for dam safety surveillance is unique to each dam but consists essentially of regular, close and systematic examination of the entire surface of the dam, where exposed, and the surrounding area.

A safety surveillance inspection includes taking appropriate measurements, perhaps from instruments, both simple and sophisticated, and making concise and accurate records of observations made.

The visual observation and monitoring would include

- reservoir level
- catchment
- the condition of all faces of the dam, whether it be an embankment dam or a concrete dam
- reservoir margin areas
- inlet works
- outlet works
- ancillary structures – towers, shafts and tunnels
- spillways
- instrumentation
- downstream areas.

Instrumentation in a dam can be regarded as part of the toolbox designed to achieve reservoir safety. Readings from the various instruments, which might include level pins on the crest or slopes, piezometers, V-notch weirs or inclinometers, must be taken at regular intervals and the readings must be taken accurately. Readings should be related to reservoir level and rainfall. Readings should be analysed soon after they are taken, and they should, usually, be plotted against rainfall and reservoir level so that changes and trends can be identified.

Traditional methods of dam engineering have always been focused on performance integrity. But safety is not an absolute condition, and various levels of risk will always remain. Society is now asking that risks are identified and managed, and so risk assessment is one technique that seeks to ensure uncertainties are explicitly taken into account. The inclusion of risk assessment in the process of dam safety management is often referred to as a risk-informed approach which enables one to help in the decision-making process and the management of dam risks.

8.9. Past failures and the associated risks

This section reviews a number of notable failures and associated risks. The case histories of failures are used to demonstrate a number of different types of failure and reasons for failure.

A review of incidents, accidents and failures allows engineers to learn from their mistakes, and this is one of the main reasons why incident reporting is likely to become mandatory under the new UK legislation.

8.9.1 Bilberry – internal erosion driven by a managed spring

In 1836, in Holmfirth, a group of master clothiers and other manufacturers met to try to resolve their problems of a shortage of water to operate their waterwheels. They undertook to construct eight reservoirs to regulate the flows of the Rivers Holme and Colne and to seek parliamentary powers to enable the dams to be built.

In 1839, the first contractor started work at Bilberry, and brought the dam to full height. The dam had a maximum height of 67 ft (20.4 m) and comprised a clay core and a clay-filled cut-off. The core was 8 ft (2.4 m) high at the top and 16 ft (4.8 m) wide at the bottom. The dam was built from the Kinderscout grit (of the millstone grit series of rock), and it is likely that weathered shale was used for the puddle clay. The spillway was formed as a 15 ft (4.5 m) diameter shaft connecting with a culvert under the embankment.

On filling, the dam leaked very badly. The contractor made attempts to seal the leak, but then a dispute arose and a second contractor was engaged to try to repair the reservoir between 1843 and 1845. On the 5 February 1852, shortly after 1 a.m., and some seven years after the second contractor had left site, the dam failed. The resulting flood wave killed 81 people and caused significant damage and destruction in the Digley valley and Holmfirth. The disaster caused widespread unemployment, as most of the mills in the area were out of commission or destroyed.

An inquest was held after the failure and it was found that a spring was encountered in the bottom of the cut-off trench during construction. It appears that, rather than isolating and diverting the spring, it was just covered with puddle clay. In addition, it appears that there was a bad leak into the masonry culvert, with muddy water entering the culvert. At the inquest the engineer was quoted as saying that he had dug down to the culvert and replaced the poor masonry, but the leak was not stopped.

Later the leak got worse and burst through the walls of the culvert. Settlement of the top of the embankment was severe, being some 2–3 in. (50–75 mm) a day, and was not corrected. Settlement continued to more than 10 ft (3 m), by which time the top of the embankment was lower than the crest of the overflow, and there were also reports of a sinkhole on the upstream side above the line of the culvert.

In late January and early February there was heavy rainfall. Eye-witness reports state that water was seen running into the sinkhole and then the dam started to overtop. Within hours the dam had failed, causing death and destruction downstream (Binnie, 1981).

In conclusion, it appears that a process of internal erosion caused by a spring that was not bottomed out led to fill being removed, and settlement of the crest then occurred until the level of the crest was below that of the spillway weir. The dam was then overtopped and failed.

8.9.2 Dale Dyke – insufficient engineering skill and attention

In 1853, the Water Company obtained parliamentary powers to construct Dale Dyke and Agden Reservoirs. Dale Dyke embankment was situated some 9 miles west of Sheffield on the River Loxley, a tributary of the River Don. The dam had a maximum height of 95 ft (29 m) and a crest length of 1254 ft (382 m). It had a top width of 12 ft (3.6 m) and slopes on both sides of 2.5 to 1. It included a narrow puddle-clay core, which was 4 ft (1.2 m) wide at the top and 16 ft (4.8 m) at the base, with side slopes of 1 in 16. The puddle-filled cut-off trench was 60 ft (18 m) deep. The shoulders were constructed from shale fill, apart from in the upper portions where it was described as 'rubble, stone and shale', and there was a rockfill zone in the downstream toe (Figure 8.13).

Figure 8.13 Cross-section of Dale Dyke reservoir (Binnie, 1981)

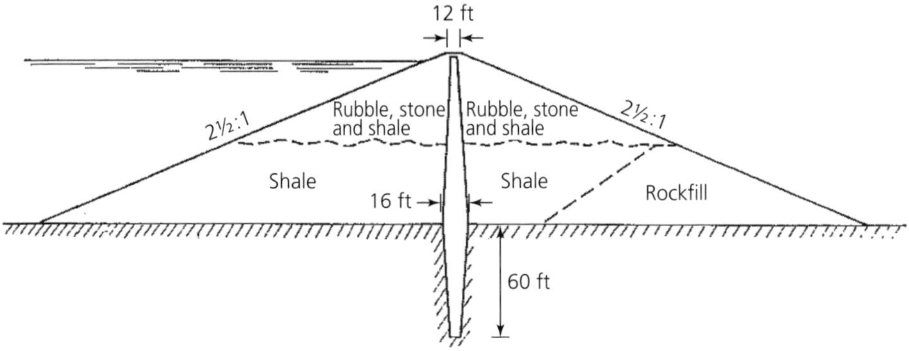

A bottom outlet chamber was constructed, from which twin outlet pipes were laid in a straight line diagonally beneath the embankment. They were cast-iron spigot and socket pipes 1.5 in. (40 mm) thick, 18 in. (450 mm) in diameter, in 9 ft (2.7 m) lengths.

The reservoir started filling when a heavy flood destroyed the diversion channel in June 1863. Filling proceeded intermittently, but by 10 March 1864 the water level was within 2 ft 3 in. (685 mm) of the crest of the weir, and still rising. During the latter part of the afternoon of the 11 March, a horizontal crack along the downstream slope near the crest was observed, and at 11.30 p.m. the same day, by which time the reservoir was more or less full, the embankment suddenly collapsed with no warning. Nearly 250 people lost their lives, and there was considerable loss of life and damage to property in Sheffield.

A number of reasons were suggested for the failure. One was that the shoulders were not compacted properly and that differential settlement between the shoulders and the clay core caused the core to crack, which led to leakage and erosion, and then failure, due to instability of the downstream face. Another theory was that the pipes had cracked due to the weight of the fill or had pulled apart, causing leakage, erosion and then instability of the downstream face. Binnie (1981) felt that there was also evidence that a spring was encountered in the puddle trench which may not have been dealt with properly, or that abrupt steps in the puddle clay core trench would have caused the core to tear and therefore lead to erosion and subsequent instability. On balance, the evidence would seem to suggest that springs in the foundation were again the cause of the failure. The jury at the coroner's enquiry expressed the opinion that there had been insufficient engineering skill and attention to the construction of the works.

8.9.3 Skelmorlie – cascade failure

In April 1925, the lower of two reservoirs that supplied the town of Skelmorlie in Ayrshire failed. The failure caused the death of a woman and four children.

It is said that heavy rainfall filled a quarry situated just upstream of the reservoir, which suddenly emptied and the spillway to the main dam was unable to cope. As a result, water overtopped the embankment and caused failure and the attendant loss of life (Binnie, 1981).

8.9.4 Eigiau and Coedty (Dolgarrog) – foundation failure leading to a cascade of collapse

On 2 November 1925, the failure of two dams caused a flood that swamped the village of Dolgarrog, killing 16 people. The disaster was started by the failure of the Eigiau Dam, a gravity dam owned by the Aluminium Corporation. The upper dam failed by water blowing a hole through the foundation beneath the dam. The water released from the reservoir flooded downstream, and overtopped the Coedty Dam, an earthfill embankment dam with a concrete core. This dam also subsequently failed as a result of the overtopping, releasing the combined volume of the two reservoirs onto the village of Dolgarrog. The breach in the upper dam is shown below in Figure 8.14.

Figure 8.14 The breach in Llyn Eigiau Dam, showing the gully cut by the flood water. A second, different breach was later deliberately made in the main wall to prevent it happening again
Sourced from Wikimedia Commons. Reproduced under the terms of the Creative Commons Attributions-ShareAlike 2.0 Generic License

Many more villagers could have been killed had they not been in the local theatre watching a film that night. This disaster and the earlier failure of the dam at Skelmorlie generated sufficient pressure from the civil engineering profession and the public to lead the British Parliament to pass the Reservoirs (Safety Provisions) Act 1930, which introduced laws on the safety of reservoirs.

8.9.5 Carsington – failure to properly prepare the foundation

Carsington Dam was under construction and nearing completion when a slide of the upstream shoulder occurred in June 1984. The slip occurred over a length of some 500 m. The dam was 1225 m long and had a maximum height of 35 m. It had a central, boot-shaped core, with shale shoulders.

The failure was caused by a layer of soft yellow clay that was left in the foundation and through which a slip surface formed as the embankment was constructed. The embankment was rebuilt after removal of some of the clay and by adopting shallower slopes (Skimpton and Vaughan, 1993).

8.9.6 Malpasset – insufficient site investigation leading to poor design choices

The Malpasset Dam was an arch dam constructed on the Reyran River, sited approximately 7 km north of Fréju, in Southern France. The dam failed on the 2 December 1959, killing 421 people in the area downstream. The cost of the damages was estimated as $68 million.

The entire wall of the arch collapsed, with only a few blocks remaining on the right bank. Pieces of the dam are still scattered throughout the area. The breach created a massive wall of water, 40 m high and moving at 70 km/h, which destroyed two small villages (Malpasset and Bozon) and a highway construction site, and took only 20 minutes to reach the village of Fréju, where the flood wave was still 3 m (10 ft) high. Various small roads and railway tracks were destroyed as the flood wave made its way to the sea.

Geological and hydrological studies were conducted in 1946 and the dam location was considered suitable for an arch dam. Due to a lack of proper funding, however, the geological study of the region was not thorough. The rock underlying the dam is known to be relatively impermeable, meaning that there is no significant groundwater flow within the rock unit and it does not allow water to penetrate the ground.

Weeks before the breach, some cracking noises were heard in the area, but they were not investigated. It was also known that in November 1959 some leaks had appeared on the right abutment. A tectonic fault was later found to be the most likely cause of the disaster. Water collected under the dam and was unable to escape through the rock. Water pressure built up and was directed along a joint towards the abutment. In addition, an engineer local to the area suggested that explosions carried out during building of the adjacent highway might have caused shifting of the rock at the base of the dam.

Between 19 November and 2 December there was 50 cm of rainfall, with 13 cm of this falling in the 24 hours before the breach. The water level in the dam was only 28 cm away from the overflow when the owner wanted to open the discharge valves to lower the level, but the authorities refused, claiming the highway construction site would be in danger of flooding from the water release. However, five hours before the breach, at 6 p.m., the valves were opened, but only with a discharge rate of 40 m^3/s, and this was insufficient to empty the reservoir in time.

Investigations after the failure found that the left abutment had moved into the hillside and downstream as the load was applied to the arch. Movement of more than 500 mm had occurred, which would have been enough to lose the arching action of the structure, after which failure was inevitable. The remains of the structure are shown in Figure 8.15.

8.9.7 Vajont – failure to consider all possible loading conditions

The Vajont Dam (or Vaiont Dam), now disused, was completed in 1959 in the valley of the Vajont River beneath the Monte Toc mountain, some 100 km north of Venice, Italy. In 1963, a huge landslide involving more than 200 million m^3 of fill caused the overtopping of the dam by more than 100 m and around 2000 deaths in the villages on the other side of the valley. The wave was caused as the landslide displaced the water in the reservoir. Surprisingly, the dam did not fail, and still stands today.

Vajont is a tall concrete arch dam some 262 m high, 27 m thick at the base and 3.4 m wide at the top, where piers carried a road bridge across the valley. It seems that the overtopping incident occurred because the stability of the side of Monte Toc seems not to

Figure 8.15 Malpasset Dam after the failure
Sourced from Wikimedia commons. Reproduced under the terms of the Creative Commons Attribution-Share Alike 3.0 Unported license

have been taken into account. During first filling, warning signs in the form of cracking and movements were seemingly not acted on quickly enough, and this eventually led to a massive landslide, which caused the overtopping event and catastrophic flooding and destruction to the Piave valley below, wiping out several villages completely. When those monitoring the slope decided to open the valves to empty the reservoir, they were too late. The mountain was moving into the basin quicker than water was being released from the reservoir. It is said that, just before the failure of the slope, all the animals and birds moved off the hillside.

The dam still stands today (Figure 8.16) but, instead of a reservoir existing behind the dam, there is a hillside with mature trees, It has been calculated that the dam was loaded to eight times its design load, and yet there appears to be no damage. The only damage sustained by the dam itself was that the crest bridge was torn off. The picture in Figure 8.16 shows the walkway the owners have installed on top of the damaged bridge piers to provide access across the top of the dam.

8.9.8 Taum Sauk

The Taum Sauk pumped storage plant is located in Missouri, USA. The pumped storage hydroelectric plant was designed to help meet peak power demands during the day, and involves a lower reservoir and an upper reservoir. Construction of the Taum Sauk plant began in 1960, and operation began in 1963.

Figure 8.16 Vajont Dam after the failure
Sourced from Wikimedia Commons. In the public domain

There had been minor leaks in the reservoir since it was constructed. A 'pumpback' station had eventually been installed to collect and return leakage to the reservoir, and in 2004 remedial works were instigated by installing a membrane. On the morning of 14 December 2005 the upper reservoir failed catastrophically when a triangular section on the northwest side of the reservoir washed away, releasing 4 million m^3 of water in 12 minutes and sending a 7 m high wall of water down the Black River. The breach is shown in Figure 8.17. The reservoir was operated by pumping water up to the reservoir in periods of low demand for electricity, and using the water to generate power when there was a high demand for power. In normal operating conditions, alarms and trip mechanisms would be in place to stop pumping to the upper reservoir before it became too full, and certainly before overtopping of the embankment took place. On this occasion it appears that the alarm levels, and the actual water level, were not being recorded properly. As a result, the pumps continued to run even though the reservoir was full. The water overtopped the walls, washing out the downstream shoulder, and eventually leading to failure.

Figure 8.17 Taum Sauk with the upper reservoir breached
Sourced from Wikimedia Commons. In the public domain

There was no overflow spillway in the original reservoir, and the reservoir would have overflowed in approximately 72 minutes once the maximum level was exceeded. It was likely that the reservoir failed once water started to overtop. Fortunately, no fatalities were caused by the failure, even though an employee of the firm lived close by. Aerial photographs before and after the failure are shown in Figure 8.18.

8.9.9 Summary

It is clear that we can learn from incidents and accidents that happen at dams. There are a number of publications that give information that will enable people to learn from the mistakes of the past. In the UK, an incident reporting system has been set up by the Environment Agency, which has had some success in gathering information and disseminating it in the form of briefing sheets to owners and Panel Engineers so that all become aware of the incident and the lessons that can be learned. However, not all incidents are reported, and thus there is a proposal currently under consideration to make the reporting system a mandatory requirement under the Flood and Water Management Act 2010.

8.10. Failure modes and analysis methods

The failure of a dam is normally considered to be a major uncontrolled and unintended release of retained water, or an event where a dam is rendered unfit to safely retain water because of a total loss of structural integrity.

Figure 8.18 Aerial photographs of Taum Sauk showing the upper reservoir before and after failure
Sourced from Wikimedia Commons. In the public domain

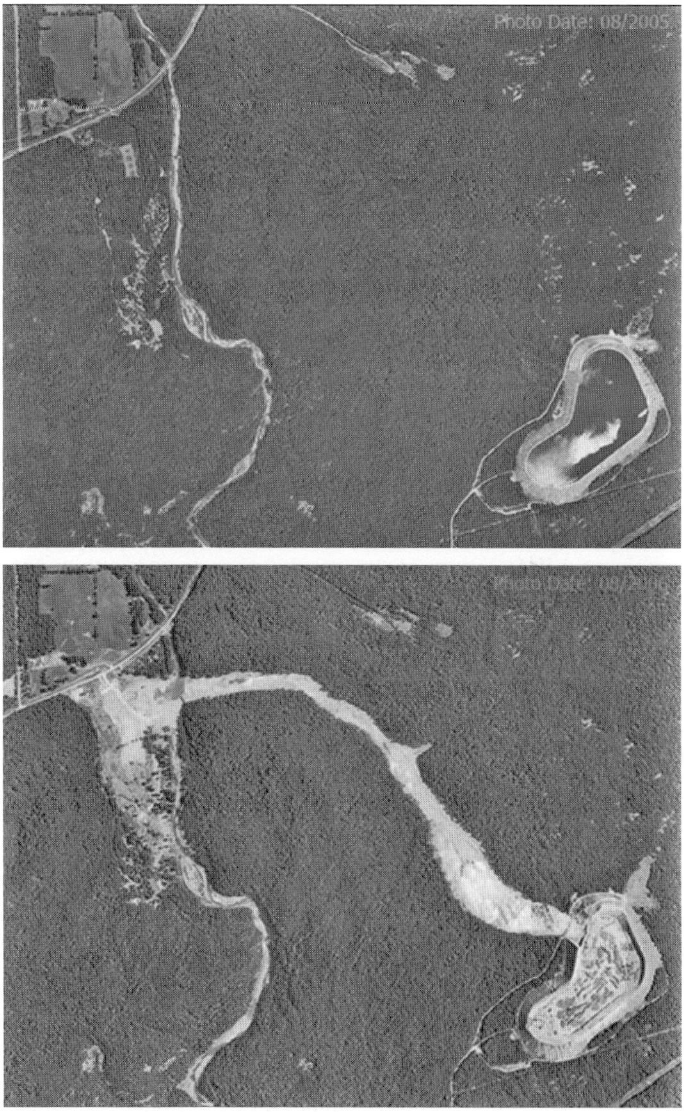

The main failure modes associated with embankment dams are: overtopping, instability of the upstream or downstream faces, excessive seepage or leakage, and piping failure. The failure modes are described in the Environment Agency (2011) document *Modes of Dam Failure and Monitoring and Measuring Techniques*.

Techniques are available with which to review and consider failure modes that a particular dam may experience. A useful aid to the understanding, piping failure, for example, is the USACE (2008) publication *Risk Analysis for Dam Safety, A Unified Method for Estimating Probabilities of Failure of Embankment Dams by Internal Erosion and Piping.*

Traditionally, methods of analysis have been based on performance integrity, producing a stock of dams that have performed generally well according to a series of defined standards. However, society now often demands that risks are explicitly identified, assessed, kept under review, and controlled. The use of probabilistic methods of risk assessment enables an improved understanding of the level of risk faced and the effectiveness of risk-reduction measures. Risk assessment allows the evaluation, usually at a screening level, of the absolute risk posed by an individual reservoir to the public downstream. The risk so determined can then be assessed for tolerability against current standards for public safety in other high hazard industries and/or other reservoirs.

8.11. Design and assessment aids

The design of dams has changed significantly over time as our understanding of materials and geology, and analytical techniques, have changed. Historically, embankment dams were designed as homogeneous dams with no drainage, and in some cases not even a cut-off. As time has moved on dams have become more sophisticated. Early developments were the inclusion of a puddle-clay core and a cut-off supported by shoulders of coarser material to provide support. Modern construction now seeks to use all materials available on a site and to form ‘zoned’ structures.

In designing a structure, it is the practice to consider the failure modes likely to occur, and to design the structure and its elements to safeguard against the principal modes of failure. A prerequisite would be an information study and site survey to find a suitable site for the dam, and to assess that sufficient suitable material is available for construction. The construction materials should be suitable for the fill of the embankment, be fairly impermeable when compacted, be available in sufficient quantity for the selected dam, be relatively easy to win, transport and place, and to be close to the dam.

The basic principle of design is to provide a functioning and safe structure, but often this needs to be achieved at minimum cost. Any design is site specific, and must take into account all aspects of the site. The principal requirement is to ensure that the structure is safe and stable during all phases of construction and operation. Criteria for design include the following.

- The dam must not impose excessive stresses on the foundation.
- The embankment slope must be stable under all conditions.
- Seepage beneath and through the dam and abutments must be controlled so as not to affect the safety or function of the dam.
- The dam must be made safe against overtopping by provision of adequate spillway capacity and freeboard for a defined level of risk.
- The upstream slope needs to be protected against wave erosion.
- The crest and the downstream slope need to be protected against erosion and damage from rainfall and wave spray.
- The embankment must be protected against livestock and animal damage.

Table 8.4 Factors of safety

Design loading condition	Factor of safety	
	Downstream slope	Upstream slope
Under construction	1.25	1.25
Long term – reservoir full	1.5	1.5
Rapid drawdown		1.2
Seismic loading	1.1	1.1

8.12. Stability analysis

The stability of the embankment must be assessed in relation to the changing conditions of loading and seepage that develop from construction through first filling and into operational service, including rapid drawdown of the reservoir. The methods of analysing slope stability are detailed in any soil mechanics textbook, and range from the analysis of circular slip planes to the analysis of non-circular slip planes. All methods are based on the consideration of static equilibrium of the potentially 'active' mass of soil overlying a possible failure surface. The stability of an embankment is highly sensitive to the water pressures in the fill, and these can be predicted from pore pressure parameters, field data, or a seepage flownet. Load conditions at the end of construction – steady state (reservoir full), rapid drawdown and under seismic loading – would normally be analysed. The analyses are usually carried out using effective stress parameters, rather than total stress parameters, which are only suitable for short-term situations.

The factors of safety that are normally used are shown in Table 8.4.

8.13. Seepage

Seepage through the embankment and foundation is of concern because excessive seepage could cause a process of internal erosion and possible failure of the embankment. Erosion usually proceeds slowly at first, with slow-flowing water carrying away the finer particles of the soil. However, at some point the rate of erosion will increase rapidly and perhaps lead to a collapse of a section of embankment. This process is often called 'piping failure'.

Seepage will always occur as it is impossible to build a completely watertight embankment. However, to ensure the stability of the embankment, the foundation and the embankment should be designed together to limit the seepage to acceptable and safe amounts, perhaps by providing a cut-off and suitable drainage and filtering features.

The fill material must be adequately compacted in order to maximise its density and reduce the amount of air and air/water filled voids. Compaction will also increase strength, decrease permeability and minimise post-construction settlement.

When considering concrete dams, the primary modes of failure are overturning and sliding on the foundation, overstress and material failure, and failure processes associated with the erosion of the foundation. The analysis of modes of overturning and sliding failure involves determining a number of forces acting on the structure. These include the water load, the self-weight of the structure, seepage and uplift loads, sediment load, hydrodynamic wave loads, ice load, thermal-effect loadings and seismic loads. The seepage and uplift forces are established because of water penetration along discontinuities in the foundation. It is usual to assume a linear or, if drains are installed, a bilinear pressure distribution.

8.14. Emergency planning and assessment aids

Emergency planning and emergency preparedness is about having a plan of what to do in the event that failure of the dam appears imminent or that the dam fails. The Water Act 2003 introduced the concept of a Flood Plan for a dam, and as part of the new legislation dam owners will be directed to assemble parts of a Flood Plan.

A Flood Plan involves three elements: an on-site plan, an inundation map, and an off-site plan. The philosophy of the approach is that pre-planning for an emergency is an important means by which to minimise the impact of a dam failure.

Owners can play an important role in ensuring the safety of their dams through adequate surveillance, monitoring, inspection and maintenance, but they should also have a plan of action in case the dam fails or is threatening to do so. It is imperative to issue an effective and timely warning to downstream residents if the dam is about to fail. The better the quality of the warning and the earlier the warning is given, the less likely it is that injury, death and damage will occur.

8.14.1 On-site plans

On-site plans are prepared by the reservoir owner. They provide useful information about the dam's location, construction and capacity, and the operation of its valves and other key equipment should there be a problem at the site. They also provide information on access to the site, which mobile phone networks work at the site and places where pumps might be set up, and give information on where vital resources (pumps, sandbags, etc.) might be obtained. Information on the rate of drawdown is also presented so that it can be gauged how quickly the hazard can be reduced.

In general, the on-site plan also

- provides clear definitions of the roles, responsibilities and actions of each agency at particular stages of the on-site response
- provides a response escalation procedure and the actions to be taken as part of the incident management, from the initial alert, to full plan implementation through to stand-down
- sets out the links to the off-site plan and the coordination and control arrangements for each level of response, which should be agreed with the relevant agencies from the Local Resilience Forum

- specifies the manner in which information should be communicated to staff and partners in an accessible and consistent fashion
- provides contact details to facilitate an efficient call-out of resources.

It should be assumed that

- the plan will be used by people who are unfamiliar with the reservoir (e.g. a temporary supervising engineer, a newly appointed inspecting engineer, the emergency services and the Environment Agency)
- the Prescribed Form of Record is available
- any existing drawings of the dam(s) are available
- information about catchment diversions may be available
- information on valves and other means of reducing the water level is available.

The on-site plan should state where all the statutory information (water level observations, leakages, settlement, etc.; names and addresses of the Inspecting Engineer, Supervising Engineer; certificates and reports and details of works, including direct and indirect catchment area details; other relevant information, reports, drawings, records, etc.) is kept.

8.14.2 Inundation mapping

Inundation mapping is an essential part of any emergency planning process. The maps provide outline areas of flooding, the depth and velocity of the water at any point, and the speed of travel.

The modelling techniques available have improved dramatically over the years, and there are now a number of one- and two-dimensional models available to assist in identifying the areas affected and the characteristics of the flow. As with any type of analysis there are areas of uncertainty, and different models have different strengths and weaknesses (Environment Agency, 2009). Uncertainty can exist in the actual information relating to the terrain and the valley downstream, and any constraints within it such as roads, motorways or rail embankments. It may also be associated with the physical properties of the dam and its foundations, and in particular how the dam fails and how quickly it fails. The assumptions about how the dam fails and how quickly the dam fails are fundamental to the results obtained, and so it is important that dam engineering professionals are involved in the choice of parameters used in the analysis.

Clearly one of the benefits of having an inundation map for the particular dam is that the area affected can be identified. Inundation mapping can show the area which will actually be flooded, but this will vary from areas where the water is very shallow and things just get wet, through to areas of deep slow-moving water, but also fast moving waters. The response by the emergency services will vary, and therefore the emergency services have been supplied with maps that give details of the depth of water and the depth of water at any particular point. This enables estimates of the extent of damage to be made, whether it involves structural damage to housing, and whether, for example, roads are washed away. Other valuable information includes the time of travel, which obviously gives an emergency service the ability to target its efforts where it has time to react.

8.14.3 Off-site plan

These plans are prepared by Local Resilience Forum (LRF). The LRF is a group of organisations including the police, the fire brigade, the ambulance services, the local authority, the Environment Agency, etc., all of whom would have a part to play in an emergency. Off-site emergency plans detail how emergency responders will respond to a potential or real reservoir failure. In the UK, the police will act as the coordinator, but obviously other emergency organisations and disaster relief organisations will have to be involved.

The emergency services are also interested in whether their vehicles are able to pass through the waters where a road is inundated.

8.15. Conclusions

At the World Summit of the United Nations General Assembly in 2005, eight Millennium Goals were presented for achievement by 2015 to eradicate poverty, improve education, improve health, combat disease and ensure environment sustainability. Clearly, dams and reservoirs will continue to play a major role in the management of the world's water resources.

Effective management of this resource by dams and reservoirs is essential to sustaining both the existing and future populations of the world. Dams are able to provide water for water supply, irrigation, hydropower generation, river regulation and navigation, as well as playing a role in flood control and alleviation. However, dam engineers do have to be sensitive to the impacts of dams on their surroundings, including resettlement and relocation of the affected population, environmental concerns, socio-economic impacts, sedimentation issues and safety issues.

Dams and reservoirs will continue to be needed to supply water in large quantities and of a sufficient quality to meet the needs of the world. The challenge for dam engineers is to plan, design, construct and operate cost-effective and efficient dams that are safe and socially and environmentally sensitive elements of our infrastructure.

REFERENCES

Alberta Government (2000) *Alberta Water Legislation, Regulations and Guidelines*. Available at: http://www.albertawater.com/index.php?option=com_content&view=article&id=54&Itemid=60 (accessed 22 March 2012).

ANCOLD (Australian National Committee on Large Dams) (1994) *Dam Safety Management Guidelines*. ANCOLD, Hobart, Tasmania, Australia.

Binnie GM (1981) *Early Victorian Water Engineers*. Thomas Telford, London, UK.

British Columbia Government (2000) *British Columbia Dam Safety Regulation*. Reg 44/2000. Available at: http://www.bclaws.ca/EPLibraries/bclaws_new/document/ID/freeside/10_44_2000 (accessed 22 March 2012).

British Columbia Ministry of Environment (2008) *Dam Audit Checklist Sheet*.

CDA (Canadian Dam Association) (2003) *Canadian Dam Association Bulletin.*

CDA (2007) *Dam Safety Guidelines.*

Defra (Department for Environment, Food and Rural Affairs) (1975) The Reservoirs Act 1975. HMSO, London, UK.

Defra (2010) *Flood and Water Management Act*. HMSO, London, UK.

Environment Agency (2009) *Benchmarking of 2D Hydraulic Modelling Packages.* SC080035/SR2 Environment Agency/Defra. Available at: http://publications.environment-agency.gov.uk/PDF/SCHO0510BSNO-E-E.pdf (accessed 22 March 2012).

Environment Agency (2011) *Modes of Dam Failure and Monitoring and Measuring Techniques.* SCO80048 Scoping Study. Available at: http://evidence.environment-agency.gov.uk/FCERM/en/Default/HomeAndLeisure/Floods/WhatWereDoing/IntoTheFuture/ScienceProgramme/ResearchAndDevelopment/FCRM/Project.aspx?ProjectID=E1ADD755-9DD1-46BE-89A5-06907A3DED00&PageId=a0fe6dfc-506a-452c-9bff-a7ec06b4e6b (accessed 22 March 2012).

FEMA (Federal Emergency Management Agency) (2007) *Annual Report Fiscal Year 2007.*

FEMA (2004) *FEMA 333 Federal Guidelines For Dam Safety: Hazard Potential Classification System For Dams.*

FERC (Federal Energy Regulation Commission) (2004) *Preparations for Handling Emergencies and Potential Emergencies at Projects.*

Finnish Environment Institute (2001) *Development of Rescue Actions Based on Dam-Break Flood Analysis (RESCDAM).*

ICOLD (International Commission on Large Dams) (2007) *Dams and the World's Water.* ICOLD.

Norwegian Government (2000a) Water Resources Act. Available at: http://www.nve.no/en/About-NVE/Acts-and-regulations (accessed 22 March 2012).

Norwegian Government (2000b) Regulations Governing the Safety and Supervision of Watercourse Structures. Available at: http://www.nve.no/en/About-NVE/Acts-and-regulations (accessed 22 March 2012).

Portuguese Government (1990) *The Portuguese Code of Practice for Observation and Inspection of Dams.*

Quebec Government (2002) *Dam Safety Act 2002.*

Queensland Government (2000) The Water Resources Act of 2000. Available at: http://www.legislation.qld.gov.au/LEGISLTN/CURRENT/W/WaterA00.pdf (accessed 22 March 2012).

State of Washington (1991) *Chapter 173–175 WAC: Dam Safety.*

USACE (United States Corps of Engineers) (2008) *Risk Analysis for Dam Safety, A Unified Method for Estimating Probabilities of Failure of Embankment Dams by Internal Erosion and Piping.*

Wieland M and Mueller R (2009) *Dam safety, emergency action plans, and water alarm systems.* International Water Power and Dam Construction.

FURTHER READING

Binnie GM (1978) The collapse of the Dale Dyke dam in retrospect. *Quarterly Journal of Engineering Geology* **11**: 305–324.

Binnie and Partners (1986) *Modes of Dam Failure and Flooding and Flood Damage Following Dam Failure.* Department of the Environment, London, UK. Contract PECD 7/7/184.

Binnie and Partners (1991) *Estimation of Flood Damage Following Potential Dam Failure: Guidelines*. Foundation for Water Research, Marlow, UK. Department of the Environment Contract PECD 7/7/259.

Charles JA (1997) Acceptable risk and reservoir safety. *Dams & Reservoirs* **7(2)**: 7–12.

Charles JA (1998) Lives of embankment dams – construction to old age. *Dams & Reservoirs* **8(3)**: 11–23.

Charles JA and Tedd P (1996) *Bibliography of British Dams*. Building Research Establishment, Watford, UK. Report 303.

Charles JA and Wright CE (1993) Reservoir safety research in the United Kingdom. *Maintenance of Older Dams: Accidents, Safety Assessment, Repairs. Proceedings of Technical Symposium,* Vol. 1, pp. 307–323. Comité Francais des Grands Barrages, le Bourget du Lac, France.

Charles JA, Abbiss CP, Gosschalk EM and Hinks JL (1991) *An Engineering Guide to Seismic Risk to Dams in the United Kingdom*. Building Research Establishment, Watford, UK. Report BR 210.

Charles JA, Tedd P, Hughes AK and Lovenbury HT (1996) *Investigating Embankment Dams: A Guide to the Identification and Repair of Defects*. Building Research Establishment, Watford, UK. Report BR 303.

Hall MJ, Hockin DL and Ellis JB (1993) *Design of Flood Storage Reservoirs*. CIRIA/ Butterworth-Heinemann, London, UK.

Herbert DM, Lovenbury HT, Allsop NWH and Reader RA (1995) *Performance of Blockwork and Slabbing Protection for Dam Faces*. HR Wallingford, Wallingford, UK. Report SR 345.

Hewlett HWM, Boorman LA and Bramley ME (1987) *Design of Reinforced Grass Waterways*. CIRIA Report 116. Available at: http://www.thenbs.com/PublicationIndex/DocumentSummary.aspx?PubID = 122&DocID = 264342 (accessed 22 March 2012).

Institution of Civil Engineers (1995) *Information for Reservoir Panel Engineers*. ICE, London, UK.

Institution of Civil Engineers (1996) *Floods and Reservoir Safety*, 3rd edn. Thomas Telford, London, UK.

Institution of Civil Engineers (1998) *An Application Note to an Engineering Guide to Seismic Risk to Dams in the United Kingdom*. ICE, London, UK.

Johnston TA, Millmore JP, Charles JA and Tedd P (1999) *An Engineering Guide to the Safety of Embankment Dams in the United Kingdom*, 2nd edn. Building Research Establishment, Watford, UK. Report BR 363.

Kennard MF, Hoskins CG and Fletcher M (1996) *Small Embankment Reservoirs*. CIRIA Report 161. Available at: http://www.thenbs.com/PublicationIndex/DocumentSummary.aspx?PubID = 122&DocID = 200605 (accessed 22 March 2012).

Kennard MF, Owens CL and Reader RA (1996) *Engineering Guide to the Safety of Concrete and Masonry Dam Structures in the UK*. CIRIA Report 148. Available at: http://www.thenbs.com/PublicationIndex/DocumentSummary.aspx?PubID = 122&DocID = 200603 (accessed 22 March 2012).

Reader RA, Kennard MF and Hay J (1997) *Valves, Pipework and Associated Equipment in Dams – Guide to Condition Assessment*. CIRIA Report 170. Available at: http://www.thenbs.com/PublicationIndex/DocumentSummary.aspx?PubID = 122&DocID = 201174 (accessed 22 March 2012).

Skimpton AW and Vaughan PR (1993) The failure of Carsington Dam. *Géotechnique* **43**: 51–173.

Tedd P and Hoton IR (1994) *Register of British Dams*. Building Research Establishment, Watford, UK. Report BR 261.

Yarde AJ, Banyard LS and Allsop NWH (1996) *Reservoir Dams: Wave Conditions, Wave Overtopping and Slab Protection*. HR Wallingford, Wallingford, UK. Report SR 459.

USEFUL WEBSITES

The British Dam Society: http://www.britishdams.org (accessed 22 March 2012).

ICOLD (International Commission on Large Dams): http://www.icold-cigb.net/GB/ICOLD/icold.asp (accessed 22 March 2012).

US Bureau of Reclamation (USBR): http://www.usbr.gov (accessed 22 March 2012).

Australian National Committee on Large Dams (ANCOLD): http://www.ancold.org.au (accessed 22 March 2012).

NSWDSC: http://www.damsafety.nsw.gov.au (accessed 22 March 2012).

Flood Risk
ISBN: 978-0-7277-4156-1

http://dx.doi.org/10.1680/fr.41561.211

Chapter 9
Trash screens at culverts

Nicholas Wallerstein and Scott Arthur
School of the Built Environment, Heriot-Watt University, Edinburgh, UK

Blockage at inline structures such as bridges and culverts by natural debris or debris contributed by humans poses a significant hazard to effective flow conveyance. The focus of this issue in the UK by and large relates to blockage of culvert systems where the potential consequences may be severe. In order to manage this risk, culverts often have a trash screen fitted at their inlet to prevent the ingress of unwanted material. In this chapter, a range of factors pertinent to the operation and maintenance of trash screens are outlined. This includes the current state-of-the-art methods offered in the UK for prediction of the volume of debris that is likely to accumulate at a screen over a given time-frame, or flood event, and methods for assessing the risk of blockage to culvert structures. Current UK guidance on optimal screen design and approaches for hydraulic modelling of prospective structures is also summarised. Ongoing field and laboratory-based research into blockage characterisation is also presented. This focuses on the analysis of large empirical datasets using multiple regression techniques from which equations have been generated that predict debris-delivery probabilities.

Notation

α		Value of the constant in linear multiple regression models
β		Value of the coefficients in linear multiple regression models
AG	%	Percentage of agricultural land-use cover in a screen's contributing catchment area
Da	m^3/year	Debris amount
Dda	m^3/year	Design debris amount
f	nominal	Inspection frequency/screen design/maintenance options
H	nominal	Hazard associated with screen blockage (nominal scale 0–1)
ID	%	Income domain
MDM	index	Multiple deprivation measure
NL	m	Channel network length upstream of a screen
$P_{d(m/s/a)}$	%	Probability of a significant debris-load delivery to a screen based on monthly (m), seasonal (s) and gross average (a) datasets
Q_n	m^3/s	Flow rate with return periods of $n = 2$, 10, 50 and 100 years
R	nominal	Risk associated with debris delivery at screen I (nominal scale 0–100)
RU	%	Percentage of rural land-use cover in a screen's contributing catchment area

S_i		Screen in series $S_{i(1 \ldots n)}$
SA_b	m^2	Screen area blocked by debris
SU	%	Percentage of suburban land use in a screen's contributing catchment area
SL	mm^{-1}	Channel slope upstream of trash screen
SO	%	Percentage of suburban open land use in a screen's contributing catchment area
U	%	Percentage of urban land use in a screen's contributing catchment area

9.1. Introduction

Culverts are primarily structures that provide a route for watercourses to pass under obstacles such as roads, railways and urban developments. As such, they must be engineered and maintained effectively to minimise flood risk to property, members of the public and infrastructure. In-channel debris can have a significant impact on conveyance through culverts if it becomes lodged within the structure, especially during storm events, as this can lead to flooding and sometimes damage to the culvert itself. This problem has had real consequences in recent flooding events in the UK, and was highlighted in the Foresight (2004) report in which it was recognised that: 'debris within the urban system may also increase the potential for blockage and impair performance of outlets to the urban drainage system' (p. 309). This blockage hazard is often combated by fitting a trash screen at the inlet of the structure, but these can become a hazard in themselves, as they are, by their very nature, inherently designed to block (also known as 'blinding'), and may therefore cause flooding if an effective maintenance regime is not implemented. The problems associated with effective screen operational practices are widely recognised among asset managers, who are well aware that undertaking capital works through fitting a screen does not ensure a reduction in flood risk unless the maintenance regime and necessary associated budget are adequate to ensure that the structure operates effectively over its design life. Unfortunately, too much emphasis is often placed on the construction of new flood defence works, while the necessary associated maintenance budgets may be an afterthought, which, in the case of trash screens, may actually lead to an increase in flood risk. This problem has been highlighted in the Environment Agency's Strategy for Flood Risk Management (Environment Agency, 2003), which states that:

> Flood defence assets and their management are extremely important . . . The replacement value of our flood defence asset base is more than £15 billion, with only £100m (0.7% of asset value) spent on maintaining and operating the assets. The operation and maintenance of this major asset infrastructure is key to the minimisation of damage from flooding and this level of expenditure is believed to be inadequate. (p. 9)

There has, to date, been only limited research in the UK that has systematically and comprehensively explored the dynamics of debris transport to, and accumulation at, trash screens, especially with respect to the development of robust decision-support tools that could be used to help those who are responsible for screen assets to employ

Figure 9.1 (a) Example of organic debris build-up at a screen; (b) anthropogenic trash dumped at trash screens

(a)

(b)

effective screen management strategies. However, a research theme within the Flood Risk Management Research Consortium project (Phase 2) (FRMRC II) recently tackled this issue. Through this project theme, primary research was conducted in order to produce methods/models that can be used to help those asset managers who are responsible for trash screens to make informed decisions regarding the optimisation of screen inspection and maintenance frequencies and selection of appropriate screen design parameters. This work is reviewed in a section in this chapter within the context of the current knowledge base and best-practice guidelines.

Before exploring screen blockage issues in detail in the following sections, it should be noted that the term 'trash' is here used synonymously with the term 'debris'. These two terms refer to both organic material (leaves, branches, trees, grass, etc.), which enters the channel through the processes of wind transport, bank erosion and wash-in from the catchment surface (Figure 9.1(a)), and anthropogenic material (litter, general household waste, and large non-domestic waste such as shopping trolleys and pallets, etc.), which has been dumped or blown into the channel (Figure 9.1(b)). Inorganic debris, such as silt and gravel, which may collect at screens is not discussed here, as it is considered to warrant analysis as an independent topic in its own right (e.g. Federal Highways Administration, 1985, 1990; Lagasse *et al.*, 2001).

9.2. Screen design

Screens can vary widely in size, as shown in Figure 9.2(a,b), and also vary in complexity, with such features as multi-stage screens, which are designed to cope with excessive loads at high flows (Figure 9.2(c)), and bypass chambers designed to route flow behind the culvert entrance if the screen becomes severely blocked (Figure 9.2(d)).

The term 'trash screen', is synonymous with the term 'trash rack', and may also be referred to as 'trash grilles' (the latter being common parlance in Northern Ireland). These structures are considered to be different, however, from what are known as 'roughing screens' (or grilles), which are generally located upstream of the screen proper, and which have large diameter bars with a coarse spacing and are usually set

Figure 9.2 (a) Trash screen protecting the inlet to small diameter pipe; (b) screen protecting a large culvert located in Belfast city centre; (c) two-stage screen with access platforms to allow for limited hand rake lengths; (d) screen with bypass channel (entrance at right)

(a) (b) (c) (d)

perpendicular to the channel bed (Figure 9.3(a)). These are normally designed to prevent any large or heavy items from reaching the screen proper, which could otherwise cause damage to, or severely impede the cleaning of, the screen. The term 'security screen' refers to structures that are superficially similar in appearance to trash screens but which primarily serve to prevent unauthorised access to a culvert, thereby reducing risk to life through drowning. These structures are placed directly across the upstream mouth and also often over the exit of a culvert, flush with the headwall and, importantly, often have a bar spacing lower than that of trash screens in order to be effective. Such structures will obviously act in the same way as purpose-built trash screens, and must therefore be treated in the same manner with regard to maintenance. An example of a security screen is shown in Figure 9.3(b). Security screens at culvert exits are viewed as

Figure 9.3 (a) Example of a roughing screen (flow direction is into the page); (b) a security screen fitted to a culvert outlet
Contains Environment Agency information © Environment Agency and database right

(a) (b)

notably problematic, as they are particularly difficult to clear during wet weather. One design solution, which may be employed to deal with this problem, is to use shear bolts, allowing the screen to hinge open once the force acting on the screen reaches a predetermined level.

Screen clearance methods range from simple direct raking of the face of the structure by field operatives (Figure 9.4(a)), to clearance assistance using mechanical excavators and, finally, to elaborate automated clearance devices that can be activated automatically when flow depths reach a critical level (Figure 9.4(b)).

Figure 9.4 (a) Operations staff clearing a screen during flood flow; (b) large automatic screen-raking device installed on a channel in Bristol city centre
(a) Reproduced by kind permission of Dr John Riddell

(a) (b)

9.3. Modes of structural failure caused by debris

The impact of debris deposition at the upstream face of inline structures such as culverts (with and without trash screens) can be roughly grouped into three broad categories. First, and generally associated with culverts, debris can cause blockage within the structure or, more commonly, may cause significant blockage of trash screens at the upstream face of the culvert if the maintenance interval is too great and/or if large quantities of material are transported to the screen during flood flows (Figure 9.5(a)). The consequence of this process is a reduction in conveyance capacity within or upstream of the culvert, which may significantly increase the risk of localised flooding. It is this issue that will be examined in more detail is this chapter.

Second, and usually associated with bridges, debris that becomes trapped against piers and abutments can effectively both reduce the bridge opening area and increase the pier diameter. This can result in excessive pier and contraction scour. This problem has not been widely reported in the UK, with one or two notable exceptions, including destabilisation of a pier on the Lower Ashenbottom railway viaduct in Lancashire in

Figure 9.5 (a) Large screen with significant blockage by organic debris; (b) debris jam on the upstream face of a pier on the Lower Ashenbottom Viaduct, Lancashire, June 2002; (c) trash screen bars deformed by debris loading (screen section on right removed for repair)
(a) Contains Environment Agency information © Environment Agency and database right; (b) Reproduced by kind permission of Bury Council

(a) (b) (c)

1992 (Figure 9.5(b)). This problem has been more widely reported overseas, especially in the USA, New Zealand and Australia, and a significant number of papers and reports have documented this phenomenon and/or modelled it and developed protection strategies. Noteworthy among these are the publications by Laursen and Toch (1956), Jones *et al.* (1991), Melville and Dongol (1992) and Pagliara and Carnacina (2010).

Third, debris will impart a dynamic 'impact' force when it strikes in-channel structures and, if enough debris is present, may result in significant drag forces once the blockage causes the water-surface level to reach, and exceed, the deck or soffit of the blocked structure. The former phenomenon can cause significant physical damage to the structure, and this may impede its function (in the case of trash screens) or may cause deformation of piers and wingwalls, while the latter can, in some circumstances, cause a structural component such as a bridge deck to be lifted from its footings. In either instance, the damage comes in the form of structural failure rather than increased flood risk. An example of damage caused by the dynamic impact of debris on a trash screen is shown in Figure 9.5(c). Again, the incidence and severity of this problem appears to be much lower in the UK than overseas, consequently the breadth of efforts to model and mitigate against this phenomenon are far greater elsewhere (see e.g. Apelt, 1986; Parola, 2000; Haehnel and Daly, 2002).

9.4. Currently available screen design and management tools

9.4.1 Guidance manuals

At present in the UK there are two principal support manuals – the *Trash and Security Screen Guide 2009* (TSSG) (Environment Agency, 2009a) and the *Culvert Design and Operation Guide* (CDOG) (Construction Industry Research and Information Association (CIRIA), 2010) – which describe in detail best practice guidance associated with trash screens. The TSSG is free to download from the Environment Agency website (see Environment Agency, 2009a), while the CDOG can be purchased via the CIRIA website (see Useful websites at the end of this chapter).

The TSSG is dedicated to the analysis and design of trash screens, while the CDOG covers a broad range of considerations associated with all aspects of culvert design, of which one chapter is dedicated to trash screen issues (there is, in fact, considerable replication on this subject between the two documents). The TSSG offers the current state-of-the-art information associated with: site-specific risk assessment to determine whether a screen is in fact needed; evaluation of potential debris loads; current screen design and modelling techniques (hydraulic and structural); management and monitoring issues; and operational requirements. In the remainder of this chapter the detail of these key themes is distilled and examined by making reference to the TSSG and also to other UK and overseas reports, modelling tools, legislative documents and new research conducted through the Flood Risk Management Research Consortium (FRMRC) project.

The TSSG contains a blockage risk assessment table (Table 4.4 on page 20 of that document) and a flow diagram decision support tool (Figure 3.3 on page 12 of that document), which are designed to aid in the selection of an appropriate blockage mitigation strategy. The risk assessment table, which is designed to be used when considering the probability and consequence of blockage and damage to a culvert, is reproduced in Table 9.1, and should be consulted first and a risk score determined. After the score has been obtained it should be applied to stage 1 of the decision support flow diagram, which is reproduced in Figure 9.6. This process should then be followed through step by step, considering the pertinent design factors for the site under investigation so that, through an iterative approach, the optimum strategy can be ascertained.

9.4.2 Screen design tools

Magenis (1989) devised a method for estimating the required screen area for protecting culverts using empirical data from which he compared the characteristics of well-performing screens with those where problems were experienced. This method has been refined over the intervening period, appearing first as a design tool in a National Rivers Agency design guidance document (National Rivers Authority, 1995), and then in an updated form in the current TSSG. The approach requires knowledge of the total upstream river length that contributes to the screen of interest and of the upstream land-use types, and these are applied to the diagram shown in Figure 9.7 in order to obtain an annual debris amount (Da, m^3/year). This value is then converted to a design debris amount (Dda) using a slope correction factor, which is derived from knowledge of the channel slope upstream of the screen. The Dda is then used to determine the optimal screen area (in m^2) by dividing it by the product of a blinded depth factor (Bdf), which is associated with the most prominent upstream land-use type, and the notional number of significant flow events that are likely to occur in any given year (suggested to be three).

Despite being based on very limited data, and in the absence of an alternative, the method has been found to perform reasonably well, and is recommended for use in the TSSG when designing new screens where little or no pre-existing site data are available.

9.4.3 Blockage risk assessment tools

At present there is just one tool (as far as can be determined) which has been developed for UK environments that is designed to give an estimate of blockage risk at culverts. This was developed as an R&D project for the Environment Agency South West Region by Faulkner and Waller (1998), and was commissioned in response to the observation that many local flood-level estimates were being underpredicted, principally because of blockage at structures. From this research, an Excel-based tool, the *blockage risk model* (BRM), was developed, which contains risk probability values for a number of hydrological, debris and culvert parameters based on results from a questionnaire that was circulated to a wide number of organisations and individuals. In the spreadsheet (Figure 9.8), the user supplies weightings for each parameter based on his perception of the process and its risks, and these are applied to the relevant probabilities contained within the model and combined to generate a single, overall, *risk of blockage* percentage score. This score is then applied to a decision support tree, which helps the user assess the risk of blockage based on upper and lower bounding percentage limits (low ($<$20%), medium (20–40%), and high ($>$40%)). The blockage model and decision tree are referred to collectively as the *blockage risk procedure* (BRP). The model, while currently the state of the art in the UK, has had only limited uptake outside of the Environment Agency South West region, where it is currently used routinely.

Table 9.1 Guidelines for determining potential risks posed by screens
Contains Environment Agency information © Environment Agency and database right

	Score				
	5	4	3	2	1
Blockage of culvert					
Probability	**More frequently than one in 2 years** Regular recorded blockage (e.g. once or twice in the last 2 years)	**One in 2 years to one in 5 years** Some record of blockage (e.g. once or twice in the last 5 years) *or* Culvert size under 1 m^2, catchment urban or woodland	**One in 5 years to one in 10 years** Culvert size under 1 m^2 and at least 50% urban or woodland *or* Culvert size over 1 m^2 and under 3 m^2 with potential blockage points	**One in 10 years to one in 25 years** Culvert size over 1 m^2 and under 3 m^2 *or* Culvert size over 3 m^2 with no upstream public access	**Less frequently than one in 25 years** Culvert size over 3 m^2
Consequence	**Over £1 million** Cost of flooding damages *plus* Cost of removing blockage and culvert repair (per event)	**£100 000 to £1 million** Cost of flooding damages *plus* Cost of removing blockage and culvert repair (per event)	**£10 000 to £100 000** Cost of flooding damages *plus* Cost of removing blockage and culvert repair (per event)	**£1000 to £10 000** Cost of flooding damages *plus* Cost of removing blockage and culvert repair (per event)	**Under £1000** Cost of flooding damages *plus* Cost of removing blockage and culvert repair (per event)
Damage caused by debris to infrastructure of culvert					
Probability	**More frequently than one in 2 years** Impact damage to structure from debris in flow (e.g. once or twice in the last 2 years)	**One in 2 years to one in 5 years** Impact damage at frequency of once or twice in the last 5 years	**One in 5 years to one in 10 years** Some impact damage possible due to size of debris	**One in 10 years to one in 25 years** Low, but some possibility of impact damage	**Less frequently than one in 25 years** Rare likelihood of damage or no impact damage possible
Consequence	**Over £1 million** Repairs involving diversion of watercourse and works to full length of culvert	**£100 000 to £1 million** Repairs involving significant temporary works and works to more than half of the culvert length	**£10 000 to £100 000** Repairs involving some temporary works and repairs to less than half of the culvert length	**£1000 to £10 000** Repairs requiring no temporary works	**Under £1000** Minor repairs required, not in urgent need of attention. May be encompassed in general maintenance

Figure 9.6 Flow chart designed to aid in the consideration of the requirement for a new trash or security screen
Contains Environment Agency information © Environment Agency and database right

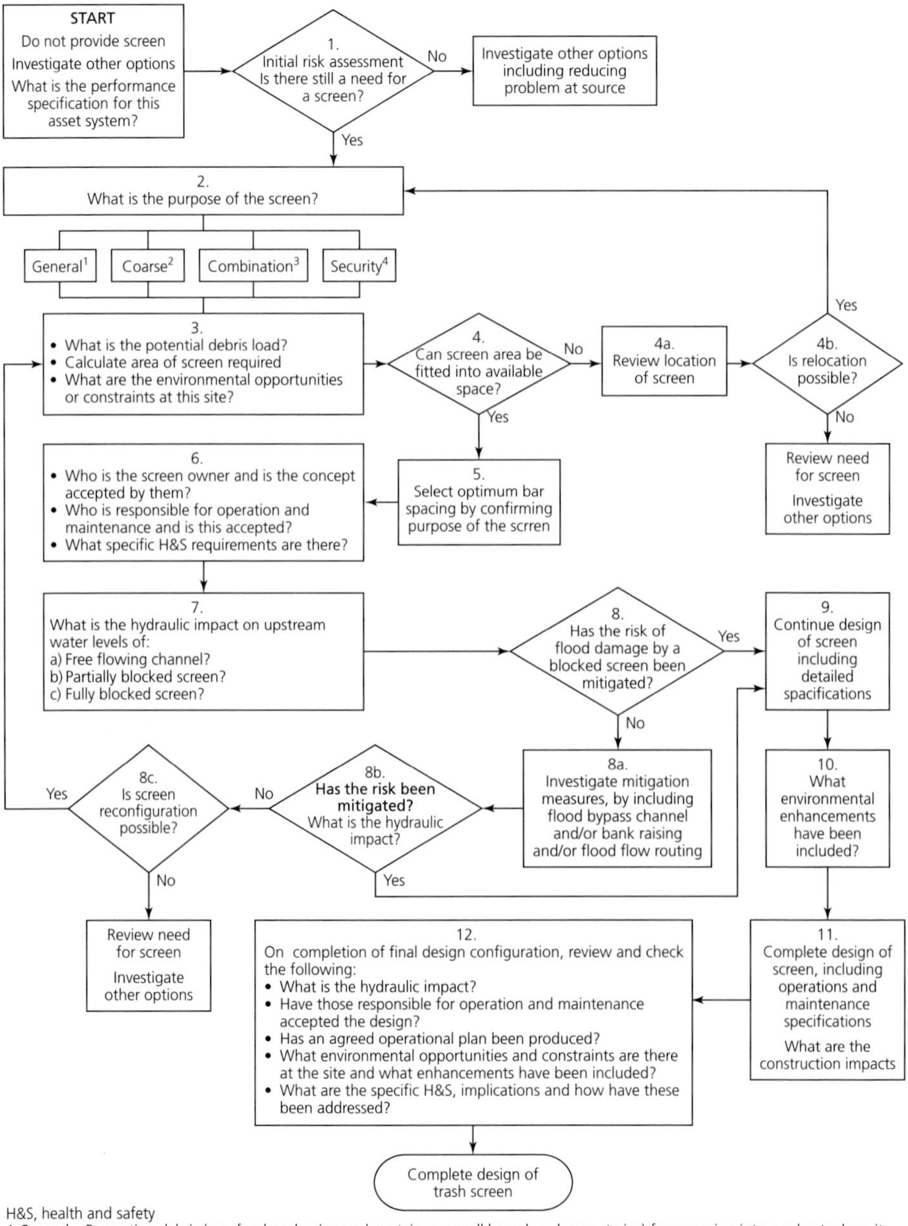

H&S, health and safety
1 General – Preventing debris (e.g. food packaging and containers, small branches, leaves, twigs) from passing into a culvert where it could accumulate and obstruct the flow
2 Coarse – Preventing large items of debris (e.g. oil drums, pallets, sofas) from passing downstream where they could become lodged in a culvert and thereby cause a blockage
3 Combination – A mixture of two or more of the other screen purposes
4 Security – Preventing people (principally children) from entering a culvert

Figure 9.7 Graph for use in Magenis' method to determine the annual debris load at a screen
Contains Environment Agency information © Environment Agency and database right

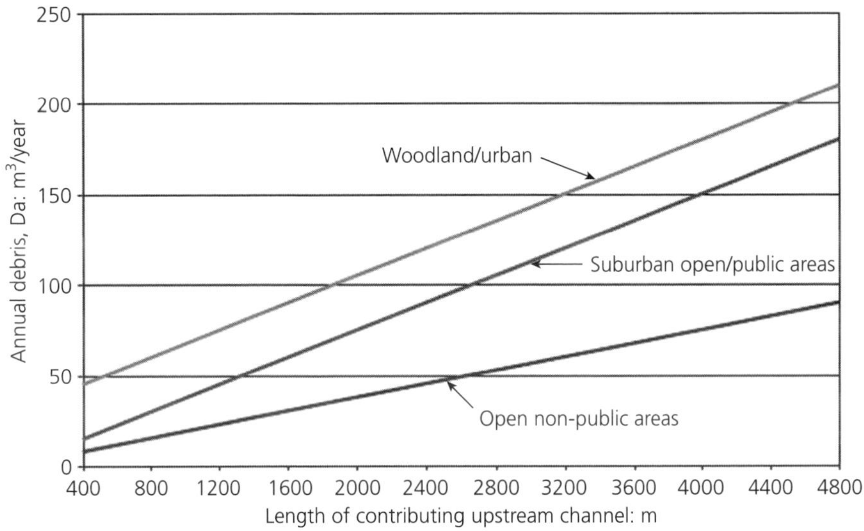

9.4.4 Approaches used outside of the UK

There are a number of other blockage design and risk assessment tools that have been developed outside the UK and reported in guidance manuals. These include the US Department of Transportation Federal Highways Administration (2005) HEC 9 manual, which represents the state-of-the-art research into blockage risks and associated design tools in the USA. This manual borrows heavily from previous reports and papers by Diehl (1997) and Wallerstein *et al.* (1997) and Abt and Thorne (1999). The manual, however, while providing some useful information, is unfortunately really only applicable to US environments, where the main concern is blockage, scour and damage at bridges, especially on large rivers. There have also been a considerable number of studies performed in Australia (Austroads, 1994; Queensland Government, 2007; Weeks *et al.*, 2009), but again the type of structures assessed, and the physical environments in which they lie, are quite different from those generally encountered in the UK. However, these reports do contain very useful and generally applicable guidelines for hydraulic modelling of afflux at screens with blockage, recommendations on serviceability associated with debris, and structural measures to cope with blockage.

9.5. Blockage management options

Where a new culvert is to be constructed, before deciding to install a screen, it is crucial to first consider as to whether this is necessary and will actually reduce flood risk because, as stated at the outset of this chapter, screens pose inherent dangers in themselves. Consequently, blockage risk management generally falls into three broad categories.

Figure 9.8 Microsoft Excel user interface for the blockage risk model
Contains Environment Agency information © Environment Agency and database right

Culvert Blockage Risk Assessment — *Version 1.04*
South West England Version — *January 1998*

Help? Version?

Name of Structure:	Type location here	Grid Ref:	EEEE NNNN
Hydrometric Code and Area :	xx	MAF :	5.00 m3/s

Flood Growth Curve		MAF	Q5	Q10	Q25	Q50	Q100
	1	5.0	6.2	7.5	9.2	10.6	12.1

Structure Details					
Upstream Invert Level	2	100.00 m AOD			
Downstream Invert Level	3	99.90 m AOD			
Barrel Length	4	15.0 m			
Bed Slope	5	0.007 m/m	1 in 150		
Shape Code	6	1			0%
Span or Diameter	7	2.000 m		*Barrel Flow Area (m2)*	2.00
Rise	8	1.000 m		*BarrelWetted Perimeter(m2)*	6.00
Number of openings	9	1 m		*Barrel Compound n*	0.014
Barrel 'n'	10	0.012		*Barrel Full Velocity (m/s)*	2.84
Bed 'n'	11	0.018		*Structure Full Capacity (m3/s)*	5.68
Trash screen fitted?	12		No		

Structural Factors		Value	Category	Potential
Multiple openings in structure	13	N/A	Single	
Poor entrance conditions	14	10%	Good	0.3%
Poor exit conditions	15	10%	Good	0.1%
Poor barrel condition	16	10%	Good	0.1%
Poor river approach	17	10%	Good	0.4%
Aspect ratio (span < rise)	18	N/A	Broad	

Flood Hydrology Factors		Value	Category	Potential
Capacity relative to target flood	19	10%	Low	9.5%
Proximity of design flow to soffit	20	100%	High	9.7%
Frequency of flash flooding	21	50%	Medium	3.3%
Frequency of long slow rise floods	22		None	

River Condition Factors		Value	Category	Potential
Level of urbanisation near structure	23	20%	Low	1.6%
Evidence of tipping into river	24	15%	Low	0.8%
Density of bankside growth (large)	25	15%	Sparse	0.4%
Density of bankside growth (small)	26	50%	Medium	1.3%
Amount of loose bankside material	27	15%	Low	1.9%

Maintenance Factors		Value	Category	
Frequency of maintenance visits	28	50%	Medium	3.7%

Risk of Blockage in Flood Event	29	Medium	33%

- Do nothing – a baseline case, which is actually the most desirable in many instances.
- Reduce the debris problem at source – this represents an off-site management approach, which is a potentially sustainable solution.
- Design, install and maintain a screen.

9.5.1 Do nothing

This option is the most realistic when both the probability and consequences of blockage within the culvert are low. The option has the greatest chance of success when a new or retro-fitted culvert is constructed according to a number of key design features. The recommended construction rules, as set out in the TSSG are to

- design the culvert with the largest cross-sectional area possible
- avoid bends, steps, changes in cross-section and sighting utility installations within the structure
- be aware that the longer the culvert is the greater the risk of blockage
- recognise that culverts designed to accommodate free surface flow are less likely to block, even in floods, than are culverts, which surcharge in such events.

9.5.2 Reduce the risk at source

This represents an off-site management approach, and is attractive as it may potentially represent the most sustainable long-term solution. Methods include the following.

- Enforcing fly-tipping legislation (see UK Government (1990) and Webb *et al.* (2006)) to the degree that the penalties deter the general public from such actions, and also enforce the equivalent legal stipulations for businesses (see Environmental Campaigns (ENCAMS), (2006)). There are also practices that should be adhered to associated with Forestry Commission land management (Forestry Commission, 2003), although these are not legally binding.
- Stakeholder engagement, whereby residents living near watercourses are educated, and encouraged through financial aid to form stewardship schemes. A good example of this approach has been the initiation of the Sheffield River Stewardship Company (see Wild *et al.*, 2008). Also, to provide financial incentives for farmers to dispose of agricultural waste effectively (recommended in the *Pitt Review* (Pitt, 2008)).
- Day-lighting channels, where possible, to remove the risk of non-accessible blockages.
- Restriction of access to the watercourse to prevent direct fly-tipping.
- Stabilisation of river banks, removal of high-risk trees and active removal of in-channel debris material at source. These management methods were recommended to control debris delivery on the River Valency in the aftermath of the Boscastle floods (Environment Agency, 2006). Note that Beschta (1990) presents a method to aid with identification of trees that are likely to enter watercourses through bank failure.
- Enforce legislation that pertains to the responsibility that landowners must obey with respect to any reach of river that they might own (Environment Agency, 2007).

Alongside these responsibilities, it is worth noting that the Environment Agency can perform clearance work at high-risk sites and charge it back to the owner if the owner is found to be negligent in his legal control duties.

9.5.3 Design, install and maintain a screen

9.5.3.1 Structural considerations

The current state-of-the-art method for estimating the optimum screen frontal area at UK sites is, at this time, still considered to be the one developed by Magenis (1989) (see Section 9.4.2), and is the recommended approach presented in the TSSG guidance. Once the optimum screen area has been established, the particulars of the structure can be defined. Table 9.2 represents a compilation of the currently accepted best practice design considerations for UK screens.

9.5.3.2 Ongoing research into screen design

To further understand how trash screens perform, a research project, linked to FRMRC II, is now underway. In this project, Froude-scale flume experiments are being conducted in which simulated debris elements of various lengths and diameters are released into the flume flow upstream of a modular, variable geometry trash screen attached to the inlet of a simple culvert, in order to determine the consequent blockage characteristics. Initial results show that the percentage of debris pieces retained at the screen increases in a logarithmic manner as the debris length/bar spacing ratio rises (Figure 9.9). In addition, it has also been found that blockage potential increases as the approach flow velocity falls. Findings from the experiments performed to date also show that the trash-screen angle has a significant influence on blockage potential, with screens set at a an angle of 60° causing fewer blockages than screens with the same bar spacing set at 45°. Although this study is still at a formative stage, it has already demonstrated that a number of different aspects of screen design can significantly influence the rate and degree to which a screen can become blocked by debris. This research project is ongoing, and investigations are now being conducted to determine in more detail the influence that screen geometry and local hydraulic conditions have on debris blockage characteristics.

9.5.3.3 Hydraulic modelling

An integral part of the screen-design process is to determine its optimum design features through the use of hydraulic models, the ultimate goal being to estimate head loss through the screen under various blockage and discharge scenarios. The recommended design conditions that should be tested are presented in Table 10.2 in the TSSG, which also recommends that blockage sensitivity testing should be carried out for hypothetical blinding of 30%, 67% and 100% of the screen. Abt *et al.* (1992) suggested that hydraulic analysis should assume a critical blinding limit of 50% in the case of super-critical flows, and it is worth considering this scenario if such conditions might prevail. The TSSG recommends that the screen should be modelled as a discrete unit so that local energy losses can be simulated, and so that screen blinding, in most circumstances, can be adequately represented using weir and orifice hydraulic equations. The manner in which three commonly used one-dimensional hydraulic models can represent screen blockage is discussed in some detail in Benn *et al.* (2004), and a simplified form of a summary table to be found in that work is presented here in Table 9.3. Alongside these models, the Environment Agency has developed the afflux estimation system (AES) (Environment Agency, 2009b), which is designed to enable the hydraulic influence of a bridge or culvert to be modelled. The model is available as freeware in a stand-alone form (it

Table 9.2 Summary of screen design guidelines

Bar spacing	Trash screens: culverts <1 m diameter	150 mm
	Trash screens: culverts >1 m diameter	150 mm or more
	Trash screens: urban areas	300 mm
	Trash screens: pumping stations	75 mm
	Security screens	140 mm
Bar dimensions	Recommend flat bars	8 × 75 mm
	If extra strength required	Flat bars: 10 × 120 mm
	Max. unsupported bar length	500 mm
Screen angle	Optimum: 45°	
	Max.: 60°	
Screen geometry and construction material	Screen length parallel to bars	2000 mm max. (if higher screen required, stage it with intervening platforms)
	Platform width parallel to flow	Min. 1500 mm, max. 2500 mm
	Cross-bracing	Behind bars to enable efficient raking
	Alignment	90° to flow (diagonal screen if space limited)
	If debris material large	Consider roughing grille
	If consequence of flooding high	Consider bypass channel
	Consider change in future debris load (climate change + urbanisation) and screen deterioration	Future-proof design
	Preferred construction material	Galvanised mild steel
Screen cleaning	Method consideration: (1) hand raking, (2) mobile plant, (3) automatic devices	
	Vandal-proofing	
	Safe access structures for routine and non-routine events	
	Internal access hatch	
	Temporary debris storage structure	
	Hard standing	
	Lighting	

can be downloaded from the HR Wallingford CES/AES website, and has also been incorporated in the ISIS hydrodynamic model (see the Halcrow website). At present this software does not contain a specific function to model screens, but it is hoped that future updates will do so.

9.5.3.4 Monitoring

In the case of high-risk screens, where the rate of blockage may be very rapid and the consequence severe, the use of closed-circuit television (CCTV) and water-level monitoring

Figure 9.9 Percentage of debris pieces retained at a model trash screen as the debris length/bar spacing ratio increases
Modified from Blanc *et al.* (2011)

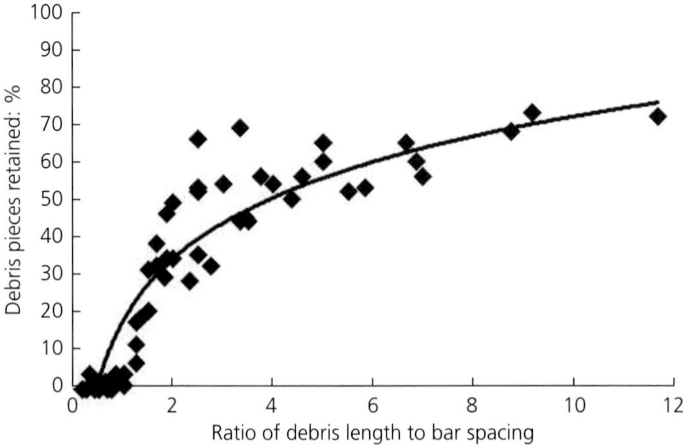

equipment mounted at the screen, incorporating telemetry to transmit the data to an operations centre, should be considered. The use of CCTV allows monitoring staff to observe conditions at a screen in real time, so that a response team can be sent out when conditions are critical. When considering the use of this technology, it is important to determine, at the outset, whether, even when critical conditions are recorded, a response team can actually reach the site to alleviate the blockage problem in time, otherwise the operational efficiency of the system will be redundant. Data transmitted

Table 9.3 Summary of blockage modelling capabilities available in HEC-RAS, ISIS and MIKE11

Blockage capability/method/procedure	HEC-RAS[2]	ISIS[3]	MIKE11[4]
Floating debris at bridge piers (applies a 'cylinder' around the pier)	✓		
Sediment in culvert invert	✓		
Manual estimation	✓	✓	✓
Transient blockage modelled as a sluice gate	✓	✓	
Blockage represented by separate hydraulic units such as weirs	✓	✓	
Model blockage as located at bridge/culvert inlet section (blockage only applied to the bounding channel section)[1]	✓		

Data taken from Benn *et al.* (2004)

[1] See page 57 in Benn *et al.* (2004)

[2] See Hydrologic Engineering Center (HEC) US Army Corps of Engineers (USACE) website HEC-RAS (USACE): www.hec.usace.army.mil/software/hec-ras

[3] See Halcrow website

[4] See Danish Hydraulic Institute website

to a publicly accessible website from a number of Environment Agency CCTV installations located in the south-west UK can be viewed at the Environment Agency website. The Environment Agency also currently has a series of actively monitored CCTV installations across the UK, which have been installed to assess the wider applicability of this technology in terms of its potential physical and financial robustness. A summary report of the findings from this project is given in *Remote Camera Pilot Scheme. Phase 3 – Post Pilot Evaluation* (Environment Agency, 2008).

9.6. Research developments: the FRMRC II Blockage Risk Assessment Project

9.6.1 Introduction

Through observation of the screen clearance, inspection and design practices employed by asset managers in the Environment Agency, Northern Ireland Rivers Agency (NIRA), private consultancies and a limited number of local authorities in the UK, it became evident that a number of different approaches have been adopted. In many cases the approaches have evolved according to the level of knowledge assimilated by asset managers at a local level, and, as such, it would be highly beneficial to have comprehensive documentation of these to enable the formation of coherent practices at a country-wide level. However, very few organisations appear to have documented their protocols, if indeed they formally have one. Typically, strategies employed, where made apparent, range from rigid site inspection regimes, in which detailed maintenance field notes are taken, through to completely *ad hoc* approaches, where no information is recorded and where sites are visited when field operatives believe the risk of blockage is likely to be high (e.g. according to CIRIA (2010), a number of Scottish local authorities make weekly superficial inspections of screens, but only during heavy rainfall). It therefore became apparent that, at present, there are no consistent, UK-wide inspection protocols that stem from simple, robust tools that can be employed to aid with the prioritisation of screen clearance, maintenance and design activities. The need for such tools has, over recent years, become increasingly important in order that screen management tasks can be performed within a rational framework, so that the cost-effectiveness of any given remediation strategy can be demonstrated in an accountable manner.

Recognition of this need led to the instigation of research into the topic through the FRMRC II project, the results of which are discussed in the following section. In this analysis, trash screen monitoring records and geographical information system (GIS) datasets, supplied by NIRA, were used to develop empirical equations, through the use of multiple regression analysis, that have been designed to enable the estimation of the probability of delivery of significant debris loads to screens. The initial results obtained from this analysis have been published in the work of Wallerstein and Arthur (2010a,b).

The datasets were associated with trash screens located around the city of Belfast, Northern Ireland (see Figure 9.10). The data consisted of observations made by asset maintenance staff at 140 trash screen sites between April 2002 and October 2008. Maintenance staff inspected each screen approximately once every seven days (the modal

Figure 9.10 Location map of the study area in and around Belfast, Northern Ireland
Modified from Wallerstein and Arthur (2012). John Wiley & Sons Ltd

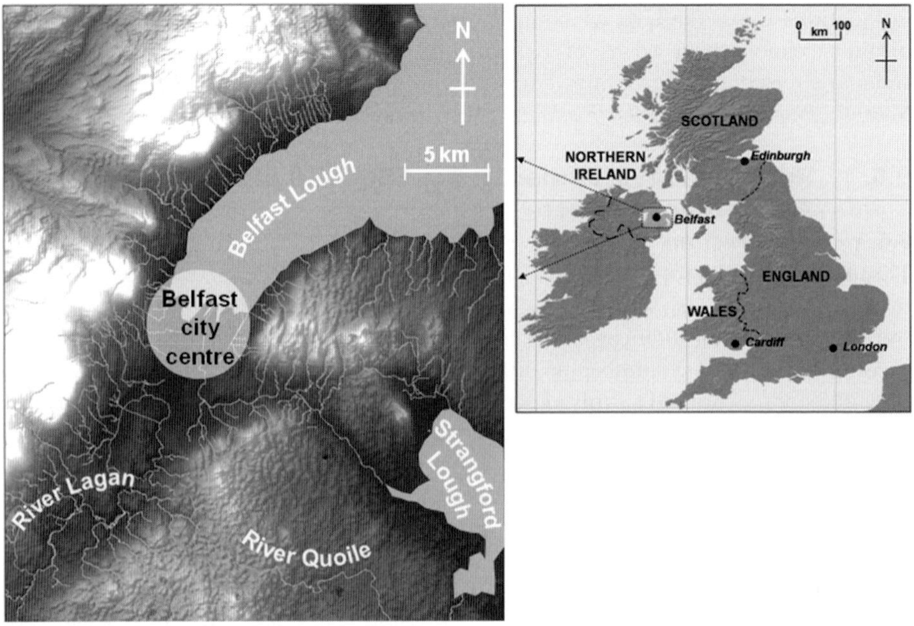

value) throughout this period in order to remove all material blocking the screen. There was some variation in this interval, with shorter periods between inspections (as little as 1 day) associated with clearance during storm events, and intervals of up to 2–3 weeks during dry conditions. During each inspection, staff also recorded an estimate of the percentage of the screen area blocked by debris and a breakdown of the composition of that material. A total of 25 265 observations were made at the 140 sites over the study period. The 140 screens used represented a subset of the total number maintained by NIRA (459 screens), and were selected on the basis that they had consistently reliable observations, and were also considered 'critical' screens which had, at the very minimum, an annual average percentage blocked screen area of 10%. This may seem a very low value in terms of selecting a critical blockage threshold, but it was used by necessity because a threshold of, say, 50% would have reduced the number of sites to the extent that statistically significant results could not be obtained.

For each screen, the raw data were initially processed by ascribing a value of one to each of the inspection records where a debris accumulation had been observed and a value of zero to those records where no debris accumulation had been observed. From the processed data the probability of delivery of significant debris loads to each screen was calculated for a range of different time-frames (seasonal and monthly, and a gross average of all the observations). This was achieved by invoking a generalised empirical probability, where the frequency of significant delivery events $\approx$ probability of delivery

(P_d), such that P_d = number of observations where a screen was observed to be blocked/ total number of observations. This assumption was considered to be justifiable because all debris accumulations were removed by the maintenance teams during each site visit, meaning that the screen blockage status observed at any given inspection depended solely on the delivery conditions between that inspection and the one immediately prior to it. Thus, the screen blockage status observed during any given site visit could be treated as a semi-independent probability.

It is important to understand that the probability of delivery of significant debris loads to trash screens, as assessed in the research presented here, depends solely on upstream catchment factors, and not on the characteristics of the screen itself. Through this reasoning, it is obvious that the inclusion of screen parameters as independent driving variables when examining the probability of trash delivery is a fallacy. However, there was a potential methodological problem associated with capturing debris-delivery data, in that the screen itself represented the device used to detect the arrival of material. Therefore, if a screen did not retain at least some proportion of the material delivered in any given event, the actual number of registered significant delivery events would be under-represented. Fortunately, however, this is rarely a problem, because screens, especially those that have been designed correctly, will, during the majority of debris-delivery events, retain at least some proportion of the imposed load, even fine material (grass, leaves and small litter, etc.) which one might think would easily pass through all but the narrowest bar spacings. This assertion is based on field observations made be the authors and findings presented in the TSSG. The probability and extent of blockage of screens by debris depends on the process of delivery and also on the complex structural and hydraulic conditions at the screen itself. While, due to space limitations, this issue is not discussed here, it is noted that a range of physical modelling experiments were conducted within the framework of the FRMRC II project to characterise this component of the subject.

9.6.2 Predictive driving variables

In order to generate predictive equations through the use of regression analysis, it was necessary to select a number of independent driving variables. These were chosen on the basis of reasoning, and observations made in the field and from the (limited) literature (including the Magenis (1989) approach). Four parameter groups were selected: channel, flow, land use and social deprivation factors.

The channel factors were extracted from GIS shape file layers supplied by NIRA, which included the trash screen and culvert locations with associated meta-data, the river network, and a 5 m horizontal resolution digital elevation model (DEM). From these data an upstream network length parameter was determined, which was calculated as the long-stream channel distance upstream of the screen in question, including all tributaries extending to the next upstream screen or to the channel headwaters if no intervening screens were present. Culverted channel lengths were removed from the total network length because debris cannot enter the channel directly along these sections (see Figure 9.11a). An upstream channel slope parameter was also calculated, this being the average gradient between the 10% and 85% distance along the river main stem,

Figure 9.11 (a) Schematic representation of GIS data from which the upstream network length, channel slope and income domain (ID) scores were extracted; (b) river network overlaid onto a Google Earth image from which contributing land-use types were estimated
Map © Google Earth. Figure reproduced from Wallerstein and Arthur (2012). John Wiley & Sons Ltd

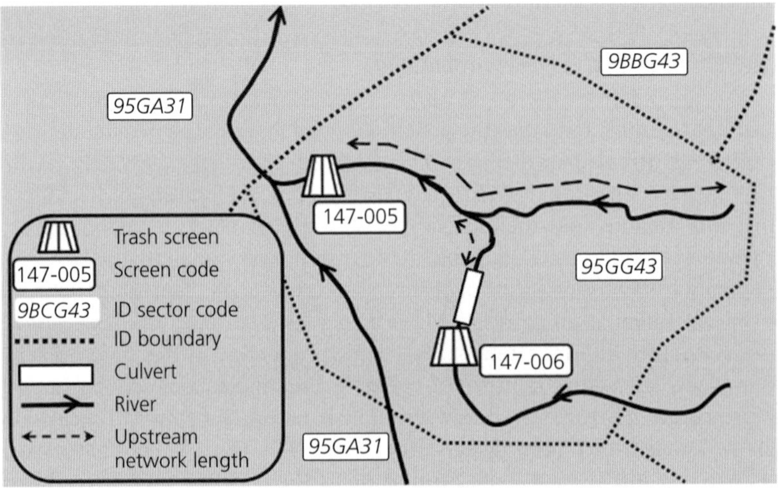

(a)

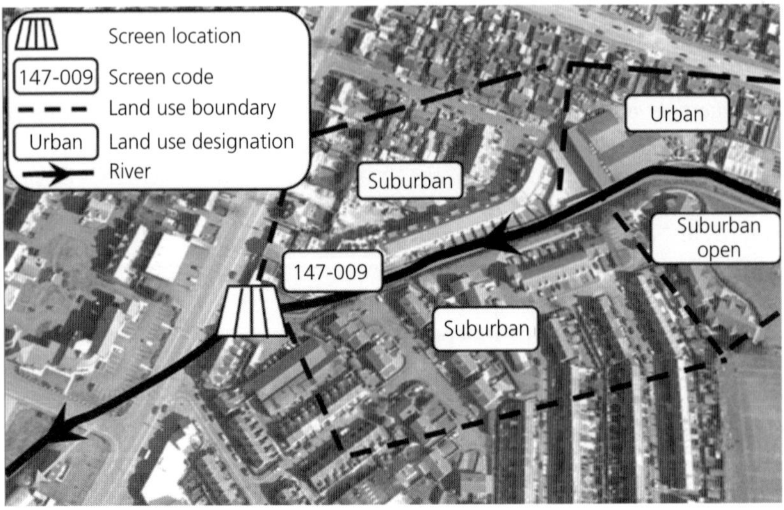

(b)

upstream of the screen. This variable is known as the S1085 slope, and was selected because it was used as a predictive variable in Magenis (1989).

Discharges with a 2-, 10-, 50- and 100-year return periods were used to characterise the flow parameter. The values for each return period were tested independently in the analysis. These discharges were calculated for each screen using the *Flood Estimation*

Handbook (FEH) method (Centre for Ecology and Hydrology, 2008), with appropriate input parameters obtained from the GIS and land-use data.

Land-use cover was determined visually by overlaying the river network shape files, exported from the GIS, onto Google Earth images, and the percentage contributing area for each type calculated as the long-stream extent of that type multiplied by an arbitrary lateral extent of 100 m divided by the total catchment area contributing to the screen (see Figure 9.11b). Land use was divided into five categories: rural (non-agricultural grassland, woodland and moors); agricultural (arable and pastoral); suburban (cover with dwellings); suburban-open (public access land such as parks and golf courses); and urban (industrial land, urban wasteland, city centres and shopping precincts, etc.).

A social deprivation factor was included because relevant literature, site-based observations and discussions with Scottish local authorities and Environment Agency personnel gave a clear indication that the degree of fly-tipping into, and near streams in the UK is strongly correlated with the level of urban and suburban poverty (see Webb *et al.*, 2006; Environment Agency, 2009a). The social parameter was originally represented in the work of Wallerstein and Arthur (2010a,b) by the Northern Ireland Statistics and Research Agency (NISRA) multiple deprivation measure (MDM) (Northern Ireland Statistics and Research Agency, 2005). However, this variable was replaced in later work by the income domain (ID) score, which represents the number of households within ward-level spatial units that receive financial support from the government (a percentage score, where higher values represent wards where a greater number of households need support). This variable was used because it is applicable across all regions of the UK, whereas the MDM score is only applicable to Northern Ireland. The ID score associated with each screen was determined by overlaying a GIS layer, which delineated each super output area (SOA) region (equivalent to ward-level regional units in Northern Ireland), onto the channel network GIS layer, and assigning the deprivation score for each screen as that associated with the most spatially prevalent upstream SOA (see Figure 9.11a).

9.6.3 Regression analysis

Once the dependent variable, the probability of significant debris-load delivery, and the independent variables (hypothesised driving variables) had been extracted, they were related using linear stepwise multiple regression analysis to develop equations for estimating the probability of debris delivery. The hypothetical relationship was formalised as

$$P_{\mathrm{d}} = f(\mathrm{NL}, \mathrm{SL}, Q_n, \mathrm{RU}, \mathrm{AG}, \mathrm{SU}, \mathrm{SO}, \mathrm{U}, \mathrm{ID}) \qquad (9.1)$$

where NL is the upstream network length (m), SL is the upstream main-stem channel slope (mm^{-1}), Q_n is the flow with return periods of $n = 2$, 10, 50 and 100 years (m^3/s), RU is rural land use (%), AG is agricultural land use (%), SU is suburban land use (%), SO is suburban-open land use (%), U is urban land use (%) (whole number percentages) and ID is the income domain score (decimal %).

Before performing the analysis, the network length and land-use classification variables were log-transformed to account for non-normal data skew. P_d also had to be logistically transformed, as it represents a probability variable (see e.g. Shaw and Wheeler, 1985). The logistic (logit) function is defined as

$$\text{logit } P_d = \ln[P_d/(1 - P_d)]$$

and the reciprocal, the actual probability of debris delivery, is defined as

$$P_d = \frac{e^{\text{logit } P_d}}{1 + e^{\text{logit } P_d}}$$

Asset managers are likely to be interested in critical screen blockages over a range of different temporal scales, so analysis was performed using monthly ($P_{d(m)} = P_{d(\text{Jan,Feb},\ldots)}$) and a seasonal ($P_{d(s)} = P_{d(\text{spring,summer},\ldots)}$) blockage sample subsets. The seasons were defined as: spring (March–May), summer (June–August), autumn (September–November) and winter (December–February). A gross average probability was also computed using all site observations for each screen ($P_{d(a)}$). This was computed in order to create a baseline equation that might be used by the asset manager if a time-independent measure were considered suitable. After transformation of the data, linear stepwise multiple regression analysis was performed using a 95% probability significance level for variable exclusion. The form of the model tested was:

$$\begin{aligned}\text{logit } P_{d(m,s,a)} = \alpha &+ (\text{NL}_{\log} \cdot \beta_1) + (\text{SL} \cdot \beta_2) + (Q_n \cdot \beta_3) + (\text{RU}_{\log} \cdot \beta_4) + (\text{AG}_{\log} \cdot \beta_5) \\ &+ (\text{SU}_{\log} \cdot \beta_6) + (\text{SO}_{\log} \cdot \beta_7) + (\text{U}_{\log} \cdot \beta_8) + (\text{ID} \cdot \beta_9)\end{aligned} \tag{9.2}$$

where α is a constant, log indicates the $\log_{10}$ transformation of an independent variable, and $\beta_{(1,2,\ldots,n)}$ are coefficients.

9.6.4 Results

9.6.4.1 Properties of the probability distribution

The distribution of the number of screens that lay within the 10% class intervals of debris-delivery probability (using the gross average probability calculations) for the 140-screen dataset is presented in Figure 9.12. The data spread was found to have a near normal distribution, meaning that, while the gross average probability of significant debris-load delivery at a high proportion of screens appeared to be around 50–50, some sites were likely to have a very high chance of receiving significant debris loads, while others appeared to have a very much lower likelihood. This outcome demonstrates that the optimisation of inspection and maintenance frequencies on a site-by-site basis is potentially an important exercise in order to maintain an acceptable level of flood risk as set against maintenance costs.

9.6.4.2 Multiple regression analysis results

The constants and coefficients that were determined through the use of multiple regression analysis for the gross average, monthly and seasonal equations are presented in Table 9.4. Before analysing the equations, it was thought to be important to consider the *a priori* expectations about how each independent variable might affect the

Figure 9.12 Number of screens falling within the 10% probability of delivery classes
Reproduced from Wallerstein and Arthur (2012). John Wiley & Sons Ltd

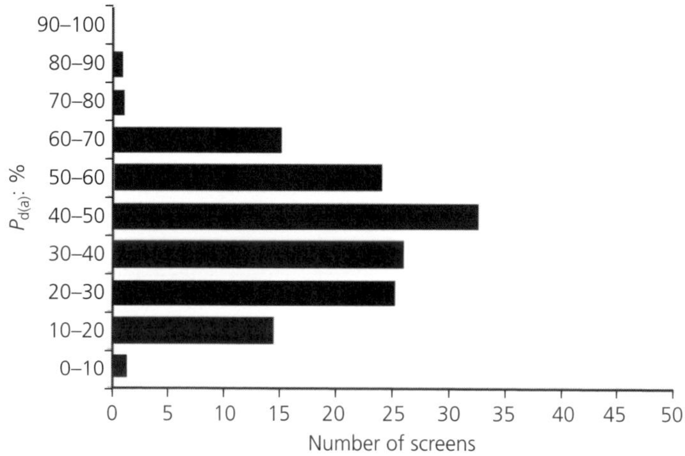

probability of significant delivered load delivery to trash screens, in order to judge whether the magnitude and influence of the coefficients appeared reasonable.

First, it was reasoned that the relationship between upstream network length and probability of delivery of significant debris loads would most likely be positive, the argument being that the longer the contributing river length, the greater the length along which trash may enter the channel. This assumption was borne out in the equations, as all coefficients were found to be positive, as shown in Table 9.4.

Second, it was reasoned that steeper upstream channel slopes were likely to positively influence debris-load delivery, the argument being that slope represents a proxy for flow energy and, therefore, the ability for a river to transport material to screens. On the other hand, however, it was recognised that this relationship might be negative, because steeper streams tend to occur towards the headwaters of catchments where channel widths are low, and therefore represent locations where it is more likely for debris to become entangled in trees and jam across the channel itself, thus impeding its downstream transport. The slope parameter did not appear as a significant driving variable in any of the equations developed in this analysis, however, and it therefore appeared to exert no significant influence.

Third, it was considered reasonable to assume that screens that experienced high discharges as compared with those that experienced relatively lower discharges would be likely to have greater debris-delivery probabilities due to greater flow depths and velocities. However, none of the flows associated with each of the four return periods used appeared as significant driving variables in any of the equations (the column headed $Q_{2,10\ldots}$ represents coefficients for flows of all four return periods, as none of these were statistically significant).

Table 9.4 Constants and coefficients for multiple regression equations that estimate the probability of significant debris-load delivery to trash screens, and the model F-test significance statistics (F), critical F-test values (F_{crit}) and the coefficient of determination values (R^2)

Timeframe	Constant: α	Coefficients: β									F-test and R^2 stats		
		NL_{log}	SL	$Q_{2,10...}$	RU_{log}	AG_{log}	SU_{log}	SO_{log}	U_{log}	ID	F	F_{crit}	R^2
Gross average	−1.14	0.28			−0.19	−0.24				1.12	13.0	2.4	0.26
Monthly													
January	−0.06						0.31				8.4	3.9	0.06
February	−0.71					−0.18					5.2	3.9	0.03
March	−1.03	0.33				−0.32					10.4	3.0	0.12
April	−0.49				−0.35	−0.29					10.6	3.0	0.12
May	−1.56						0.24			2.68	13.9	2.6	0.16
June	−2.30	0.32				−0.22	0.31			1.92	21.9	2.4	0.38
July	−1.54					−0.28	0.31			3.00	25.9	2.6	0.35
August	−1.90	0.25				−0.22	0.25			2.30	17.2	2.4	0.32
September	−1.62	0.28			−0.28	−0.33				1.62	12.6	2.4	0.25
October	−0.44						0.30				12.6	3.9	0.08
November	−0.62	0.44				−0.48					19.9	3.0	0.21
December	−0.81	0.28				−0.20				1.57	9.1	2.6	0.15
Seasonal													
Spring	−1.10	0.23					0.17				6.7	3.0	0.08
Summer	−1.45	0.26			−0.24	−0.25				1.12	12.4	2.4	0.25
Autumn	−1.60				−0.19	−0.24	0.22			2.12	21.5	2.4	0.37
Winter	−0.52	0.30				−0.31					12.4	3.0	0.14

Fourth, it was argued that, as the percentage of rural land use increased relative to the total land-use area for any screen, the probability of delivery of significant debris loads would fall, because debris input from rural land is likely to be limited to organic debris that is relatively small in comparison with anthropogenic waste. It was also reasonable to make the same assumption about agricultural land use, and the sign of the coefficients for both of these variables was found to be negative in all cases, which confirmed this assumption (see Table 9.4). In contrast to this, it was considered likely that, if the proportions of suburban, suburban-open and urban land-use types were to rise relative to the total land-use area, the probability of debris delivery would also be likely to rise as a result of the increased input of anthropogenic waste. This assumption was borne out in the case of the suburban land-use category (Table 9.4), because all regression coefficients for this variable were positive. However, the suburban-open and urban land-use types did not appear as significant driving variables in any of the equations. It was interesting to find that, in the detailed breakdown of blockage material types recorded by the maintenance teams for each screen, slightly over 45% of the screens that had a suburban contributing area of 50% or more received debris containing garden waste such as hedge and grass cuttings. This demonstrated a causal link between a particular type of anthropogenic waste and the land-use type from which one would expect this waste to emanate.

Finally, it was reasoned that, as the ID percentage score increased, so the probability of significant debris-load delivery would also be likely to rise, because higher ID scores represent a greater degree of social deprivation. Again, this assumption appeared to be correct, because all associated regression coefficients were found to be positive. It was interesting to find that the coefficients associated with the ID variable were all relatively large, meaning that this parameter had a strong influence on debris delivery. This outcome has implications for debris management activities, as source control through stakeholder engagement, enforcement of fly-tipping regulations and managed access to channels may well significantly reduce the risk of blockage.

9.6.4.3 Statistical significance of the equations

The coefficient of determination (R^2) for the gross average equation, as shown in Table 9.4, was calculated to have a value of 0.26, which is rather low. However, the associated significance of the equation, determined using the F-test statistic (F), had a value of 13.0, which was greater than the associated critical value (F_{crit}) of 2.4 (95% probability level). This indicated that, while the explanatory power of the equation was weak, it did in fact represent a statistically significant expression for estimating the probability of delivery of significant debris loads. The null hypothesis, that of the equation having no explanatory capability, was therefore rejected.

The R^2 values for each monthly equation (see Table 9.4) were also found to be low. In relative terms, the R^2 values were high for the June, July and August equations as compared with those for the other months, which were extremely weak, especially those for January, February, October and December. However, again, the F-test values for these equations were all greater than the F_{crit} values (95% probability level) and, consequently, although the explanatory power of the equations was weak, they did in fact represent statistically significant estimates of the probability of delivery of significant debris loads.

The R^2 values of the seasonal equations (see Table 9.4) showed that the value for August, while still weak, was the highest of all other monthly and seasonal equations and that of the gross average equation. However, the R^2 values for the other seasonal equations were very low, although all associated F-test values were greater than their corresponding F_{crit} values (95% probability level). It was, therefore, reasonable to conclude that these equations, as in all other cases, were statistically significant.

9.6.5 Decision support

Based on the analysis presented here, a scheme was devised to determine blockage risk associated with any given screen. This scheme is presented in Figure 9.13. The first step in the process is to calculate the probability of significant debris-load delivery to each screen under consideration, using the multiple regression equation(s) appropriate to the timeframe of interest to the user (see Table 9.4). This can be approached in three ways: (a) calculate delivery using the equation for ($P_{d(a)}$) if it is gross long-term average conditions that are thought to best characterise the debris-delivery risk;

Figure 9.13 A suggested scheme for quantifying risk associated with delivery of significant debris loads to trash screens
Reproduced from Wallerstein and Arthur (2012). John Wiley & Sons Ltd

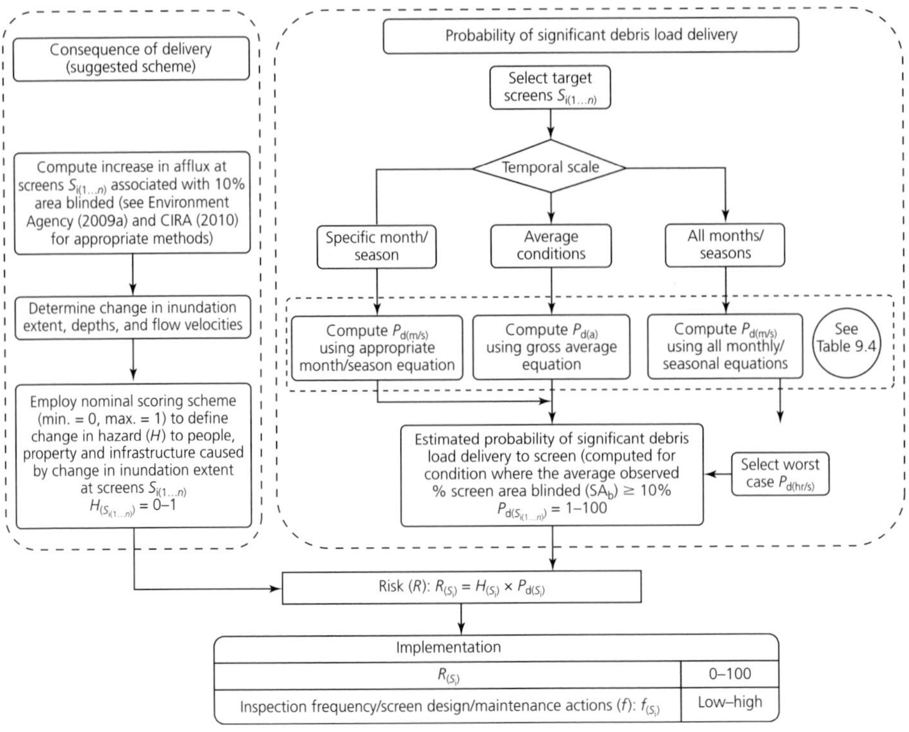

(b) calculate probabilities for a specific month or season if a specific time-frame is of greatest concern; or (c) calculate probabilities using all the month or season equations and select that equation that has the greatest associated delivery risk so that temporal 'worst-case' conditions can be pinpointed. Whichever equation is chosen, a percentage value (0–100) will be extracted that represents the screen's probability of significant debris delivery. It is important to note that these probabilities are based on a dataset in which the screens selected for analysis all had an average annual screen area blocked of 10% or more.

The second step is to calculate the consequence of debris-load delivery. The immediate consequence is blinding of the receptor screen, which increases upstream afflux and, therefore, potentially increases the extent and depth of flooding upstream above that which would occur for the condition where the screen is not blocked. In order to perform this task, the afflux caused by 10% blinding of each screen must be determined. Approaches to doing this are discussed in CIRIA (2010) and Environment Agency (2009a). Once this step has been completed, the consequent changes in inundation extent, flow depth and velocity distributions can be determined using appropriate one- or two-dimensional models. Finally, the altered inundation properties must be linked to the potential changes in physical threat to life, psychological trauma, damage to infrastructure and property, and financial losses. In the scheme presented here, it is recommended that, in order to do this, one should: (a) overlay GIS layers of land-use characteristics and other pertinent receptor properties onto an inundation map that covers the area of the screen under investigation; and (b) apply a nominal hazard score to the screen to quantify the degree to which the increase in inundation extent it causes when blocked will affect the most sensitive local receptor (people, infrastructure, etc.). For the sake of simplicity, the hazard scheme is envisaged to be best represented using a nominal scoring approach, with values ranging from 0 (no impact, such as the case where the flows remain in-bank) to 1 (very high impact, such as potential for loss of life and damage to crucial infrastructure).

The risk of debris delivery associated with any given screen is, therefore, defined as the product of the debris-delivery probability and the consequence of delivery in terms of increased flood hazard. This risk score will, therefore, take a value that lies between 0 and 100. The risk values calculated for the screens that the asset manager is responsible for can then be used as a tool to prioritise screen inspection frequencies, screen design (new or retro-fit) and maintenance activities, by ranking them according to their risk scores. These risk values are obviously on a relative scale, so the actual time between inspections in days must be determined by the asset manager. An example of an approach for achieving this is to: (a) establish from the present maintenance regime used for the screens in question the highest inspection frequency, and ascribe this value to the screen that has the highest risk score; and (b) develop the maintenance frequencies for all the other screens by multiplying their computed risk scores, as decimal percentages, by the highest inspection frequency.

9.6.6 Conclusions from FRMRC II research

The research presented here represents a method for estimating the probability of delivery of significant debris loads to culverts that have trash screens in place. Based on the results obtained through this research, it is possible to draw the following conclusions.

- Gross average, seasonal and monthly debris-delivery probabilities, calculated from the field data for each screen, were related to pertinent channel, flow, land use and social deprivation driving variables using stepwise linear multiple regression analysis in order to generate predictive equations. The results of the multiple regression modelling indicate that the upstream channel length contributing to screens, three land-use types (rural, agricultural and suburban) and the degree of social deprivation all have a statistically significant influence on debris delivery (see Table 9.4). The independent variables used were selected partly on the basis that they could, to a reasonable degree, be acquired through secondary data sources, because it was considered important that the equations should be relatively easy to implement by the asset manager, both in terms of the data acquisition and the processing resource burden.
- Although the analysis identified several key influencing variables, and the F-test values for the equations demonstrated that they were all statistically significant, their explanatory power as defined by the associated R^2 values was low. However, discovery of the significant driving parameters and their relative measures of influence does represent a step change in our understanding of this complex subject.
- Furthermore, despite the weak explanatory capabilities of the equations, this research represents interim findings of an ongoing project, and factors such as incorrect selection or selection of too few or too many independent variables must be assessed in greater detail in order to improve the results.
- A scheme, outlined in Figure 9.13, has been presented in which the predictive regression equations can be used by the asset manager to rank screens that are of concern to them, according to the risk that they may pose in terms of flood inundation. This scheme, therefore, offers an approach by which rational, and therefore accountable, decision-making can be put into effect when prioritising screen inspection and design needs.
- It is recognised that there is uncertainty associated with the use of the equations derived here because, although the NIRA dataset has enabled useful relationships to be generated, they are based solely on rivers located around Belfast, and therefore cannot fully encompass the characteristics of all environments in the UK. Therefore, in order to make the equations more robust, similar data are required from other regions of the UK, although it has not, to date, been possible to source the comprehensive datasets required.

9.7 Case study: failure of a trash screen on the River Sheaf, Sheffield

An example of the potentially catastrophic consequences of severe screen blockage is that which occurred at a large screen on the River Sheaf in Sheffield during a storm event in December 1991. This blockage occurred when an automatic screen-raking device failed, resulting in an almost complete blockage of the culvert entrance, and subsequently causing flooding of a large area of the lower city centre, including the city's railway

Figure 9.14 (a) Flood water overtopping a trash screen on the River Sheaf, Sheffield, in December 1991 (screen submerged at left). (b) Screen bypass channel (right of centre) and hydraulic grab fitted in response to the failure of the original raking device
(a) Reproduced by kind permission of Dr John Riddell

(a) (b)

station. A picture taken at the screen during this event is shown in Figure 9.14a (flow from left to right). The screen was originally installed to mitigate against the potential for internal blockage of the culvert after a severe storm event in 1990 raised concerns about the high potential for blockage as a consequence of such an event. After the 1991 flood event the screen was redesigned, with a large bypass channel and both a screen-raking device and a hydraulic grab (Figure 9.14b). This example highlights the need to determine the potential loading of debris that may arrive at screens during extreme conditions, so that both the screen design and its maintenance schedule can be optimised to cope with such events.

9.8. Conclusions

The potential for blockage within culverts by debris, both natural and anthropogenic, is often mitigated by installing a trash screen over the culvert inlet. These are designed to prevent the ingress of that portion of the imposed debris load, which might otherwise block within the culvert and therefore threaten conveyance capacity. However, trash screens may pose a threat in themselves because their function is, by design, to trap debris and thus, if they are not cleared regularly, can cause an increase in afflux upstream to the point where out-of-bank flood flows occur. It is, therefore, imperative that screens are well designed, and have an appropriate inspection regime in order to ensure that the risks associated with their blockage are minimised.

Current best-practice guidance regarding trash screens in the UK is to be found in the TSSG (Environment Agency, 2009a). Two tools are presented in this guide to aid with screen management, one which is designed to determine optimum screen areas, developed by Magenis (1989), and the other devised specifically for the TSSG, which offers a methodology for choosing whether a screen should be installed at all. A further tool, developed by Faulkner and Waller (1998), which is not included in the

TSSG, is designed to help determine blockage risk at screens based on a nominal scoring system.

When considering how to avert blockage at culverts, three options should be considered. First, and the most desirable, is not to install a screen and thus eliminate the problems associated with a screen altogether. This option can be employed where the imposed debris load is highly unlikely to block inside the structure. Second, is to reduce the risk at source, using land and channel management methods, fly-tipping legislation and public engagement approaches. Third, but least desirable, is to install and maintain a screen.

If a screen is to be constructed, the design properties must be carefully determined using current best-practice guidance (a summary is presented in Table 9.2) and appropriate hydraulic modelling, along with consideration of whether CCTV and water-level monitoring equipment might substantially reduced flood risk.

To date, it appears that no tools have been developed that can be used to predict the probability of significant debris-load delivery to screens. However, this issue has been tackled through a research theme within the FRMRC II project, the results of which are presented in Section 9.6. In the research analysis, screen blockage inspection records were processed to calculate debris-delivery probabilities for 140 screens located in Belfast, Northern Ireland. These were related, using multiple regression analysis, to a series of hypothesised driving factors, including catchment, flow, land use and social deprivation variables, from which a series of regression equations were generated (see Table 9.4). The input of appropriate data into these equations enables the user to determine the probability of debris-load delivery to screens, and, through a flow diagram, to obtain the blockage risk associated with each screen (see Figure 9.13). This methodology, therefore, allows asset managers to rank the screens under their control in order of necessary relative inspection frequency using a rational, scientifically based approach.

Acknowledgements

The research reported here was conducted as part of the Flood Risk Management Research Consortium (FRMRC). The FRMRC is supported by grant EP/F020511/1 from the Engineering and Physical Sciences Research Council, in partnership with the DEFRA/Environment Agency Joint Research Programme on Flood and Coastal Defence, UKWIR, OPW (Ireland) and the Rivers Agency (Northern Ireland). This financial support is gratefully acknowledged. The research is also part of the SAWA Interreg IVB project. The authors would like to acknowledge Bury Council for permission to reproduce the photograph presented in Figure 9.5b, and Dr John Riddell for permission to reproduce the photographs presented in Figures 9.4a and 9.14a. Special thanks are extended to Jim Martin and associated staff at the Northern Ireland River Agency for supplying the datasets used in the analysis presented here. The authors would also like to thank Caroline McGahey and Paul Sayers from HR Wallingford, and Adebayo Adeloye, Gavin Gibson and George Streftaris from Heriot-Watt University for their valuable technical contributions to this research. The authors are also grateful to Janice Blanc (jb201@hw.ac.uk) for providing Figure 9.9.

REFERENCES

Abt SR and Thorne CR (1999) Debris management techniques at hydraulic structures. In *Proceedings of the International Water Resources Engineering Conference* (Walton R and Nece RE (eds)), Seattle, WA, USA. EWRI-ASCE (CD ROM).

Abt SR, Brisbane TE, Frick DM and McKnight CA (1992) Trash rack blockage in supercritical flow. *Journal of Hydraulic Engineering, ASCE* **118(12)**: 1692–1696.

Apelt CJ (1986) *Flow Loads on Submerged Bridges*. Institution of Engineers Australia (Queensland Division), Queensland, AU. Queensland Division Technical Papers 27(19). Publication No. QBG 175612–22.

Austroads (1994) *Waterway Design: A Guide to the Hydraulic Design of Bridges, Culverts and Floodways*. Austroads National Office, Sydney, Australia.

Benn JR, Mantz P, Lamb R, Riddell J and Nalluri C (2004) *Afflux at Bridges and Culverts: Review of Current Knowledge and Practice*. Environment Agency, Bristol, UK. Defra/Environment Agency Flood and Coastal Defence R&D Programme. R&D Technical Report W5A-061/TR1.

Beschta GE (1990) Identifying trees in riparian areas that can provide coarse woody debris to streams. *Forest Science* **36(3)**: 790–801.

Blanc J, Wallerstein NP, Arthur S and Wright GB (2011) Modelling the impact of trash screen design on debris related blockage at culvert inlets. In *Proceedings Rivers 2011*, Penang, Malaysia.

Centre for Ecology and Hydrology (2008) *Flood Estimation Handbook*, vols 1–5 and FEH CD ROM. Available at: http://www.ceh.ac.uk/Feh2/fehintro.html (accessed 10 February 2012).

CIRIA (Construction Industry Research and Information Association) (2010) *Culvert Design and Operation Guide*. CIRIA, London, UK. CIRIA Report C689.

Diehl TH (1997) *Potential Drift Accumulation at Bridges*. US Department of Transportation, Federal Highways Administration, Washington, DC, USA. Publication No. FHWA-RD-97–28.

Environment Agency (2003) *Strategy for Flood Risk Management: 2003/2004–2007/2008*. ver. 1.2. Environment Agency, Bristol, UK.

Environment Agency (2006) *River Valency Catchment Management Strategy*. Environment Agency, Birmingham, UK.

Environment Agency (2007) *Living on the Edge: A Guide to the Rights and Responsibilities of Riverside Occupation*, 3rd edn. Environment Agency, Bristol, UK. Available at: http://www.environment-agency.gov.uk/homeandleisure/floods/31626.aspx (accessed 10 February 2010).

Environment Agency (2008) *Remote Camera Pilot Scheme. Phase 3 – Post Pilot Evaluation. Final Report*. Environment Agency, Bristol, UK. Project No. 9T0223.

Environment Agency (2009a) *Trash and Security Screen Guide 2009*. Flood and Coastal Erosion Risk Management Research and Development Programme. Bristol, UK: Environment Agency. November 2009. Available at: http://tinyurl.com/EA-TSSG (accessed 10 February 2010).

Environment Agency (2009b) *Recommendations for Maintaining the Scientific Relevance of the Conveyance and Afflux Estimation System*. Environment Agency, Bristol, UK. Flood and Coastal Erosion Risk Management Research and Development Programme. Report: SC070032/R1.

Environmental Campaigns (ENCAMS) (2006) *Your Rubbish and the Law: A Guide for Business*. Report No. ENC/04/06. (ENCAMS is now Keep Britain Tidy). Available at: http://kb.keepbritaintidy.org/smokinglitter/publications/rubbish.pdf (accessed 10 February 2010). Full legislation is available at: http://www.legislation.gov.uk (accessed 10 February 2010).

Faulkner BL and Waller SJ (1998) A risk assessment approach to the prediction and modelling of structure blockage during flood flows. In *Proceedings of the 33rd MAFF Conference of River and Coastal Engineering*. Keele University, Keele, UK.

Federal Highways Administrations (1985) *Hydraulic Design of Highway Culverts*. Department of Transportation, Federal Highways Administration (FHWA), Washington, DC, USA. Hydraulic Design Series No. 5. Report No. FHWA-IP-85-15.

Federal Highways Administration (1990) *Highways in the River Environment*. Department of Transportation, Federal Highways Administration (FHWA), Washington, DC, USA. Report No. FHWA-HI-90-016.

Foresight (2004) *Future Flooding*. Office of Science and Technology, London, UK. Available at: http://www.foresight.gov.uk (accessed 10 February 2012).

Forestry Commission (2003) *Forest and Water: Guidelines*, 4th edn. Forestry Commission, Edinburgh, UK. Available at: http://www.forestry.gov.uk/pdf/FCGL007.pdf/$FILE/FCGL007.pdf (accessed 12 February 2012).

Haehnel RB and Daly SF (2002) *Maximum Impact Force of Woody Debris on Floodplain Structures*. Cold Regions Research Engineering Laboratory, Engineering and Research Development Centre, US Army Corps of Engineers, Washington, DC, USA. Technical Report ERDC/CRRELL TR-02–2.

Jones JS, Trent RE and Potter DL (1991) Bridge scour research needs. In *Proceedings Hydraulic Engineering '91* (Shane RM (ed.)), pp. 323–328.

Lagasse PF, Schall JD and Richardson EV (2001) *Stream Stability at Highway Structures*. Department of Transportation, Federal Highway Administration (FHWA), Washington, DC, USA. Hydraulic Engineering Circular No. 20, 3rd ed. FHWA NHI 01-0022.

Laursen EM and Toch A (1956) *Scour around Bridge Piers*. Iowa Institute of Hydraulic Research/Iowa State Highway Commission and Bureau of Public Roads, Iowa City/Ames, IA, USA. Bulletin No. 4.

Magenis SE (1989) *Efficiency of Debris Screens in Urban Watercourses*. Masters of Philosophy thesis, School of Civil Engineering, Kingston Polytechnic, Kingston, UK.

Melville BW and Dongol DM (1992) Bridge pier scour with debris accumulation. *Journal of Hydraulic Engineering, ASCE* **118(8)**: 1306–1310.

National Rivers Authority (1995) *Design and Operation of Trash Screens: Interim Guidance Notes*. National Rivers Authority, Bristol, UK. Flood Defence Publication P-126 (first published 1993, revised 1995).

Northern Ireland Statistics and Research Agency (2005) *Northern Ireland Multiple Deprivation Measure: 2005*. Northern Ireland Statistics and Research Agency, Belfast, Northern Ireland. **doi**: 10.1080/00221680903568592. Available at: http://www.ninis.nisra.gov.uk (accessed 10 February 2012).

Pagliara S and Carnacina L (2010) Temporal scour evolution at bridge piers: effect of wood debris roughness and porosity. *Journal of Hydraulic Research* **48(1)**: 3–13. **doi**: 10.1080/00221680903568592.

Parola AC (2000) *Debris Force on Highway Bridges*. Transportation Research Board/National Academy Press, Washington, DC, USA. NCHRP Report 445.

Pitt M Sir (2008) *The Pitt Review: Learning Lessons from the 2007 Floods*. Cabinet Office, London, UK.

Queensland Government (2007) *Queensland Urban Drainage Manual,* Vol. 1, 2nd edn. Department of Natural Resources and Water, Brisbane, Australia.

Shaw G and Wheeler D (1985) *Statistical Techniques in Geographical Analysis*. Wiley, Norwich, UK.

UK Government (1990) *The Code of Practice on Litter and Refuse: Issued under Section 89 of the Environmental Protection Act*. Her Majesty's Stationary Office (HMSO), London, UK.

US Department of Transportation Federal Highway Administration (2005) *Debris Control Structures: Evaluation and Countermeasures,* 3rd edn. Hydraulic Engineering Circular No. 9. Publication No. FHWA-IF-04-016. Available at: http://www.fhwa.dot.gov/engineering/hydraulics/pubs/04016/hec09.pdf (accessed 10 February 2012).

Wallerstein NP, Thorne CR and Abt SR (1997) *Debris Control at Hydraulic Structures in Selected Areas of the United States and Europe*. US Army Corps of Engineers Army Research Development and Standardisation Group – UK. Contract Report No. CHL-97-4. Available at: http://chl.erdc.usace.army.mil/Media/7/5/6/CHL-CR-97-4.pdf (accessed 10 February 2012).

Wallerstein NP and Arthur S (2010a) Improved methods for predicting trash loading at culverts and trash screens. In *Proceedings of the Flood and Coastal Risk Management Conference 2010*. Environment Agency and Defra, Telford, UK.

Wallerstein NP and Arthur S (2010b) Development of equations for prediction of blockage at trash screens. In *Proceedings of the First European Congress of the IAHR*, Edinburgh, UK.

Webb B, Marshall B, Czarnomski S and Tilley N (2006) *Fly-tipping: Causes, Incentives and Solutions*. UCL Jill Dando Institute of Crime Science, London, UK.

Weeks W, Barthelmess A, Rigby E, Witheridge G and Adamson R (2009) *Australian Rainfall and Runoff Revision projects: Project 11. Blockage of Hydraulic Structures*. Engineers Australia, Barton, Australia. Stage 1 Report. Report No. PII/SI/007.

Wild TC, Ogden S and Lerner DN (2008) An innovative partnership response to the management of urban river corridors – Sheffield's River Stewardship Company. In *Proceedings of the 11th International Conference on Urban Drainage*, Edinburgh (Ashley RM (ed.)).

FURTHER READING

Technical reports

Environment Agency (2004) *Afflux at Bridges and Culverts: Review of current knowledge and practice. Annex 4: A Review of Current Practice for Afflux and Blockage Estimation in UK, Europe and Asia*. Environment Agency, Bristol, UK. Defra/Environment Agency Flood and Coastal Defence R&D Programme. R&D Project Record W5A-061/PR4.

Environment Agency (2004) *Afflux at Bridges and Culverts: Review of current knowledge and practice. Annex 5: Review of the Environment Agency S. W. Region Study 'Risk Assessment of Structure Blockage During Flood Flows'*. Environment Agency, Bristol, UK. Defra/Environment Agency Flood and Coastal Defence R&D Programme. R&D Project Record W5A-061/PR5.

Highways Agency Network Service (2009) *Priority Culverts: Vulnerability of the Network to Flooding*. Prepared for the Highways Agency Network Services through Research and Development Framework by Highways Group/Highways Research Group (HCG/HRG).

Scottish Environment Protection Agency (SEPA) (2010) *Engineering in the Water Environment: Good Practice Guide. River Crossings*. 2nd edn. SEPA, Galashiels, UK.

SEPA (2006) *Conceptual Design Guidelines: Application of Engineered Logjams*. SEPA, Galashiels, UK. Prepared by Herrera Environmental Consultants, Inc., Seattle, WA, USA.

Academic works

Braudrick CA and Grant GE (2000) When do logs move in rivers? *Water Resources Research* **36**: 571–583.

Braudrick CA, Grant GE, Ishikawa Y and Ikeda H (1997) Dynamics of wood transport in streams: a flume experiment. *Earth Surface Processes and Landforms* **22**: 669–683.

Clark P, Tsikata JM and Haresign M (2010) Experimental study of energy loss through submerged trashracks. *Journal of Hydraulic Research* **48(1)**: 113–118. **doi**: 10.1080/00211680903566026.

Gippel CJ, O'Neill IC, Finlayson BL and Schnatz S (1996) Hydraulic guidelines for the reintroduction and management of large woody debris in lowland rivers. *Regulated Rivers: Research & Management* **12**: 223–236.

Tsikata JM, Katopodis C and Tachie MF (2009) Experimental study of turbulent flow near model trashracks. *Journal of Hydraulic Research* **47(2)**: 275–280. **doi**:10.3826/jhr.2009.3381.

USEFUL WEBSITES

Construction Industry Research and Information Association (CIRIA): http://www.ciria.org.uk (accessed 10 February 2012).

Danish Hydraulic Institute, MIKE11 software: http://www.dhigroup.com/software/waterresources/MIKE11.aspx (accessed 10 February 2012).

Environment Agency, Thorverton Gauging Station webcams: http://tinyurl.com/trashcam (to view all cameras and archived images go to the 'Click Here' link located at the bottom left-hand side of this page) (accessed 10 February 2012).

Flood Risk Management Research Consortium project (Phase 2) (FRMRC II): http://www.floodrisk.org.uk (accessed 10 February 2012).

Halcrow, ISIS hydrodynamic model: http://www.halcrow.com/isis (accessed 10 February 2012).

Highways Agency: http://www.highways.gov.uk (accessed 10 February 2012).

HR Wallingford, CES/AES software: http://www.river-conveyance.net (accessed 10 February 2012).

Hydrologic Engineering Center (HEC), US Army Corps of Engineers (USACE), HEC-RAS software: http://www.hec.usace.army.mil/software/hec-ras (accessed 10 February 2012).

SEPA position statement on regulatory approach to culverting watercourses: http://www.sepa.org.uk/water/water_regulation/guidance/engineering.aspx (accessed 10 February 2012).

Flood Risk
ISBN: 978-0-7277-4156-1

http://dx.doi.org/10.1680/fr.41561.245

Chapter 10
Design for environmental improvements

Julian Francis
Environment Agency, UK

The pertinent question is perhaps not whether we value our environment, but why there should be a continual need for design teams to have to justify undertaking integrated environmental design work related to engineering projects.

Good engineering design within a good landscape setting is something design professionals aspire to. 'Functionality', 'innovation' and 'environmentally sustainable' solutions are words that we like to use. But how can we still achieve this outcome on publicly funded flood defence projects, against a backdrop of public sector spending reviews, financial cuts and a drive for greater efficiency?

Good design should not be an additional cost, so the answer, in theory, is relatively straightforward. Firstly, a well-integrated design team can positively embrace the environmental constraints and opportunities. Secondly, skilled and experienced landscape architects and environmental professionals should be involved from the outset and work alongside the engineering team, rather than being engaged too late in some instances in the project cycle to make a real difference, and perhaps being briefed to 'prettify' an engineering design that might be so advanced along the design process that the resulting outcome might already be a foregone conclusion. Readers who seek to redress similar situations should find this chapter useful.

10.1. Introduction

This chapter explores the environmental design principles and options available to project teams working on flood defence projects, and seeks to raise awareness for the acceptance of early involvement of environmental design professionals in an integrated team role. This will enable them to play a key role towards agreeing priorities and producing environmentally and technically sustainable solutions, while working within the framework of legislation and guidance.

A number of design and assessment aids are outlined which, if carefully applied, can assist landscape and environmental design professionals to work in an integrated fashion with engineering designers. A key result of this close working is inevitably the generation of clear aims and priorities from the outset, with the production of well-engineered, sympathetic designs that will both reduce environmental impact and enhance and improve the landscape. Case studies are used to give a broad sample of the ways in which this methodology, together with the application of innovation, can achieve good, and indeed best, practice.

10.2. Principles

Human beings create landscapes, and many of the changes to the environment in which we live and work do not just happen by chance. Landscapes are, in fact, the big picture, containing buildings, natural features such as rivers, the horizon and the places where town turns into country or the land meets the sea. The external environment is affected by how we think about landscapes, use them, and change them over time. They can be designed and managed to look and function in a certain way. Rivers, for example, may be diverted or be subject to engineered defences or flood storage areas to control and to help alleviate the effects of flooding.

Environmental professionals such as landscape architects are trained to deal with the many ways in which our landscapes can change and be changed. Some landscapes – such as national parks – are protected by law, because, as a nation, we feel very strongly about conserving their beauty. Environmental professionals are also able to promote the environmental dimension of sustainability and to promote the best practice standards in environmental management, auditing and assessment. This is particularly important given underlying concerns of climate change and an increasing call for a more efficient use of funding for a more robust cost–benefit analysis of flood defence projects.

Nowadays, landscape architects are playing a big part working alongside architects and the engineering professions in regeneration and public realm improvements. Landscape architects restore land for new uses and develop plans for how different landscapes should be managed, whether they are for wildlife, people or industrial development. The profession also has an important role in dealing with modern-day problems such as: making it easier to walk or cycle instead of using a car; and tackling the effects of climate change by, for example, planting more trees to create shade and cool buildings down. Importantly, the profession is dealing with flooding by studying water volumes and flow, and then designing landscapes that can cope better with heavy rainfall. It is also playing an increasingly important part in providing soft and hard engineered defence solutions to protect homes and businesses from the effects of flooding (Landscape Institute, 2008).

Fluvial and tidal flood defences, by their very nature, are hard engineered defences. Experience has shown that the full environmental potential of a site is far from straightforward, and requires early input by environmental professionals as part of the project team to help develop clear aims and priorities from the outset. For example, the benefits of river restoration and wetland creation can be applied to flood defence projects so that the environmental benefits complement the improved protection. There are a number of organisations, such as the River Restoration Centre (1995) in their *Manual of River Restoration Techniques*, that adequately detail methods. The techniques described in this chapter tend to be linked to improved environmental function in 'hard' engineered defences.

10.2.1 Key drivers

The diversity of landscapes is a key environmental, social and economic resource for the nation. By their very nature, landscapes are the settings for everyday life, and are a reminder of how our culture interacts with the physical world. This is because there is a close affinity between people and place.

In planning new construction projects, there is almost certainly going to be a resultant change on the physical landscape. The impact is often a measure of the degree to which the impact is either short term, or a residual one with lasting effects. It is often the case that projects are screened at an early stage as part of their feasibility for their impact on the environment in which they will be located or have a wider effect. One of the common criteria is whether there is a designation attached to that built or natural landscape. While this is an important factor, it should also be remembered that among some of the best-known landscapes are those that do not hold a formal recognition. This issue was identified at a European level, and eventually led to the development of an international agreement devoted to the protection, planning and management of all landscapes, whether they are of outstanding beauty, degraded or ordinary landscapes. It applies to rural, urban, coastal and marine areas. This convention is The European Landscape Convention (the Florence Convention the (ELC)) (Council of Europe, 2000). This is an important factor in the future planning and implementation of developments, including flood defence provision, and is the first international instrument devoted exclusively to the protection, management and planning of all landscapes in Europe in their entirety. It was published in 2000, signed by the UK in February 2006, and came into effect in England in March 2007. Article 1 of the ELC, defines landscape as: 'an area, as perceived by people, whose character is the result of the action and interaction of natural and/or human factors' (Council of Europe, 2000).

10.2.2 Growth

The forward-looking approach of the ELC is concerned with providing the highest quality landscape for future generations, by protecting special landscapes and by managing and enhancing all landscape everywhere. National implementation is led by the Department for Environment, Food and Rural Affairs (Defra). A workshop on the ELC held in London and York (RECEP/ENELC, 2009) indicated that there is a need for regional planners, policy makers and stakeholders to work together and develop action plans that will result in the effective implementation of the ELC. This will be important for future flood defence provision, particularly as implementation at all levels of government is expected to capture the cultural and historical, as well as the natural, richness of the landscape.

The ELC states that the importance, qualities and functions of all landscape should be further integrated into spatial planning and land management sectors as well as other sectoral policies. It also states that landscape provides a spatial framework, applicable at different geographic scales, where ecology, landscape character, cultural and aesthetic values, and the ecosystem goods and services that landscapes provide, can be brought together and support an integrated approach to decision-making.

The ELC has recognised the requirement for legal intervention to develop landscape policy that can protect, manage and create quality landscape. The legislation incorporates the knowledge that the participation of stakeholders at all levels, including national government, regional policy makers, local authorities, businesses and the general public, will be vital for the successful design and implementation of policy. The ELC is the first international convention to focus specifically on landscape.

Natural England is leading on the implementation of the ELC in England. An implementation framework has been prepared with Defra and English Heritage to guide the development of action plans by other stakeholders and organisations. Natural England has prepared its own ELC action plan to embed the convention's measures through its own work and through that of others.

Raising awareness in the value of landscape in all sectors of society and the role that these stakeholders play in shaping our landscapes is a main theme in the ELC. Specific measures include

- the identification and assessment of landscapes, including the analysis of landscape change, with the active participation of the stakeholder
- the setting of objectives for landscape quality with the involvement of the public
- the implementation of landscape policies through the establishment of plans and practical programmes.

The UK has already put many of the principles of the ELC into practice. This is shown through the joint character area map of England, and also through the well-established practice of using landscape character assessment (LCA) to inform local policy making. LCA is a tool to identify what makes a place unique, and can serve as a framework for decision-making that respects local distinctiveness. It also provides a mechanism by which local communities and other stakeholders can contribute towards the decisions that affect their local surroundings. A number of key Environment Agency Flood and Coastal Risk Management (FCRM) strategies, including the Lower Lea, and the Dee and Clwyd Strategies, have already used LCA successfully. For these, LCA provided a framework for spatial planning and targeting of green infrastructure and land management initiatives. Furthermore, through a shared understanding of the distinctive characters of the landscapes, it provided a framework to support decisions about strategic-scale landscape change.

A significant project that exemplifies the crux of the principles underpinning this chapter is the Thames Estuary 2100 (TE2100). This Environment Agency project embraces climate change as a key driver in developing a tidal flood risk management plan for the Thames Estuary through to the end of the twenty-first century. The project is the first major flood risk project in the UK to have put climate change adaptation at its core. It demonstrates the Environment Agency's desire to create a better place when designing and implementing flood management responses across a full range of different situations that exist along the River Thames, through urban London and eastwards into the more rural areas of the estuary for a distance of approximately 68 miles.

The landscape architect took the leadership role, heading a multidisciplinary team of engineers, ecologists, environmental professionals and planners. The team produced award-winning guidance (Highly Commended in the Landscape Policy category of the 2009 Landscape Institute Awards) to inspire and inform those involved in flood risk design to think and act creatively to create quality urban landscape, green infrastructure and open spaces resilient to climate change. A tool kit was devised in the development of environmental enhancement measures for flood risk management. The primary objective of the guidance is to inspire and inform those involved in flood risk design to think and act creatively in order to successfully integrate environmental and ecological enhancement into future flood risk management schemes in the estuary. A critical factor was that a strategic approach was made which meets the requirements of raised defences, while also including a primary consideration of the broader environmental objectives of retaining and enhancing the visual and physical connections of river communities with the Thames itself in order to create a better place (Environment Agency, 2011).

LCA is a significant driver in the above project developments, which also underpins Natural England's 'All Landscapes Matter' approach (Natural England, 2009). All Landscapes Matter sets out the position that Natural England will adopt when addressing the conservation and enhancement of England's landscape. It outlines how others should consider and work with landscapes everywhere, at all scales and in all contexts. This statement also forms part of a suite of detailed landscape positions that addresses future landscapes, protected landscapes, and the historic environment, all of which apply to the design of flood defences.

As the legislation and policy with respect to landscape is fairly well developed and the level of linkage that exists between landscape and other environmental resources is intertwined, there is potential to place regional policy for landscape as a central core upon which to build policy for other areas such as biodiversity, water, and marine environments, air quality, soil and climate. Discussions and engagement with Natural England with regard to strategic- and project-level flood risk management will continue to see this play a key role in decision-making. It is likely that engagement with stakeholders on FCRM work will be expected to embrace LCA more commonly than is being practiced at present.

10.3. Planning matters relevant to flood defence design

Not all projects will require an application for planning permission, which is largely centred around whether the works are an 'improvement' or 'development', or are largely maintenance works. Permitted development for FCRM works is defined under Sections 14 and 15 of the Town and Country Planning (General Permitted Development) Order 1995 (GDPO) (SI 1995 No. 418) as: 'Development in or under any watercourse or land drainage works, and required in connection with the improvement, maintenance or repair of that watercourse or those works'. Such development does not require planning permission from the local planning authority.

Under the terms of the Land Drainage Act 1991, the prior written consent of the Environment Agency is required for any mill, dam, weir or like obstruction, or for any culvert (or alteration to a culvert) likely to affect flow in an 'ordinary watercourse'. In the case of an ordinary watercourse in an internal drainage district, the consent of the internal drainage board (IDB) is required. Where there is no IDB, the local authority is also a drainage authority, and may have its own bylaws.

Under the EIA (Land Drainage Improvement Works) Regulations 1999 as amended, 'improvement works' are defined as works that are

- the subject of a project to deepen, widen, straighten, or otherwise improve or alter any existing watercourse, or remove or alter mill dams, weirs or other obstruction to watercourses, or raise, widen or otherwise improve or alter any existing drainage work
- permitted development by virtue of Part 14 or 15 of Schedule 2 to the Town and Country Planning (General Permitted Development) Order 1995.

This work applies to improvements undertaken by land drainage bodies, including the Environment Agency. Such works are permitted development. Other projects will fall under the planning system and will require planning permission.

Therefore, a key early consideration is whether the project development is likely to fall within the remit of the Town and Country Planning Environmental Impact Assessment (EIA) Regulations, or if it is likely to be permitted development under the Land Drainage EIA Regulations or the Town and Country Planning (Development Management Procedure) (England) Order (the DMPO) from October 2010, which is a consolidated version of the former (General Development Procedure) Order (GDPO) 1995.

An early indication as a result of discussions with the local planning authority will often direct the way in which the project team will undertake early project planning aspects. There are a number of key planning legislation and policy documents relating to the environment that play a significant role in flood defence design and implementation. Of particular relevance to land-use planning and biodiversity are Flood Defence Consents (formerly known as Land Drainage Consents). The flood defence provisions under the Water Resources Act 1991 (section 109 specifically) and associated bylaws require formal consent for works in, over, under or adjacent to main rivers. This is to ensure that such activities do not cause or make worse an existing flooding problem, interfere with our work, and do not adversely affect the local environment, fisheries, wildlife and flood defences.

Planning authorities, in assessing applications for flood defences, are likely to take into account the following environmental aspects of projects

- to protect and enhance biodiversity, taking a 'net gain' approach
- focus on designated sites but also biodiversity in general
- link to biodiversity action plans
- develop, protect and enhance green/river corridors and wildlife networks, linking to flood plain protection, water-based recreation, access to watercourses, navigation, water quality and fisheries
- a landscape assessment relating to the visual impact of the proposed scheme.

In addition to obtaining planning permission, developers may need to get consents or licences for which the Environment Agency is responsible before they can proceed with development of a site. These consents are issued under separate legislation to the Town and Country Planning Act (TCPA) 1990 and can sometimes lead to difficulties and misunderstandings, as the Environment Agency may wish to refuse consent for an outfall or culvert associated with a development that already has planning permission. It is important to avoid this situation by addressing all the relevant issues at the planning-application stage. Information requested at this stage can often also be used for licensing.

The two key planning policies that are taken into consideration are discussed in Chapter 1, notably Planning Policy Statement 25: *Development and Flood Risk* (PPS 25) (Communities and Local Government (CLG), 2009) in England and Technical Advice Note 15: *Development and Flood Risk* (Welsh Government, 2004) in Wales. In addition, in England, Planning Policy Statement 9 (PPS 9: *Biodiversity and Geological Conservation* (CLG, 2005a) and the related Circular 6/05: *Biodiversity and Geological Conservation – Statutory Obligations and their Impact within the Planning System* (CLG, 2005b), sets out the UK government's expectations of the planning system in relation to nature conservation within England. PPS 9 includes government objectives, key principles and a series of policies to ensure that the potential impact of planning decisions on biodiversity and geological diversity are carefully considered. These principles place emphasis on the importance of sustaining and enhancing environmental resources and on promoting mitigation or compensation measures where appropriate. It also advises, in paragraph 28, on the importance of networks of natural habitats, and specifically states that canals and rivers are part of such networks. The accompanying *Planning for Biodiversity and Geological Conservation: A Guide to Good Practice* (CLG, 2006) complements PPS 9 and Circular 6/05, and provides good practice guidance on ways in which local planning authorities can help deliver the national policies and legal requirements.

In Wales, *Planning Policy Wales* (Welsh Government, 2011) establishes the national planning policy framework, including the Welsh Assembly's objectives for the conservation and improvement of the natural heritage. Technical Advice Note 5 (TAN 5): *Nature Conservation and Planning* (Welsh Government, 2009) (currently being revised, with consultation on a draft replacement TAN 5 completed) provides detailed guidance on nature conservation and includes guidance regarding wildlife corridors.

10.4. Design options

In response to the threat of climate change affecting our coastline and coastal communities, and riverine flooding affecting our inland communities, design professionals need to find ways of working with our partners in a smarter and more innovative manner. This includes examining the ways in which projects are funded, reviewing the costly consequences of flooding and erosion to householders, and also looking at the effects on local communities in terms of local amenity, tourism, jobs and opportunities for economic redevelopment. This aspiration is being increasingly seen in a less than positive light, given the certainty of central government reductions in public sector spending in the current political and economic climate.

Local communities are increasingly being given greater scope for engaging with project teams in the design of flood defences. These communities generally have a good knowledge of the local characteristics of the environment in which they live and work. Evidence of this can be seen by the publication of good-quality Parish Plans and Village Design Statements in England. This process also meets the aspirations of local planning authorities in helping to develop their spatial plans.

One such example is partnership funding of projects to supplement central funding with other sources of funding that reflect the range of public and private benefits that can be delivered. This can be an extremely successful way to deliver projects, and often assists in 'ownership' of schemes by stakeholders such as community groups, developers, businesses and landowners.

Statutory Guidance for the Environment Agency (Code of Practice on Conservation, Access and Recreation (Department for Environment, Food & Rural Affairs (Defra, 2000)) issued under Section 4 of the 1995 Environment Act states that the Agency should: 'take a holistic approach to the protection and enhancement of the environment. It should therefore strive through its actions to optimise benefit to the environment as a whole'. The guidance makes it explicit on the first page that the Agency's duties involve 'conservation and enhancement'.

Referring specifically to the conservation of natural beauty, the guidance acknowledges the fact that the operations of relevant bodies (including the construction of new schemes) can have significant environmental effects that may result in 'direct visual effects', and goes on to provide clarity on how we must address such impacts: 'relevant bodies should avoid damage arising from any works and land use changes which could have an adverse effect on the character of the landscape. Projects should therefore be designed to conserve and enhance the landscape character of an area, using local materials and building styles where possible and, if possible, plant native species which are appropriate to the site and of local provenance'.

10.4.1 Design and access statements

Another aspect that directly relates to flood defence projects requiring planning permission is the requirement for Design and Access Statements to accompany the planning application. These statements should explain the design principles and concepts that have informed the development and how access issues have been dealt with. The design and access components of the statement are separate, but are inextricably interrelated and influence each other.

One of the key purposes of the statement is to allow local communities, access groups, amenity groups and other stakeholders to involve themselves more directly in the planning process without needing to interpret plans that can often be technically challenging. This, in turn, helps to increase an understanding for people affected by development, and helps to improve trust between communities, developers and the planning authorities. This is primarily because the statement enables the rationale for the proposal to be more transparent to stakeholders and the local planning authority.

As set out in the GDPO, statements would not normally be required for pure engineering (or mining) operations, but, due to the broader nature of flood defence projects, most planning authorities will expect a statement to accompany both outline and full planning permission applications.

The Commission for Architecture and Built Environment (2006) has produced a good practice guide entitled *Design and Access Statements, How to Write, Read and use Them*.

It is recommended that the project design team produces a document in the early stages of the scheme development. The four key points that should be covered are: assessment, involvement, evaluation and design. Further elements will then need to be developed for design and access, as outlined below.

10.4.2 Design

There are two key aspects.

- Use.
- Amount.

For flood defence projects, these first two points are relatively simple to explain.

- *Layout*: the description will need to explain, for flood defence projects, how the line of the floodwall or embankment has been modified or placed to integrate with the local environment setting. This would include consideration of how it relates to its surroundings, and how existing views of the river or other important views have been considered and retained.
- *Scale*: this should include consideration of reducing the height of walls or embankments, and any other measures used, to ensure that they fit within the environment.
- *Landscaping*: this should include an explanation of how we treat land that may be severed by a new flood wall or embankment, with the rationale for any tree planting, or reasons for not planting trees.
- *Appearance*: the report should explain how the cladding for the walls integrates with the character of the existing setting, and the rationale for the proposed materials.

The guidance makes it clear that design includes the site layout, scale of development and landscaping. It is not simply a matter of appearance or how cladding has been applied to permanent wall structures.

10.4.3 Access

- *Vehicle access*: a description of why access has been chosen and how the site responds to local road layout and public transport provision.
- *Pedestrian access*: this section is extremely important in the design of flood defences. It encompasses the design and how people can get to and move through the space on equal terms, irrelevant of age, disability, ethnicity or social group. Flood defence walls and embankments can present very real physical barriers to disabled and less physically able users. Even relatively small changes in height or the inclusion of steps can cause real problems for users. The designers have a duty to integrate the walls and embankments in order that access is not being limited. It should also highlight opportunities for improved access to the waterside or countryside. Consideration also must be given to improved access provision in relation to impacts, particularly on people's privacy and any vulnerability to crime.

It should be remembered that the Disability Discrimination Act 1995 strengthens people's right to equal access. The legislation makes it unlawful for providers of goods, facilities and services to discriminate against a person on grounds of disability and requires them to improve access by making reasonable adjustments to assist disabled people. The Disability Discrimination Act 2005 (DDA 2005) brought further changes, requiring all public bodies to positively promote disability equality. Currently there are no statutory benchmarks for reasonable practice in the outdoors with regard to DDA 2005 – which is not surprising given the enormous diversity of green space and countryside. This makes it even more challenging for designers and flood risk managers to know what standards to apply when planning flood defences and access.

In the case of the Environment Agency as the largest provider of flood defences in England and Wales, in addition to internal approval processes to ensure that the project proposal is well founded, projects must all be achievable and represent sensible and cost beneficial use of public money. The Environment Agency will generally be required to apply for formal planning permission where permitted development powers do not apply, which clearly means that the local planning authority is likely to assign planning conditions to the consented proposal. In order to ensure smooth progress through any approval body, it is important to set out how any impacts identified will be mitigated and how the Environment Agency will enhance the environment for people and wildlife. Where any environmental assessment has stated that there will be adequate mitigation and enhancement, it is vital that this statement is followed through at the delivery stage.

Under the UK planning system, development must be good enough to approve as opposed to being poor enough to reject. As this applies to all development considered through the planning system, it is reasonable to infer that the Environment Agency (with the clear responsibilities that it has under legislation) should be doing a better job than the average developer.

10.5. Performance appraisal

Landscape and environmental design brings together the engineering proposals with the environmental impact assessment process. As an inseparable part of both processes, it helps to

- reduce or avoid negative impacts on the environment by design
- design and integrate effective mitigation for negative impacts
- identify and realise the potential for environmental enhancement within projects
- achieve creative and effective management of landscape and habitat creation schemes.

Well-founded environmental design requires a good understanding of an area's ecology, history, pedology, geology, landscape character, current land use and the requirements of the proposed development. To ensure this, rigorous analysis and site planning are needed. The process then becomes iterative, from the development of options, through selection of preferred options, to the development and then refining of the design.

There is a very clear symmetry between landscape and environmental design and the engineering design process. Landscape is largely concerned with the relationship between people and place. It is important to note that it does not merely apply to special or designated landscapes, and applies equally to urban, rural and coastal projects.

Due to the nature of many flood alleviation schemes, the urban landscape is an important aspect of many projects. Improving the more neglected urban environment is a key challenge requiring careful design. Where planning consent is required, local planning authorities expect developers to integrate their proposals in order to minimise impacts and to help improve the standard of the local environment. Flood defence designs are no exception.

Landscape character is the pattern arising from particular combinations of different components of the environment, which provides a sense of place to our surroundings. A good understanding of landscape character is a key factor in good environmental design, and thus engineering design. Landscape character assessment (LCA) (The Countryside Agency/Scottish Natural Heritage, 2002) and landscape and visual impact assessment (LVIA) (Landscape Institute and Institute of Environmental Management and Assessment, 2002) are both processes used to consider the impact of a landscape setting receiving a new flood wall. A review of LCA would be undertaken to consider the character of a landscape setting. This would help to inform a LVIA which would then be used to review the likely visual changes a flood wall would have on that setting, the key being whether the construction of a flood defence will destroy the character of the setting (Figure 10.1). Guidelines issued by the Landscape Institute and the Institute of Environmental Management and Assessment (IEMA) have classified landscape character using a five-point system. The five points place a value indicating the level of quality that a landscape might possess.

Figure 10.1 Shrewsbury Frankwell flood alleviation scheme. Note the two-tier wall structure and planter which, now established with shrubs, assists visually to help soften and screen the high wall from observers on the opposite bank
© Julian Francis

With regard to work being undertaken by the Environment Agency, the Environment Act 1995 states that the principal aim of the Environment Agency is *to protect and enhance the environment and contribute to achieving sustainable development.* Under this legislation, there is an inherent duty to promote

- the conservation and enhancement of natural beauty and amenity when developing projects or consenting others
- to conserve flora and fauna
- to promote the use of water and associated land for recreational purposes
- to conserve buildings and sites of historic interest.

There is often an anomaly with regard to the level of input required to adequately assess the impacts of new projects. Landscape and environmental design professionals from the Environment Agency's National Environmental Assessment Service (NEAS) have successfully produced guidance notes on how landscape and environmental design issues should be handled in the development of flood defence projects for the Environment Agency. The aptly entitled *Landscape and Environmental Design Guidance* (Defra/Environment Agency, n.d.), is aimed primarily at staff within NEAS managing the strategic environmental assessment (SEA)/environmental impact assessment (EIA) process and for consultants undertaking and writing SEA/EIA's and undertaking associated landscape assessment and design tasks. The guidance centres around procedures which identify the steps that need to be taken and the programming of these to ensure full consideration of the environmental opportunities and constraints when developing a project. It also gives guidance on how to manage these issues, and includes a list of design quality standards to inform designers of the approach they should adopt.

The guidance encourages good environmental design that identifies and realises the potential for environmental benefits within projects and that reduces or avoids negative impacts on the environment. It contains a series of advice notes on how to interpret standard methodologies, such as the landscape and visual impact assessment methodology, when working on Environment Agency flood defence projects. It also outlines how to deal with trees, the design of walls and embankments to fit within a landscape, the use of landscape protection zones and planting maintenance.

10.6. Design and assessment aids

A summary of relevant LEDG guidance is outlined in this section for flood wall and flood embankment design, together with the environmental benefits brought to a flood defence scheme by sympathetically designed hard and soft landscaping.

10.6.1 Flood wall design

Flood walls are likely to be the preferred option in urban or built-up areas where the available space to construct a defence is limited. While requiring a narrower strip of land compared to an embankment providing a similar level of defence height, the construction of such a structure can still require a reasonably wide corridor during the construction phase, and is often more complex to design and construct.

- The alignment of flood walls should minimise damage to landscape features (particularly mature trees), maintain access to the river and preserve existing views (see Figure 10.1).
- Good and innovative design should be applied to make flood walls a positive feature in the landscape.
- Landscape Character Assessment should be used to define the appropriate scale, form and finish of a wall.
- Particular attention should be made to the transition between different wall heights and widths, materials, changes in direction, or where a wall adjoins to an existing riverside structure (Figure 10.2).
- In the Environment Agency's case, it has statutory duties to enhance the environment, so there is a general presumption that sheet-pile and concrete walls will need to be clad or treated appropriately to ensure that they build on an area's character and not detract from it.
- On the grounds of aesthetic quality, sustainability and longevity, natural stone is preferred to reconstituted stone. Natural stone should, however, be sourced locally, and should not be purchased from overseas.

Figure 10.2 The variation of cladding materials along this section of flood wall shows the attention to detail by the design team in order to try to mirror the type of typical local boundary wall
© Julian Francis

- Consideration of health and safety aspects should be made early in the design process for items such as railings along the top of a wall.
- Expansion joints in walls should extend not only to cladding but also through any capping or coping units. Differential movement will otherwise cause cracking of joints or even the units themselves.

As a result of landscape of assessment in towns such as Bewdley and in Shrewsbury on the River Severn, proposed flood defences moved away from a solution using permanently constructed walls in their entirety, and partly took the form of other forms of defence on the grounds of amenity and visual impact. While permanent flood walls were indeed installed in these towns, they were used in conjunction with demountable flood barriers, which provided the perfect solution for the specific settings (Figure 10.3). There is continuing debate regarding the labour-intensive requirements in the installing and dismantling of this form of barrier (Figure 10.4). It has always been the case, however, that demountables are used as a last measure where other forms of defence cannot to provide a suitable solution on environmental, technical or economic grounds.

Figure 10.3 Demountable aluminium barriers erected along the quayside in Bewdley, Worcestershire
Contains Environment Agency information © Environment Agency and database right

Figure 10.4 A routine exercise by the Environment Agency to test the installation of a demountable defence
© Julian Francis

There is a variety of alternative materials for flood defence walls, including

- concrete, cast in situ or pre-cast units
- brick or stone
- steel sheet piles
- timber
- glass
- metal
- plastic.

Finishes to concrete walls can include

- textured
- coloured
- exposed aggregate
- imprinted.

Cladding materials include

- brick:
  - brick trims and patterns
  - brick corbelling
  - brick panels
  - brick specials
- natural stone
- reconstituted (artificial) stone:
  - heritage-type finish
  - split stone finish
  - architectural stone finish.

As discussed earlier, the Code of Practice on Conservation, Access and Recreation states that projects should be designed to 'conserve and *enhance* the landscape character of an area', and 'to use local materials and building styles wherever possible'. The term 'enhance' has been highlighted because the Environment Agency on its projects has to do more than just provide mitigation or replace what was there – it has to actually improve, or enhance the environment. In many cases, plain concrete or sheet-pile walls will detract from an area's landscape character, and therefore such structures will need to be clad or treated in some way.

This approach has been challenged and upheld by an independent arbitrator in an important case between the Environment Agency and a major retail outlet in the north west of England. A supermarket had its land drainage consent application for an unclad sheet-pile wall refused by the Environment Agency despite the development already having planning permission. The Agency insisted the wall should be clad in order to fulfil its duties under the Environment Act 1995. The developer contested that decision, the case went to arbitration, and the arbitrator upheld the refusal of consent saying that the condition was reasonable and so should stand.

This case set clear precedent and demonstrated that the Environment Agency had made a correct interpretation of its obligations under the Environment Act with regard to the duty to further the conservation and enhancement of natural beauty, both as a water management operator and as a regulator. Obviously, for several key reasons, the Environment Agency cannot apply lesser standards to its own work than it applies to developers. In essence, the Agency should always be looking for opportunities to improve the appearance of a setting, and certainly not diminish the visual appearance of an area.

As part of the environmental impact assessment (EIA) process, proposals are evaluated to establish their environmental impacts and assess whether those impacts can be avoided or reduced to acceptable levels. Reduction of impacts through an action of some description is known as 'mitigation', and is a fundamental element of environmental impact assessment. In other situations, through policy, aspiration or opportunity, where a developer is planning to leave an improved environment through intervention, this would be termed an 'enhancement'. Cladding, for example, is generally seen as a mitigation to offset an adverse visual impact, often of a steel sheet-pile or concrete wall (Figure 10.5). On Environment Agency projects, for example, the delivery of mitigation measures during the construction phase is crucial to ensuring that the Agency adheres to its statutory duties, as mitigation is non-negotiable for the Agency.

This case does not, of course, lead to a presumption that every sheet-pile wall and concrete wall will require cladding or some other treatment to enhance the environment. Nevertheless, the costs of cladding are normally relatively small in comparison to the cost of a project. Stripping out cladding in order to save costs will achieve very little in terms of savings, but will mean that major negative risks materialise. For example, in the case of an Environment Agency scheme, the reputation of the Agency will be significantly damaged where the assessment process has derived that a certain finish should be provided, but the product delivered is completely different, particularly in cases where the finish was not conditioned as a result of a well-founded assessment. A consequence of this would be in relation to future work in the locality, which is most likely to involve dealing with the same local planning authority and local communities. It would also mean that, as a developer, it would be open to the imposition of onerous planning conditions based around remediation and mitigation, and difficulty in taking future initiatives and partnership forward. In essence, the cost of dealing with a reputational risk may far exceed the perceived cost savings of stripping it out.

The Environment Act makes no distinction whether or not walls that are not generally visible by people need to be addressed. By way of guidance, and to ensure that it stays within the duties of the Environment Act, if the Agency is proposing anything that would reduce the appearance of the environment, it would have to ensure that it is addressed in such a manner as to prevent deterioration of the environment.

It is important to note that the legislation on the 'conservation and enhancement of natural beauty' derives from the National Parks and Access to the Countryside Act, which enabled the designation of places because they were still in a relatively wild and

Figure 10.5 The Princess Royal at the official opening ceremony of the Shrewsbury Frankwell Flood Alleviation Scheme. Provision was made for protection against more frequent flood events through the use of a raised embankment and walkway. Note the use of stone cladding and the use of a demountable barrier at a higher level to provide protection against larger events
© Julian Francis

natural state. To argue that sites that are inaccessible do not deserve conservation of natural beauty is to specifically argue against the purpose of the major legislation for protection of landscape and the countryside, including that in the Environment Act.

In the spirit of the duties conferred on the Agency by the Environment Act 1995, by the Agency's aspiration to Create a Better Place for people and wildlife, and precedent set by the arbitration case described earlier, it should be ensured that new walls of any structures do not create a visual impact and are in keeping with their setting. In effect, this means that there should be a presumption to treat the visual appearance of any sheet-pile or concrete wall so it does not prevent deterioration of the environment. Any deviation from the presumption to treat the visual appearance must be assessed and reviewed on a case-by-case basis in order to find a balanced solution.

In reaching a decision on treating the visual appearance of an Environment Agency structure, it is important to address the wider issues involved. These will generally include

- the landscape character and the landscape and visual impact assessment
- the views of the local planning authority, stakeholders and local community
- the amount of waste that will be generated
- the pollution risk to watercourses during construction
- health and safety issues associated with the structure and treatment proposals
- cost – whether it is reasonable in the context of the cost of the whole scheme
- construction, design and management (CDM) and maintenance requirements – whether the treatment will require excessive and difficult maintenance
- time – whether this will be disproportionate to the scheme
- resource efficiency – whether the measure is sensible in its use of finite resources
- whether the carbon footprint of materials and transport is acceptable.

The health and safety of employees and contractors is rightly of paramount importance and a legal requirement. Employees or contractors should not unnecessarily be placed in positions of risk, such as working over/in water, or handling dangerous materials unnecessarily. The Construction (Design and Management) Regulations 2007 are, therefore, a fundamental consideration for project managers on flood defence projects. Application of the Regulations in a balanced manner is the role of the project manager and the CDM coordinator. In finding a solution to the visual appearance, it is vital to ensure that the structure is safe to construct and will be safe during operation. A key role of the designer, for example, is to eliminate hazards and to reduce risks during design. Elimination of risk is likely to prove impossible; a solution that allows for inspection and will not compromise the CDM Regulations in any way is always possible, and it is equally important to engender a balanced 'can do' attitude in these circumstances in order to achieve the best outcome.

10.6.2 Flood embankment design

The design of flood embankments has its own unique issues and identifiers. The scale of embankments and their footprint is normally directly related to the design height, which in turn is related to the design height of the level of protection to be provided. An embankment is typically formed by the construction of an impermeable core to its centre, with layers of compacted fill material built up to the required height and profile, and finished with topsoil over the entire construction, normally for seeding or turfing with grass.

Embankments are often large structures and can run for considerable distances, and require considerable quantities of material in their construction. Locally won material can be excavated from borrow pits adjacent to the line of the defence, but sometimes there is no suitable material in the locality, meaning that it has to be transported to site.

The resulting environmental impact of this work is likely, as a minimum, to entail structures that are not natural features being constructed within or at the edge of a flood plain. There may also be an impact on the local landscape scene of borrow areas, temporary storage of materials and/or trafficking of imported material.

However, where material can be won locally there may be a double benefit. Subject to agreement with the landowner, excavations may be turned into ponds, lakes or wetland habitats, bringing local environmental benefits. The impact of transporting materials may also be minimised, and so there is less disruption and noise to local communities. In some instances, the borrow areas are re-topsoiled and returned to agricultural or recreational use.

Typical considerations in flood embankment design include the following.

- Careful alignment – to reduce the impact on existing features, through careful integration, but in most cases the structure should still be identifiable as a flood bank.
- Embankment height, which may include for a freeboard height – low spots along a length of embankment are likely to lead to a significant risk of erosion and bank failure in events where overtopping occurs.
- Gradients of the side-slope profiles – these can comprise an angular form, a rounded profile with convex side slopes, and a rounded profile with concave side slopes. Side slopes may also be symmetrical or asymmetrical, depending on how they best blend into the existing landscape. Overwidened embankments will allow more effective land for stock grazing and for silage cuts, and in some cases mitigation tree planting can be planned where a sacrificial shoulder is sufficiently wide in design. This may be an aspect arising from the LVIA, and may be a condition imposed by the planning authority. On maintenance grounds, the wider the embankment, the greater the amount of material for construction and the greater the maintenance requirement. Reduced side-slope gradients will benefit sympathetic integration and mitigate negative visual effects.
- Crest widths – typically 3.5–4.0 m wide, to allow safe passage of maintenance vehicles and machinery. Surfacing design is also an important factor. Inappropriate surfacing will be intrusive. Where spillways are required across the crest and down one side, this can normally be designed using ground reinforcing measures that are then grassed, rather than using surface-based materials.
- Construction access and temporary working areas along the length of embankments – these should be agreed with the contractor during the design phase in order that landowner consent can be arranged, and mitigation and reinstatement can be planned and agreed. Temporary stock fencing is often required to separate the construction working area, and this may be planned to be left in place, following discussion with the landowner, until the grass has established. In some circumstances, a management regime can be agreed with the landowner for reduced stock grazing to allow longer term rejuvenation of wild flora, in which case a management plan and compensation for loss of grazing would be in order.

Other issues are likely to include the following.

- Access for landowners, for example, to access parts of land split by a new defence or for passage of agricultural stock.

- Public rights of way may exist over or along the line of an embankment. These will require discussion with the local planning authority at an early stage to outline design intentions. It is likely that the project team will need to either:
  - arrange for temporary closure and re-routing during the construction phase
  - or arrange for a permanent diversion.

These will both require statutory consultation with stakeholders and interested parties by the planning authority. A formal appeal can be made by interested parties, so the best advice is to have early discussions. Suitable provision should be made in the project programme for all eventualities.

- Access way-leaves – these allow for inspections of flood embankments from the bottom of the structure side slopes. These are usually required on both sides of embankments, and are used routinely for inspection purposes, but also for machinery and equipment for routine or non-routine repair works. They are rarely surfaced except in localities known to be particularly wet, in which case stone may be laid, compacted, topsoiled and oversown with grass to reduce any visual intrusion.
- Tying into existing features – steeper embankments may need to be formed at junctions, so maintenance and turning areas for machinery and routine inspection vehicles may need to be considered at the design stage.
- Burrowing by animals – badgers and rabbits frequently inhabit embankments, and can cause a greater likelihood of premature failure unless scheduled inspections and maintenance are carried out.

Vegetation cover on embankments can be provided using a number of methods, normally over a minimum cover of 150 mm depth of prepared topsoil.

- Topsoil preparation and seeding by small machine/by hand.
- Hydra seeding using a specialist landscape contractor, which is particularly effective on very extensive and/or steeper sided profile embankments.
- Turfing is more expensive than seeding, but can be an advantage where rapid establishment is required due to the lateness in the year of engineering operations. Temperatures and growing conditions in the autumn might otherwise mean that grass seed would not germinate and establish before the 'flood season', leading to an increased risk of bank failure.
- Wild flora seed may be incorporated in a grass-seed mix, but is unlikely to be sown on its own as it would be unlikely for the roots to ensure a sufficiently dense sward to prevent soil erosion.

Methods sometimes required to reinforce sections of embankment include the following.

- Plastic geotextile netting laid over the surface of the soil prior to seeding.
- Proprietary products incorporating plastic or coir-based geotextile netting, but pre-seeded and contract grown to form a turf, which is then transported and laid on site.

- Plastic or coir-based grid mats, often using recycled products, which are laid into the topsoil layer, joined together and pegged, and then seeded to stabilise grass root growth. This can also be a useful method where infrequent vehicle access is required across the profile of a bank. Experience has shown that these products have advantages over the equivalent concrete counterpart, as the soil in the pockets does not tend to dry out as quickly during drier weather. Hence the grass cover is less prone to dying back. This die-back is often a 'trademark' of some of the concrete-based products, which store and radiate the heat to the soil in the pockets, leading to high temperatures.

10.6.3 Environmental benefits brought to a flood defence scheme by the integration of soft landscaping

It is widely accepted that there is a requirement by local planning authorities for retention of existing trees and shrubs and the planting or natural regeneration of new trees where this is consistent with flood risk management objectives, the setting's landscape character and ecology and other environmental factors. Flood risk management will normally impose restrictions on planting on the flood plain in the vicinity of, or on, flood defence embankments and associated structures.

Recognition is also made by the Environment Agency that trees and shrubs make an essential contribution to the natural beauty of England and Wales, and that they are part of the country's flora and provide habitat for much of its fauna. It is also accepted that trees make a significant contribution to improving people's quality of life by providing attractive, healthy, wildlife-rich environments, giving shelter, absorbing development, providing shade, acting as a source of inspiration, and so on. The Environment Agency also has a duty to report biodiversity outcome measures to Defra and to the Welsh Assembly Government.

Although trees generally contribute to and enhance an area's character, there are certain landscapes where tree planting may not be entirely suitable. A landscape character assessment should be used to determine the appropriateness of trees within a landscape. There may also be other factors, such as avoiding trees within open habitats where there are populations of ground-nesting birds. Trees can also potentially cause problems for people if inappropriate species are planted in small gardens, which can become overgrown. In areas of shrinkable clay, tree roots can also be a contributory cause of ground subsidence. It is, therefore, important for appropriately qualified professionals to provide technical knowledge and to pay particular attention to the locations and types of trees being proposed.

The benefits of trees are as follows.

- Trees are valuable habitats for wildlife, and riverside trees are particularly important in providing otters with holt sites and cover for fish, and help keep water temperatures lower, which helps create ideal habitat for freshwater pearl mussels.
- Trees are popular with people from an amenity aspect, who develop strong ties to them, especially when they are under threat.

- Trees and shrubs provide people with shelter, inspiration, a healthy environment, privacy, and lessen the effects of urban development by providing green infrastructure and screening.
- Environments with trees are known to have a proven beneficial effect on people's health recovery.
- Woodland is a valued recreational resource, and can absorb far more people than open landscapes.
- Specimen trees may have a particular heritage value, for example, they may be a reminder of a historical event or place name.
- Trees can help improve air quality and act as a buffer to reduce noise.
- Trees reduce the effects of climate change and provide shade.

Trees can also help to reduce the risk of flooding.

- Trees in headwaters can help to reduce soil compaction and erosion, and therefore reduce the amount of sediment entering watercourses. This reduces maintenance costs and reduces the likelihood of sediment obstructions causing flooding.
- Larger stands of trees and woodlands intercept rainfall, gradually releasing floodwaters, and thereby lowering peak flows.
- Tree roots along embankments help to bind soil together, which prevents erosion.
- The shade created by riverside trees reduces the growth of in-channel aquatic plants, thereby maintaining channel capacity.

It is accepted that trees can also be of concern for the routine flood risk management of rivers.

- Trees can obstruct routine maintenance access.
- Trees can obstruct access in emergency situations.
- Trees growing within a river channel can reduce the overall capacity of the channel. Furthermore, trees within a flood storage area may reduce the capacity significantly, and lower the protection afforded downstream.
- Trees and shrubs can trap litter and brash during high flows.
- Trees and shrubs can sometimes make routine inspections of flood defences more difficult and can obscure animal burrows.
- Dead trees pulled out of the channel during floods can expose the core of an embankment. Fallen trees can also block channels and culverts, which can lead to flooding and can cause backing up behind built structures such as bridges, leading to increased likelihood of failure.

Trees are often sited on or close to the line of proposed flood defences. A tree survey and evaluation based on the topographical survey is a prerequisite where there is any likelihood of felling or crown lifting, etc., in order that the team can make a value judgement regarding the alignment options. Sometimes a comprehensive tree survey might be required, and this will require the services of a qualified arboriculturalist. This is also likely to be a requirement of the local planning authority as part of its assessment of a planning application. Mature trees are irreplaceable in the short to medium term, so every effort must be made to retain them rather than fell them. This is likely to result in increased expense and a need for innovative design in some situations. However, where a tree or group of trees is of significance, this extra cost and effort will be justified.

Protection measures for trees is also a requirement for construction sites, and should remain in place for the full duration of the works. The line of any protective fencing should be agreed at the outset with the project landscape architect in order to avoid damage to trees, and should be shown on a tree protection plan. Protection should, wherever possible, extend to the extent of the tree canopy. Where this may not be feasible, root protection measures should be installed to the satisfaction of the project landscape architect and local authority tree officer, where applicable. It is not unknown for a local authority to impose emergency Tree Protection Orders on specimen trees or groups of trees where it believes that it is warranted in order to afford immediate protection against impending works. Reinstatement in the vicinity of trees should be agreed at an early stage of the project.

Establishment and maintenance of new tree planting should be carried out for as long as possible. In the case of the Environment Agency, this period is now 5 years from the date of planting.

10.7. Case studies

10.7.1 Case 1: Bewdley flood alleviation scheme, Environment Agency

10.7.1.1 Setting

Bewdley evolved during the fifteenth century as a crossing point and inland port on the River Severn. Today it is a quiet Worcestershire market town, with an attractive core consisting of Georgian and earlier buildings in mellow brick and local sandstone. The town's original function as an inland port is still apparent along Severnside, the historic quay, which is an important feature in attracting visitors to the town.

Bewdley has historically suffered from flooding from the River Severn, and consultations about suitable flood defences began in the early 1990s after flood events. Initial proposals comprised the construction of permanent walls on the quayside, but this was discontinued on the grounds of their visual impact.

Following extensive flooding of up to 175 properties in the town in November and December 2000, design work for flood defences for Severnside was accelerated, and commenced in mid-December 2000. Further consultation coincided with the introduction of a new form of demountable flood defence that could be installed for protection during floods and removed to storage at other times. On the basis of engagement with statutory stakeholders, the local authority and the local community, this method was chosen as the basis for the design of defences on Severnside. The design process was divided into two phases in order that the works required on site would have the minimum environmental impact.

- Phase 1 – Severnside North and Dog Lane car park (upstream of the Grade I listed Bewdley Bridge).

- Phase 2 – Severnside South and Gardner's Meadow car park.
  - There were 22 listed buildings identified within the study area, and it was ascertained that 18 of these were within the site that played an important element of the heritage resource and significantly contributed to the historic character of the town. These were all Grade II listed buildings.
  - Most of Severnside South was also part of the designated Bewdley Town Centre Conservation Area.
  - Bewdley Bridge is Grade I listed (it was designed and built by Thomas Telford between 1798 and 1801).
  - The existing riverbank along Severnside South is retained by a Grade II listed quay wall. The wall is constructed of large blocks of dressed sandstone, and most parts of the wall are at least 200 years old.

In addition to flood defence works, the ground investigations and a structural survey of the historic quayside wall, which is over 300 m long and stands 3–4 m high above the river bed, showed that the wall structure was so weak in places that it would be at risk of collapse if the proposed works were undertaken. Therefore, strengthening work was carried out on the wall, this involving constructing support beneath and behind the wall, and taking down and rebuilding the stone face.

Other existing features, such as a small grassed amenity area on the quayside and the remains of a Victorian bandstand were removed to allow the works to take place.

Part of the quayside was thought to contain medieval stonework and timbers, and relics of previous much earlier bridge abutments. The flood defence works involved extensive below-ground work and excavation. Intrusive archaeological studies and excavations were undertaken between June and September 2002 in advance of the works in order to investigate and record the archaeology, and to influence the engineering design. Throughout these stages of work there was close monitoring to an agreed watching brief by Worcestershire County Council's Historic Environment and Archaeology Services to ensure that significant historical artefacts were not inadvertently damaged or destroyed.

The production of an environmental statement setting out the enhancements that could be achieved, and the full integration of landscape architects into the design team were paramount to the success of the project. The flood defence works and associated strengthening works led to a reassessment of this urban riverside environment in the locality, and an opportunity for the restoration and enhancement of the quayside amenity (Figure 10.6).

The Environment Agency worked closely with English Heritage, Worcestershire County Council and Wyre Forest District Council Highways, Planning and Conservation Officers, the Bewdley Civic Society and Town Council, among a host of others who were consulted.

While the purpose of the Environment Agency project was to provide Severnside South with a flood defence with a 100-year level of protection, the Agency was

Figure 10.6 This segregated pedestrian walkway at Bewdley is separated from adjacent car parking by a hedgerow and tree planting. The photograph was taken six years after planting. The planting on the right of the path assists in screening the flood defence wall, which forms the rear boundary of properties
© Julian Francis

intrinsically aware of the confines of the historic environment in which the scheme had to be developed. During the design and development of the flood defence project, it was recognised that there was significant opportunity within the overall scope and implementation to additionally enhance and restore the landscape environment of Severnside South.

Notwithstanding this, the remodelling of Severnside South had to be undertaken within the engineering and environmental constraints of constructing the flood defence. Therefore, the project aimed to

- assist in the conservation and enhancement of Bewdley's heritage
- encourage more people to be involved in Bewdley's heritage
- help ensure that everyone can learn about and, importantly, have access to and enjoy the heritage of the quayside area.

This part of the work does not form part of the core business in the implementation of flood defences, but is covered under the Code of Practice on Conservation, Access and

Recreation (Defra, 2000) issued under Section 4 of the 1995 Environment Act. The primary Act was designed to set up the Environment Agency with powers to protect the natural environment. The Act provides legislation in relation to all facets of the environment, including landscape and those factors that affect the landscape.

The main aim of the environmental interface with the project was to significantly improve pedestrian access to the refurbished quayside promenade, providing access for all in the vicinity of the river's edge. This has maintained the character and setting of the buildings and quayside, while providing a safe environment for those using the adjacent highway, using the principle of shared public realm.

The principle of the design was that the quayside was rationalised to recreate an upper and lower terrace, the change between the two levels being denoted by the line of the flood defence. The proposed changes to the immediate quayside were significant, with the quay wall where it had previously been altered being returned to its original level and the multiple 'steps' down to the river's edge being removed. The steps would be replaced by two distinct levels, with access essentially being controlled from either end of the quayside (Figure 10.7). The bandstand was removed in order to create a new civic space using high-quality materials sympathetic to the setting, with stepped and ramped access leading down to the lower quayside. Further downstream, part of the

Figure 10.7 The lower promenade walkway at Bewdley Severnside South, against the backdrop of Thomas Telford's Grade I listed Bewdley Bridge, performs a multi functional use. The sandstone-faced wall and higher level public realm provides permanent protection against more frequent flood events. The seating area and walkway is fully accessible for those who are less able bodied, through the use of ramps and steps, and is extremely popular with locals and tourists alike
© Julian Francis

quay wall required strengthening repair work, involving constructing support beneath and behind the wall, and taking down and rebuilding the stone face.

There was considerable upgrading to the alignment of the quayside roadway, kerb lines and pavements, and relocating previously permitted roadside parking to nearby car parks. There were also hard and soft landscape improvements to the quality of the materials and street furniture along the entire length of the scheme, to complement the town's conservation area.

The design development was naturally subject to intense scrutiny by all local stakeholders. A method chosen by the Environment Agency to assist in the consultation and explanation of the design was a three-dimensional card model produced to scale, which was an ideal way to illustrate the surfaces, changes in level, access improvements and a real 'feel' for the new spaces being proposed. This was first presented to a Town Council meeting, and following its success as a delivery tool, was later exhibited at the Town Museum as part of an exhibition.

The baseline project was funded by the Environment Agency with Defra grant aid. In addition, a contribution from Advantage West Midlands, the Regional Development Agency (RDA), through the Market Towns Initiative to fund part of the works to this section of riverside enabled the Agency to extend its brief to cover this environmental enhancement work. This covered remodelling, environmental and landscape improvements along the whole length of Severnside South, from the line of the flood defence back to the building frontages. The Defra grant aid covered the works to the quayside wall and back to the flood defence line. The design palette developed by the Environment Agency design team was further consulted and utilised by Opportunity Bewdley, an arm of the RDA within the Bewdley community to improve additional areas of the town using a consistent design philosophy and types of street furniture.

A successful application for an Arts Council Grant during the design stage allowed the commissioning of a public realm artist to engage with the community and local schools in the design of brass 'cargo plates' and in-street furniture. These were integrated into the quayside promenade paved walkway, and the cargo plates were set out to form a trail naming key materials that were traded at the former port (Figure 10.8).

Leaflets explaining the trail were distributed by the town museum to school children. Two permanently located interpretation panels, designed in collaboration with local stakeholders and the community, further assisted in demonstrating the historic aspects of the town and the nature of the flood defence scheme in affording protection to the quayside area.

10.7.2 Case 2: New Cut Wern flood alleviation scheme – Rivers Severn and Vyrnwy confluence, Powys. Environment Agency

10.7.2.1 Background

Many of the flood embankments in the Rivers Severn and Vyrnwy confluence near Welshpool, Powys, are at least 200 years old. These embankments, locally known as

Figure 10.8 One of the cargo plates set into the paving on the lower quayside promenade walkway © Julian Francis

argaes, are at least 200 years old. They are narrow and steep sided and, for this reason, have escaped intensive agricultural change, but are prone to burrowing by rabbits and badgers, and to being damaged whilst being grazed by farm stock. Extensive lengths of these argaes have, therefore, been subject to overhaul and improvement, and topping up of levels to maintain their function as flood defence assets. This has meant that, rather than retaining their original narrow, steep-sided profile, they now largely comprise wider footprints and wider crests with one-in-three gradient side slopes. The original banks often supported a wide and varied flora, and hence formed valuable wildlife corridors (Figure 10.9). The 'improved' banks can easily be used as a source of silage, and therefore are usually sown with ryegrass.

On the New Cut Wern project, a survey of the flora showed that the diversity of flora was particularly high. One field through which the argae passed was almost worthy of designation as a Site of Special Scientific Interest (SSSI). For this reason, the Countryside Council for Wales stressed that this diversity must be protected. In addition, due to the retained nature of the original profile, the Clwyd Powys Archaeological Trust was also keen to retain this profile as one of the last remaining lengths that had not been subject to improvement.

Many options were considered as part of the project development, including offline realignment and localised topping up works. These were discounted, however, on cost–benefit, design and environmental grounds.

10.7.2.2 Design principles

Following detailed consultation, a decision was made to mitigate improvement works on the existing alignment by undertaking the following measures.

Figure 10.9 New Cut Wern, Powys – wild flora sward on a section of original steep-sided flood bank prior to reconstruction
© Dave Phillipps

- The flora-rich turf was carefully stripped and stacked on site for the duration of the works.
- The original profile, which was of interest archaeologically, was delineated by introducing and spreading a layer of slate chippings over the subsoil prior to introducing and grading new subsoil to reprofile the wider and higher embankment. The chippings were cheap and plentifully available from sources in the area. They were chosen as they were not 'native' to this locality and could easily be identified as a horizon profile in the event of any future archaeological excavation.
- It was planned to re-create the meadow after the engineering works had been finished, taking care to respect the local flora gene pool by using locally harvested seed.

10.7.2.3 Methodology

Seed collection was difficult to tie into the Capital Works Programme, but collection was organised to be done 12 months ahead of the engineering works. The seed was harvested from a local SSSI at Ty-Brith hay meadows, which were owned and managed by the Montgomeryshire Wildlife Trust. The seed was collected using a specially designed harvester machine that featured a special screen to capture any invertebrates accidentally 'harvested'. These invertebrates are often rare, being associated with ancient meadows, and were released back into the meadow. An ecological survey was undertaken before work took place to ensure that there were no ground-nesting birds likely to be disturbed by the harvesting.

It is worth noting that the timing of the harvest will reflect the type of seed mix that is collected. In general, by the end of August most seed has ripened. However, some seed may have ripened during July, and could already have dropped to the ground. Particular seed may, therefore, not be represented in the collection. It is always well worth seeking advice on this aspect on a site-by-site basis.

Seed needs to be kept in cold storage to prevent it deteriorating. The length of viability in storage will therefore need to be taken into account, particularly if there is any delay to implementation of the project once the seed has been harvested. On the New Cut Wern project, the team was particularly keen to harvest yellow rattle, which is semi-parasitic on the roots of some grasses. However, as this is an annual plant, the seed matures early, and it has poor viability in storage so additional care had to be taken with regard to the timing of the harvest.

10.7.2.4 Outcome

At the end of the engineering works, the flora-rich turf and topsoil was brought out of storage, spread across the reprofiled embankment and rotovated to break up all clods of turf. This was supplemented with the harvested wildflower seed. During the summer, the weather conditions were very favourable for the operation, and grass was quickly established. Watering was undertaken to ensure that a good sward coverage was established by the autumn (Figure 10.10).

The grass forms a vital part of the embankment, and without it there would be a risk that the wild flower seeds would not develop sufficiently to form a dense sward, which is required to assist the longevity of the embankment structure through the winter months. A contingency to sow an annual ryegrass that would quickly establish a root zone was made in the original project appraisal report, but was not required due to the effective maintenance regime that was implemented.

10.7.2.5 Landowner approvals

It is obviously essential that the landowner is involved in every stage of the process. In this project, land management agreements were set up with landowners using a number of local land agents. Along the very best stretch of argae, the landowner was paid for loss of grazing for a period of 10 years.

10.7.2.6 Aftercare and management

The precise management details varied from one section to another, depending on the mix desired. In general terms, the grass and flora needed to go unchecked between February and August. Management was achieved through a combination of mechanical means and grazing animals.

Cutting can encourage a build up of thatch that prevents plants from growing, and can, in time, favour nettles, docks and other more vigorous broad-leaved weed

Figure 10.10 New Cut Wern, Powys – the re-established wild flora sward on the reconstructed flood embankment after one season
© Dave Phillipps

species. Mechanical flail cutting twice annually can reduce this thatch, as can the removal of arisings. Traditionally, this would have been used as hay, and there is a possibility that wildflower hay could have a commercial use, either for helping to establish a new site or for horse fodder (apparently race horses have a particular penchant!). Although not practised on this project, it is of course possible to further harvest the areas for seed supply, and this can generate further income. It should be noted that an annual harvest of this type has the potential to deplete the site in the long term. Therefore, while being a practical technique, it is best to wait for a few years after establishment before undertaking such a harvest.

Grazing is another traditional management technique. Sheep grazing is usually the most effective – sheep will cope with coarser dryer grass – but will cause as much disturbance and will not enrich the fertility of the soil as much as horses and cows. When horses graze they pull on the grass, and can cause considerable damage to newly established meadow by pulling up the roots. However, some plant species, such as dog daisies and even some orchids, appear to benefit from higher levels of disturbance. If the grass becomes very dry and unappetising, it may be beneficial to cut and then graze the immediate regrowth.

The entire embankment at New Cut Wern was fenced off for the initial one-year defects liability, or rectification, period. It was essential that any poisonous plants that had become established were identified where the area was to be cut for hay or grazed.

Ragwort is an example, as this is deadly to livestock. Some plants, such as knapweed and those with deep taproots, take longer to establish, and are often not present for the first few years. Public rights of way should also be managed in a manner that prevent them from becoming overgrown. The original fencing and vehicular gates were left in place, and management of the grass within the fencing was controlled by the Environment Agency.

A further ecological survey was undertaken about ten months after the work was complete, with very positive results. Evidence of the establishment of diverse flora can be seen in the Figures 10.9 and 10.10. The Countryside Council for Wales was very pleased with the outcome, and this has helped in building our professional relationship with them for consultation on other projects.

10.7.3 Case 3: Sutcliffe Park; Quaggy River flood risk management scheme, Environment Agency

10.7.3.1 Background

The Quaggy scheme adapted a number of public and private multifunctional green spaces linked by the Quaggy River in order to reduce flood risk and the potential impacts of climate change for 600 homes and businesses, and more than 4000 people living and working in the London Boroughs of Greenwich and Lewisham. The scheme was designed and implemented by the Environment Agency over a period of 15 years up until 2007.

The core part of the project is Sutcliffe Park, a district park of 35 ha serving the surrounding area of Greenwich, on land that had been levelled in the middle of the twentieth century to provide an extensive area of featureless playing fields containing the culverted and realigned river. Following redevelopment, Sutcliffe Park has been transformed into a high-quality natural environment featuring

- a meandering river
- a fully operational flood storage area
- a lake and ponds (Figure 10.11)
- wildflower meadows and avenues of trees
- widely accessible boardwalks, paths, seating areas and viewing points (Figure 10.12)
- new and enhanced wildlife habitats, including biodiversity action plan habitat.

The project also incorporated adaptation and regeneration of green spaces at Manor Park and Weignall Road to accommodate flood storage, and innovative environmental and flood risk management measures along the Quaggy River itself, including the restoration of natural features within the channel near Manor Park and set back flood defences that were incorporated in private gardens facing the river.

10.8. Conclusions

Engineering projects with well-integrated, good-value environmental outcomes are not achieved without foresight, careful planning and the use of experienced practitioners.

Figure 10.11 Sutcliffe Park
Contains Environment Agency information © Environment Agency and database right

Figure 10.12 Sutcliffe Park – aerial view
Contains Environment Agency information © Environment Agency and database right

The projects described here prove that it is completely possible to produce examples that meet these criteria. However, such projects cannot be achieved by taking the easy-option route, and ultimately require a relatively higher level of resource input and commitment to get good results, first time, every time!

Author's note

The views expressed in this chapter do not necessarily represent those of the Environment Agency.

REFERENCES

CLG (Department for Communities and Local Government) (2005a) *Planning Policy Statement 9: Biodiversity and Geological Conservation*. Available at: http://www.communities.gov.uk/publications/planningandbuilding/pps9 (accessed 10 February 2012).

CLG (2005b) *Circular 06/05: Biodiversity and Geological Conservation – Statutory Obligations and Their Impact Within the Planning System*. Available at: http://www.communities.gov.uk/publications/planningandbuilding/circularbiodiversity (accessed 10 February 2012).

CLG (2006) *Planning for Biodiversity and Geological Conservation: A Guide to Good Practice*. Available at: http://www.communities.gov.uk/publications/planningandbuilding/planningbiodiversity (accessed 10 February 2012).

CLG (2009) *Planning Policy Statement 25: Development and Flood Risk – Practice Guide*. Available at: http://www.communities.gov.uk/publications/planningandbuilding/pps25guideupdate (accessed 2 February 2012).

Commission for Architecture and Built Environment (2006) *Design and Access Statements, How to Write, Read and use Them*. Available at: http://webarchive.nationalarchives.gov.uk/20110118095356/http:/www.cabe.org.uk/files/design-and-access-statements.pdf (accessed 10 February 2012).

Council of Europe (2000) The European Landscape Convention. Available at: http://www.coe.int/t/dg4/cultureheritage/heritage/landscape/default_en.asp (accessed 10 February 2012).

Defra (Department for Environment, Food & Rural Affairs) (2000) *Environment Act 1995. Water Industry Act 1991. Code of Practice on Conservation, Access and Recreation: Guidance for the Environment Agency and Water and Sewerage Undertakers*. Available at: http://tna.europarchive.org/20070101180421/http://www.defra.gov.uk/environment/water/industry/copcar/chapter1.htm (accessed 10 February 2012).

Defra/Environment Agency (n.d.) *Landscape and Environmental Design Guidance*. Science Summary SC020101/SS. Available at: http://publications.environment-agency.gov.uk/PDF/SCHO1006BLKQ-E-E.pdf (accessed 10 February 2012).

Environment Agency (2011) *Creating a Better Place 2010–2015*. Available at: http://www.environment-agency.gov.uk/aboutus/112323.aspx (accessed 10 February 2012).

Landscape Institute (2008) *Landscape Architecture and the Challenge of Climate Change*. Position Statement. Available at: http://conventions.coe.int/Treaty/en/Treaties/Html/176.htm (accessed 10 February 2012).

Landscape Institute and Institute of Environmental Management and Assessment (2002) *Guidelines for Landscape and Visual Impact Assessment*, 2nd edn. Spon Press, London, UK.

Natural England (2009) All Landscapes Matter. Available at: http://www.naturalengland.org.uk/Images/elc-actionplan0910_tcm6-20278.pdf (accessed 3 April 2012).

RECEP/ENELC (2009) *The European Landscape Convention. What Does it Mean for Your Organisation?* ELC Workshops, London and York. European Network of Local and Regional Authorities for the Implementation of the European Landscape Convention. Available at: http://www.recep-enelc.net/pageImg.php?idCont=632&idSez=&idlink=59&lang=en (accessed 10 February 2012).

River Restoration Centre (1995) *Manual of River Restoration Techniques*. Available at: http://www.therrc.co.uk/rrc_manual_pdf.php (accessed 10 February 2012).

The Countryside Agency/Scottish Natural Heritage (2002) *Landscape Character Assessment. Guidance for England and Scotland.* Available at: http://www.naturalengland.org.uk/Images/lcaguidance_tcm6-7460.pdf (accessed 10 February 2012).

Welsh Government (2004) *Technical Advice Note 15: Development and Flood Risk*. Available at: http://wales.gov.uk/topics/planning/policy/tans/tan15/?lang=en (accessed 2 February 2012).

Welsh Government (2009) *Technical Advice Note 5: Nature Conservation and Planning*. Available at: http://wales.gov.uk/topics/planning/policy/tans/tan5/?skip=1&lang=en (accessed 10 February 2012).

Welsh Government (2011) Planning Policy Wales, 4th edn. Available at: http://wales.gov.uk/topics/planning/policy/ppw/?lang=en (accessed 10 February 2012).

Flood Risk
ISBN: 978-0-7277-4156-1

http://dx.doi.org/10.1680/fr.41561.281

Chapter 11
Robust decision-making under uncertainty – towards adaptive and resilient flood risk management infrastructure

Paul B. Sayers,* Gerald E. Galloway† and Jim W. Hall‡
* Sayers and Partners, Oxfordshire, UK
† University of Maryland, Maryland, USA
‡ University of Oxford, Oxford, UK

It is always necessary to make decisions in the absence of perfect information. In the past, uncertainty has been implicitly accounted for in flood risk management decisions through safety factors and allowances rather than via explicit analysis. Recognising uncertainty does not, however, prevent decisions being made. In fact, recognising uncertainty is a key requirement for appropriately designing adaptive capacity and resilience into flood risk management choices. Only by quantifying and acknowledging uncertainty are we better placed to decide how best to manage it.

Perhaps the largest of these uncertainties is associated with future conditions. Climate and demographic change can have a profound influence on flood risk management and the management choices made. Infrastructure choices made today will persist for several decades if not centuries, so taking a longer term, strategic view when planning infrastructure investment is critical to making the right choice. This chapter explores various methods and approaches that have been applied in practice, as well as those emerging from research, to support good decision-making under uncertainty, including scenario development, robust decision-making and adaptive management (based on multi-stage interventions), and in particular how adaptive capacity can be appropriately embedded within infrastructure design and management using real option approaches.

11.1. Introduction

By 'uncertain' knowledge, let me explain, I do not mean merely to distinguish what is known for certain from what is only probable. The game of roulette is not subject, in this sense, to uncertainty ... Even the weather is only moderately uncertain. The sense in which I am using the term is that in which the prospect of a European war is uncertain, or the price of copper and the rate of interest twenty years hence, or the obsolescence of a new invention ... About these

> matters there is no scientific basis on which to form any calculable probability whatever. We simply do not know. Nevertheless, the necessity for action and for decision compels us as practical men to do our best. (John Maynard Keynes, 1937)

Both developed and developing countries are seeking to promote communities that are resilient (in respects that we will go on to explain) in the face of natural hazards and capable of adapting to unknown future changes. All are struggling to turn good theory into practical action. In building resilient communities and implementing adaptive management it is clear that engineered structural measures will continue to play a significant role; however, these will be increasingly working alongside a wide range of non-structural measures and instruments (Hall *et al.*, 2003; Evans *et al.*, 2004; Sayers *et al.*, 2011). As yet no blueprint is available as to what constitutes resilient design or adaptive management. However, a common understanding is starting to emerge, and this acknowledges resilient and adaptive design as a process that, as part of a wider portfolio of responses, fosters innovative approaches to the design, construction and operation of buildings and infrastructures (Bosher *et al.*, 2007; National Institute of Building Sciences (NIBS), 2010).

- Emerging principles of resilient design:
  - a resilient design is resistant to a wide range of threats, including ones that were not necessarily foreseen during the design process
  - the performance of a resilient design does not decay catastrophically when exposed to events more severe than design levels (e.g. a levee will be overtopped, but should not collapse or breach without warning; critical infrastructure such as pumping stations, bridges, gates, etc. must continue to operate)
  - a resilient design recovers rapidly from a disruptive event (supporting the rapid return to normality – avoiding the need for complex plant, highly specialised skills or difficult to source materials).
- Emerging principles of adaptive management:
  - it utilises responses that do not foreclose future options or unnecessarily constrain future choice
  - it utilises responses that are effective under the widest set of all plausible future scenarios
  - it observes change through targeted monitoring and continues to reassess scenarios of the future
  - it appropriately modifies policies, strategies and structure plans.

Delivering resilient infrastructure is much more than simply reducing the chance of damage through the provision of 'strong' structures, and adaptive management is much more than simply 'wait and see'. Both are purposeful approaches to design that are inherently risk based and, importantly, seek to actively manage uncertainty. A risk-based approach is now widely accepted (Sayers *et al.*, 2002) and is maturing in practice (Table 11.1). Accepting the future as unknown, although widely recognised as important (Evans *et al.*, 2004a,b; Hall and Solomatine, 2008), is yet to become a

Table 11.1 The change in design paradigm towards using risk to inform resilient design (Sayers *et al.*, 2011)

Management paradigm	Basis		Characteristic motivation	Example objective
Engineering design/safety standards (traditional approach)	Probability	Historical event	To prevent flooding during a repeat of a specified historical event	Design the flood defences to withstand the 1822 flood
		Single design events	To prevent flooding during a storm event of a specified return period	Design the flood defences to withstand a 1 in 100-year flood
		Multiple design events	To prevent flooding for a given design storm event set according to the nature of the land use/asset protected	In highly urbanised areas design the flood defences to withstand a 1 in 200-year flood. In rural areas design the flood defences to withstand a 1 in 20-year flood
	Consequence	Safety regulation	To limit the consequences of flooding during the probable maximum flood flow to a specified level (safety standard), regardless of the cost of doing so	During the probable maximum hydrological flow, ensure that no individual is exposed to a chance of dying in excess of 10^{-4}. Ensure that the chance of >1000 people dying is less than 10^{-7}
Risk management (modern approach)	Risk	Resource optimal and multi-criteria	To implement a range of interventions that maximise benefits (across multi-criteria) and minimise whole life resource input	Implement a portfolio of measures and instruments to reduce risk effectively and efficiently whilst achieving societal preferences for equity, safety, and ecosystem health. The increased resource inputs required to providing progressively greater reductions in risk should not be disproportionate to the additional benefits secured

routine consideration within flood risk management. Accepting this premise has a number of profound implications, both in terms of the way in which strategies are developed, and the way in which infrastructure is designed. Some of these challenges and approaches are discussed below.

11.2. Developing strategies under uncertainty

> Decisions taken today will have a profound impact on the size of flood risks that future generations will need to manage. They will also strongly influence the options available for managing those risks. (Foresight Future Flood Study, Evans *et al.*, 2004a,b)

Uncertainty is pervasive within the evidence on flood risk, both in the data and the models, and there are also profound uncertainties in how change (climate, demographics, etc.) will influence the future. As a result, uncertainty about what to do is natural. How such gross uncertainties are managed shapes the nature of the strategies and the engineering designs developed. In contrast to the linear model of strategy development, which is based on a more certain view of the future that is characteristic of traditional flood control decisions, engineers now seek to embed resilience and adaptive capacity within the choices made (Table 11.2).

Table 11.2 Comparison of the traditional linear design model and the adaptive model of strategy development and decision-making (Sayers *et al.*, 2011)
Adapted from Hutter and McFadden (2009) courtesy of the members of the Floodsite Consortium (www.floodsite.net)

Stages of strategy development	Traditional (certain) model	Adaptive (uncertain) model
Deciding what to do	Predefined system of goals, objectives and desired outcomes Defined set of activities and resource demands	Emerging pattern of goals, objectives and desired outcomes Flexible configuration of resources and priorities
Deciding how to do it	Sequential process of planning, programming and implementation Top-down strategy development	Continuous alignment of plans, programmes and implementation activities with the changing world Continuous reconciliation of the bottom-up initiatives and top-down strategies
Understanding the external and internal influences	Stable system of decision-making Predictable (deterministic) future change – climate, demographics, deterioration, preferences, etc.	Changing decision-making processes and priorities Unknown future change – climate, demographics, deterioration, preferences, etc.

Figure 11.1 Framework for adaptation decision-making proposed by Willows and Connell (2003)
© UKCIP, 2003

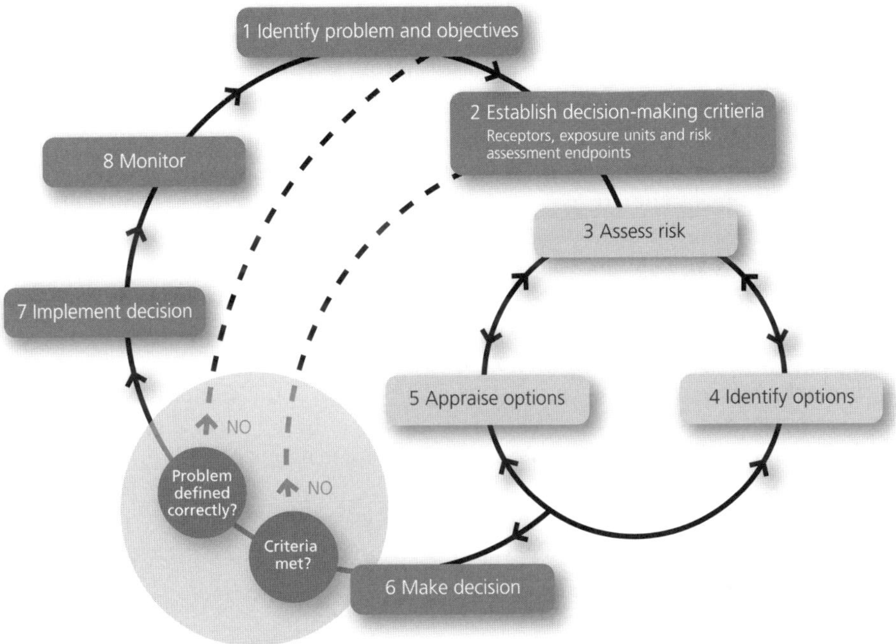

Recognition that future conditions may change (perhaps significantly) from those that exist today, or that existed when a structure was first designed, underlines the need for a continuous process of monitoring and intervention, which is essential to the success of any infrastructure project. The classical engineering control loop of data acquisition, decision-making, intervention and monitoring reappears in contemporary thinking about adaptive management. Consider, for example, the risk-based adaptation decision-making framework proposed by Willows and Connell (2003), shown in Figure 11.1, which establishes adaptive management as a continuous process of defining objectives, assessing risks, appraising options, implementation and monitoring. Conditions of uncertainty and change imply a commitment to ongoing study of and intervention in the system in question, in the context of constantly evolving objectives.

A number of important principles underpin an adaptive management approach.

- *Look beyond historical precedents.* There is no certainty about what the future holds, and increasingly a historical analogy provides limited guidance. Lack of imagination in describing possible future change can condition actions based on current knowledge and experience.
- *Look for solutions that perform well in multi-plausible futures – robust solutions.* To be meaningful, all plausible futures and possible strategies must be considered. Judgements made too early about the most likely strategic preference or possible

future can precondition the answer in an undesirable and suboptimal manner. Conversely, overcomplication must be avoided, including unnecessary detail or very localised options.

- *Take a long-term view, avoiding short-termism.* The planning and implementation of flood risk strategies is often biased towards quick wins. More progressive strategies that embed a longer term view and support progressive management are often difficult to develop and implement.
- *Look for radical solutions and challenge the status quo.* Engineers need to be brave enough to propose new or radical solutions such as land banking, forced programme integration (e.g. energy generation and flood defence, habitat creation and flood management), urban blue highways, stream buffers and land management and land-use controls.
- *Maximise the efficient and safe use of the sunk investments.* Much of the developed world, and the developing world, has already invested significant sums in an infrastructure portfolio, much of which is ageing. After detailed technical evaluation of structural integrity, it may be possible and fiscally efficient to incorporate and/or adapt existing infrastructure in future plans if such incorporation can be accomplished in a sustainable manner.

The desire for adaptive management also introduces additional decision criteria that reflect the performance of the strategy as a whole and the engineering measures it may contain.

- *Robustness*: in addition to seeking robustness of physical structures, ensuring the *strategy* also performs acceptably well in the widest set of plausible future scenarios; avoiding strategies that are tailored to a given view of the future or historical setting, and only perform well in that context.
- *Flexibility*: ensuring that the *strategy* can be changed based on monitoring and observation; avoiding, where possible, measures that foreclose future options, while promoting others that keep future options open. By considering *multi-staged decisions* rather than single trajectories, flexible strategies can be developed that have clear decision points.

11.3. Robustness

Determining what to do, given perfect information and clearly defined outcomes, would be a straightforward ranking process. In reality, however, uncertainty in both information and the outcomes to be achieved complicate this process. An underlying desire to maintain the ability of a flood risk system to perform acceptably well (i.e. avoiding catastrophic failure, limiting residual risk, maximising environmental gain, avoidance of resource waste) in the context of all plausible futures drives the need for a change in thinking – and a desire to make 'robust' choices.

Scenario analysis is a crucial first step in determining robust decisions when faced with significant uncertainty. By exploring different future scenarios, an understanding of what the future may look like and, importantly, how different decisions play out in those futures, can be developed.

Good scenario development is not straightforward, and demands a combination of expert dialogue and quantified evidence. Some of the basic rules in good scenario development are as follows.

- *Open minds to a different future* – experts must think imaginatively about change, and not simply project existing trends. A comprehensive view of the potential drivers that might influence future flood risk need to be considered and discussed. It is through this process that the *status quo* can be challenged and space given for innovation.

  > Illustrating the future by means of scenarios is a way to overcome human beings innate resistance to change. Scenarios can thus open mental horizons that allow the individual to accept and understand change, and so be able to shape the world. This approach may therefore help in seizing new opportunities ahead as well as avoiding undesirable effect or misconceived action. (Bertrand *et al.*, 1999)

- *Distinguish autonomous and purposeful actions* – potential autonomous developments (i.e. all future developments that are not purposefully influenced by flood risk management measures and related policy instruments) must be clearly identified and separated from purposeful flood risk management actions. Without this distinction, benefits can be misattributed to flood risk management activities, and resources can be unnecessarily invested, either duplicating or needlessly displacing autonomous expenditure (McGahey and Sayers, 2008; Klijn *et al.*, 2009).
- *Be internally consistent* – not every combination derived from a set of possible futures is necessarily possible or plausible. Assumptions and procedures used in the development of scenarios must be clearly identified and consistently applied in order to ensure coherence.
- *Be capable of quantified analysis* – at the core of the scenario analysis lies a system flood risk model for estimating the severity and consequence of flooding, and a cost model for computing the fiscal impacts of flood risk management options. To be meaningful, risk analysis must reflect the performance of the whole system of sources, pathways and receptors, and how each component of risk is influenced by change (Figure 11.2). By using a whole-system risk model, alongside quantified scenarios of change alternative strategies can be appraised and used to support expert selection of the preferred approach.

11.3.1 Scenario development

Scenarios can be developed as either a small number of discrete futures or as a continuous spectrum of futures. Each approach has its own advantages and disadvantages in the context of supporting flood risk management policy, strategy and engineering.

11.3.1.1 Discrete storylines

A small number (up to four or five) of contrasting scenarios are developed. This approach is widespread in the field of socio-economic scenarios (e.g. the Intergovernmental Panel on Climate Change (2000) Special Report on Emissions Scenarios and

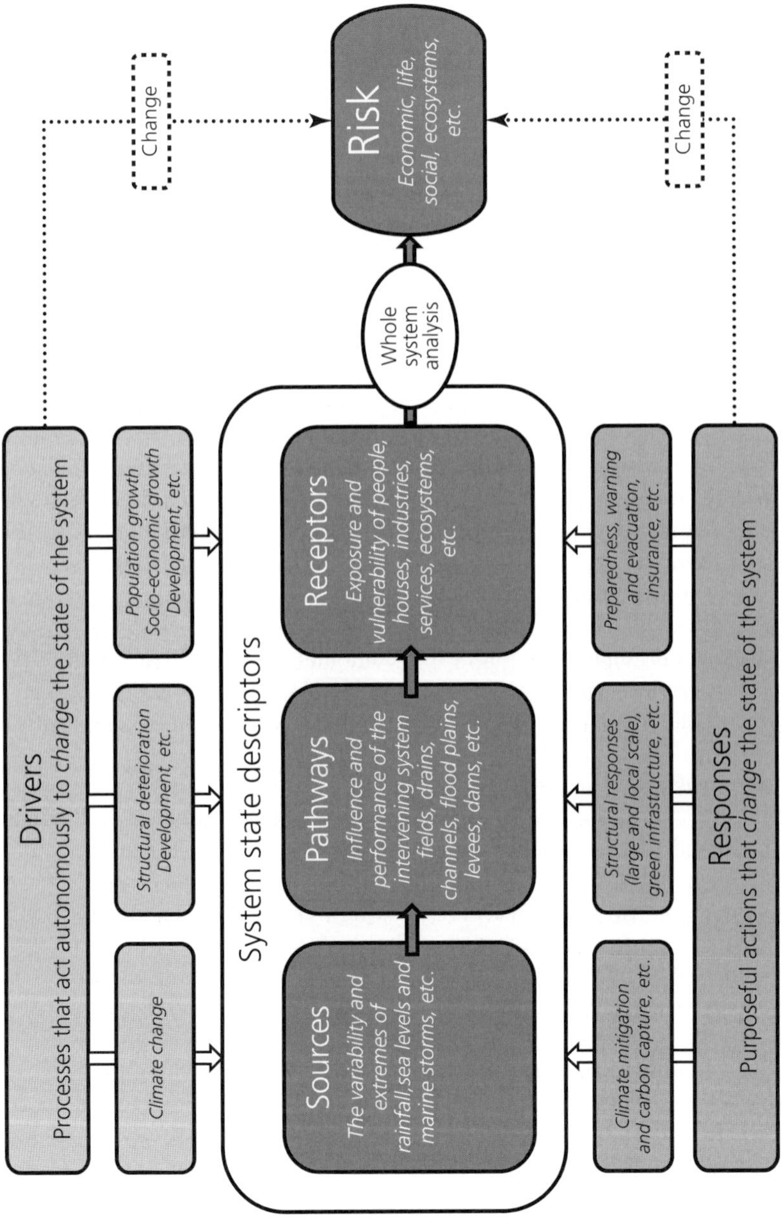

Figure 11.2 A structured framework of whole-system thinking based on understanding the sources, pathways and receptors of risk (Sayers *et al.*, 2011) Adapted from Evans *et al.* (2004)

the UK Climate Impacts Programme (2001) socio-economic scenarios), where a set of narrative storylines are developed based on a small number of distinct world views. The performance of possible management actions is then assessed in the context of each discrete future, and actions that perform well in a wide range of futures are identified. This approach is most useful for policy analysis that needs to be nuanced with respect to a wide range of attributes of the future, many of which may not be quantifiable. The approach is attractive in that it involves a small number of futures, and so is readily communicated. It has been used in the UK Foresight Future Flooding Studies (Evans *et al.*, 2004a,b) and Schelt Estuary planning (Kljin *et al.*, 2009). In both cases, scenario analysis was used to explore high-level flood risk management policies and to successfully influence national policy (e.g. in England and Wales it shaped the development of the Department for Environment Food and Rural Affairs (Defra) strategy Making Space for Water (Defra, 2005) and the subsequent Floods and Water Management Act 2008).

11.3.1.2 Continuous scenario space

The disadvantage of the discrete storylines approach is that it deals with a relatively small number of scenarios. Moreover, the narrative basis requires further elaboration before it can be used to generate quantified inputs for decision analysis. An alternative approach, promoted most effectively by Lempert *et al.* (2003) is to explore the performance of alternative policies with respect to a continuous multi-dimensional scenario space. The dimensions of this scenario space are identified to represent the main uncertain variables in a decision. Analysis of option performance with respect to this scenario space helps to identify options that perform acceptably well across a wide range of possible future conditions. This type of analysis offers advantages for engineering design in comparison to a discrete approach, as it provides the basis for quantified analysis of specific engineering alternatives and associated design characteristics (crest level, etc.).

11.3.2 Defining robustness

The issue of decision-making under severe uncertainty (i.e. where probabilities are difficult to assign) has been an ongoing subject of philosophical debate for centuries. For example, Laplace's principle of indifference or insufficient reason involves assigning an equal probability to uncertain quantities. Yet an assignment of equal probabilities represents an exact statement about the relative likelihood of the events concerned, which is hardly a reflection of severe uncertainty. Knight (1921) recognised the situation where there is no knowledge about probabilities of uncertain events, and defined it as 'decision-making under uncertainty', in contrast to the situation when probabilities are known, which he termed 'decision-making under risk'. While there are well-established rules for how a rational decision-maker should behave when taking risk decisions (Lindley, 1990), in the case of decision-making under uncertainty there are no normative rules, although several have been proposed. For example, Wald's maximin model (Wald, 1945) seeks to choose the option that maximises the reward given that the worst case materialises. This is a worst-case approach, and may be considered to be very conservative.

More generally, a solution is thought of as being robust if it performs acceptably well irrespective of what the future holds. The approach is referred to as 'satisficing', a term originally introduced by Simon (1956) to describe how decision-makers seek solutions that satisfy their decision criteria rather than optimising to find the best possible performance. Instead of seeking to maximise the expected value of an outcome (e.g. net present value), robust satisficing aims to maximise the *degree of sureness* that a satisfactory outcome will result. Robust satisficing therefore asks 'Are the outcomes good enough?', and seeks to identify options that satisfy performance thresholds across multiple criteria and under all plausible future scenarios.

A robust satisficing approach underpinned the Foresight Future Flooding studies, where the performance of different management responses was assessed against thresholds of acceptability associated with a range of criteria including cost-effectiveness, social justice, and residual flood risk (Evans *et al.*, 2004a,b). Management responses that failed to satisfy one or more of these thresholds, in one or more of the four future scenarios, were ranked lower in preference to those responses that offered acceptable performace in all futures.

There is, therefore, a trade-off between a solution that maximises performance under an assumed set of future conditions and one that maximises the robustness with respect to a wide range of possible futures (Ben-Haim, 2006). It will often be necessary to invest more in order to deliver solutions that are robust. While for decision-makers robustness is an intuitively attractive attribute, they may need some persuading to pay more (or sacrifice some performance under nominal conditions) in order to provide the investment required to provide solutions robust to uncertainty. This is where robustness analysis can contribute to engineering decision-making by highlighting the trade-offs made. Some examples are given below.

11.3.2.1 Sensitivity analysis and visualisation

The starting point for the identification of robust solutions is a process of isolating the most important uncertainties and understanding the response of decision alternatives with respect to those uncertainties. Complex systems will typically have a large number of inputs, but sensitivity analysis enables identification of a smaller number of critical uncertainties. This involves a combination of expert judgement and systematic experiments with an appropriately comprehensive system model (e.g. Gouldby *et al.*, 2008). A range of quantified methods exist for sensitivity analysis (e.g. Saltelli *et al.*, 2000; Hall *et al.*, 2009; Gouldby *et al.*, 2010). In the context of severe uncertainties, global sensitivity analysis methods are required, as local methods are sensitive to the choice of central point for the analysis. Visualisation of sensitivities is essential alongside the quantified methods of sensitivity analysis. Figure 11.3 gives an example of such a visualisation, illustrating how a key variable (flood damage in part of London) is sensitive to two uncertain hydraulic variables. McGahey and Sayers (2008) also promote the use of these types of visualisation, and further explore the sensitivity of option performance to these uncertainties by sampling from a joint probability distribution over these uncertainties. Figure 11.4 illustrates typical results from this type of analysis, based on sampling three main sources of uncertainty (sea level rise, dike deterioration and economic growth). While a probabilistic representation of these three significant epistemic uncertainties has in this case been adopted, the approach

Figure 11.3 Flood damage (in pounds) conditional upon joint loading conditions. Top: Damage within the Thames Estuary conditional on tidal level and river flow (Harvey *et al.*, 2011) with permission from IWA Publishing. Bottom: Damage at Towyn, North Wales, conditional on coastal water and water level (Sayers *et al.*, 2004)

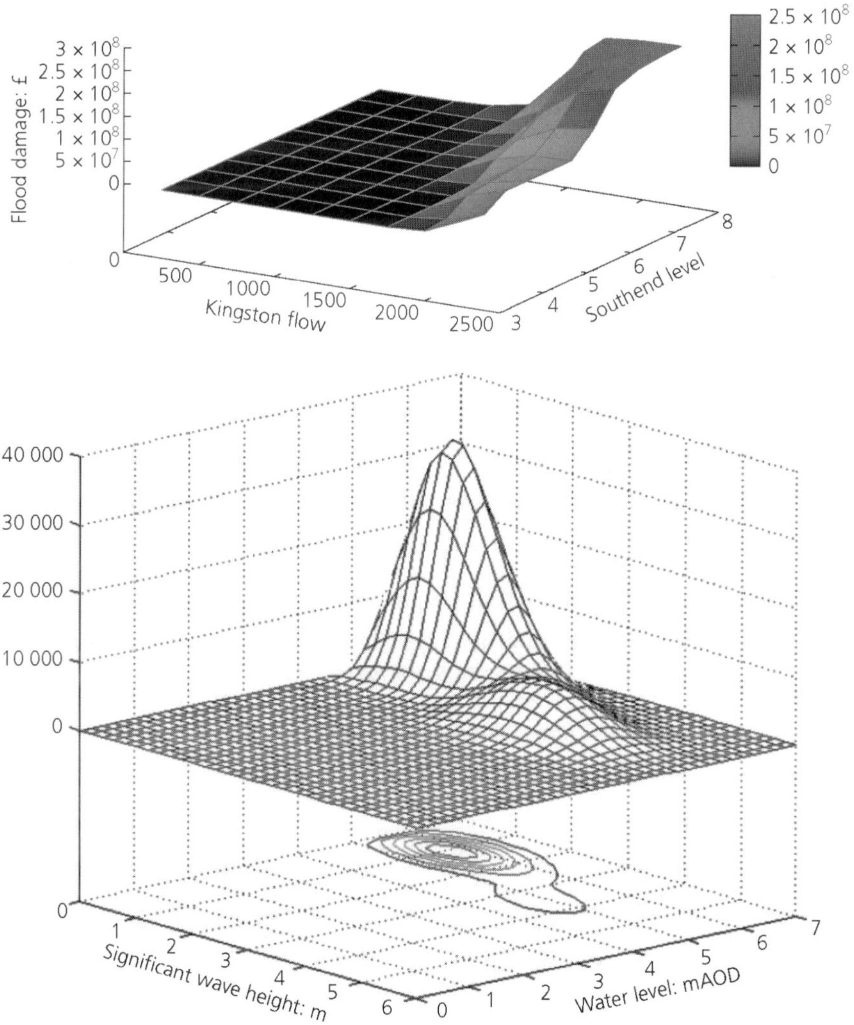

does not integrate out the uncertainties into an expectation, but illustrates the full distribution of option performance, so decision-makers can see how performance varies over a wide range of input conditions.

11.3.2.2 Robust control and robust optimisation

Recognising that the inputs to a system are often uncertain, engineers working in the fields of control and optimisation have worked since the 1960s on developing algorithms

Figure 11.4 Distribution of net present value (NPV) for four decision options based on sampling for three different sources of uncertainty (sea level rise, dike deterioration and economic growth)
From Harvey *et al.* (2011) with permission from the copyright holders, IWA Publishing

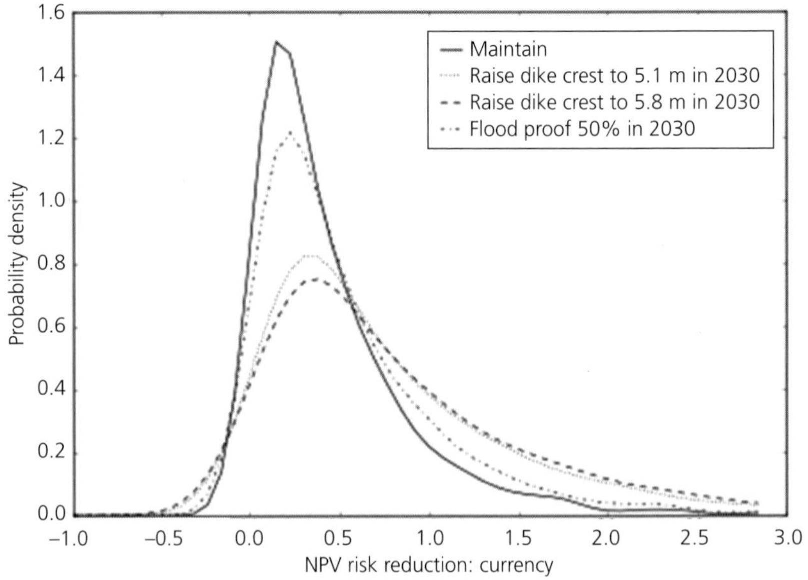

that identify solutions that perform acceptably for a range of possible inputs (see e.g. Schweppe, 1973; Kouvelis and Yu, 1997). While many of these methods have dealt with rather specific problems (e.g. control of dynamical systems), the principle that solutions can be identified that perform well with respect to uncertain system definition or subject to uncertain constraints is transferable to the flood risk management context. Moreover, these methods may be applied to identify and justify robust designs and management arrangements.

11.3.2.3 Info-gap analysis

Any approach that explores option performance over a set of possible uncertain quantities relies on the definition of that set of possibilities. Info-gap analysis (Ben-Haim, 2006) circumvents the need to precisely define the set of possible uncertain quantities by conducting a progressive sensitivity analysis with respect to an expanding set of possibilities. The approach starts at the assumed best estimate of what the future holds, and then expands the set of possibilities around that estimate in an unbounded way, so that eventually the global set of possibilities is analysed. At each progressively larger horizon of uncertainty α, the maximum and minimum performance of a set of decision alternatives is identified. Hall and Harvey (2009) applied the approach to compare four options for flood defence raising, including options that delayed intervention in order to acquire more information. The results are illustrated in Figure 11.5, in the form of 'robustness curves' that track the minimum performance at a given horizon of uncertainty α. The curves highlight the potential for rapid decay in

Figure 11.5 Results of info-gap analysis for four flood defence options. The curves illustrate the minimum level of flood defence performance (expressed as net present value (NPV) for four flood defence options) that can be expected at increasing horizons of uncertainty α, where in this case uncertainty is defined on the three-dimensional space of sea level rise, socio-economic growth and flood defence construction cost
From Hall and Harvey, 2009

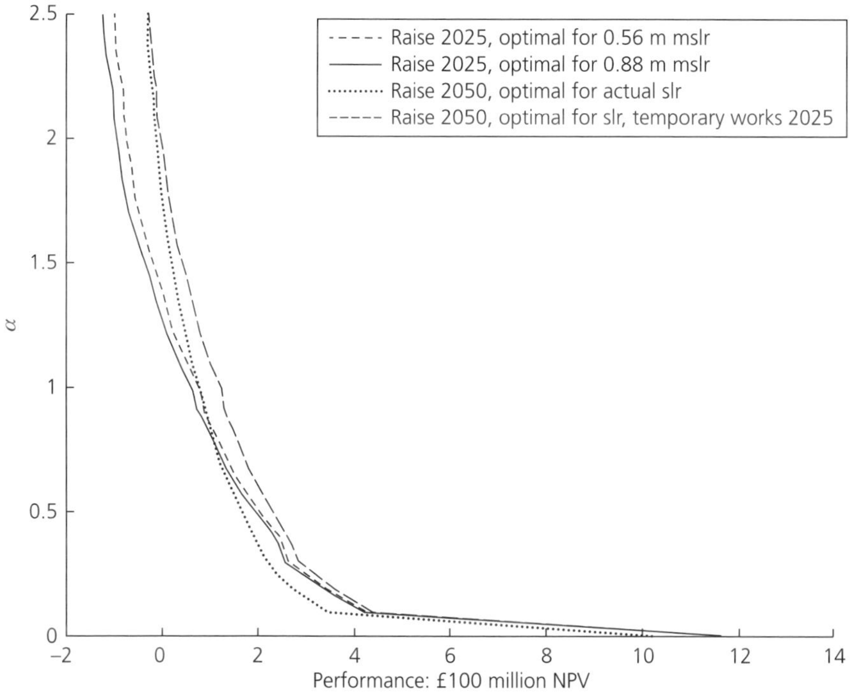

performance (in this case represented through the NPV of each option) as the reality of the future deviates, even by really small values, from the best estimate. The option that delays the commitment to structural intervention, while continuing to non-structural measures to limit the build up of risk (option 4 in Figure 11.5), is identified as being the most robust (consistently returning the greatest lower bound on NPV as the range of uncertainty in the future α increases).

11.4. Designing and evaluating flexibility/adaptability

In a changing world it makes sense to adapt solutions that can be modified if the future should turn out to be different from expectations, and adaptive management is much easier in systems that are flexible. However, designing for adaptation (by which we mean designing so that a system can, if necessary, be modified in future) will often bring some additional cost. That cost will need to be justified in terms of the lifecycle performance of the design, in a range of uncertain futures (as highlighted in Figure 11.5).

Various examples exist of an adaptable design; for example, purchasing land in the lee of an embankment to facilitate future raising or widening, or designing foundations that anticipate a heightened embankment in the future (Figure 11.6). Such options often demand greater upfront expenditure than perhaps would be the case if future change had been ignored; there are seldom true win–win situations. Flexible solutions are, however, likely to be more cost effective over the longer term. For example, beach nourishment is often promoted as a flexible solution in that the amount of fill placed on the beach can be modified from one nourishment campaign to the next, in the light of improving understanding of beach behaviour and changing objectives with respect to risk reduction.

There is, of course, a close connection between flexibility and robustness. Flexible solutions tend to be robust to uncertainty, as they are able to be adapted to perform well under a wide range of possible conditions, assuming that future adaptation decisions are taken in a timely way. Thus, the methods for robustness analysis that we have outlined in the previous section are also applicable to the analysis of multi-staged decisions that offer future choices (i.e. flexibility). Various more formal techniques are starting to emerge as practical means of constructing and analysing multi-staged decisions. Some of these are described below.

11.4.1 Decision trees

Decision trees (Figure 11.7) are a well-established method for analysis of sequential decision problems. They are very useful in the context of long-term planning problems, where processes of long-term change trigger particular system management decisions. Each decision point is constrained by previous actions and each is more or less suited to different future states that may exist. The performance of each decision pathway – the set of decisions that constitutes a single route through the decision tree – under each future can then be assessed against a range of future scenarios and the most robust strategy identified (either through a robust-satisficing, robust-optimising or a combined approach). The performance evaluation is over the whole lifetime of the strategy. By taking this whole-life view, flexible options are often highlighted as preferred, as they tend to perform better over a wider range of possible future conditions; this is despite the additional cost that is typically associated with flexible strategies at certain stages during the lifecycle.

Analysis with decision trees provides an intuitively appealing means of developing flood management strategies and identifying those that offer maximum flexibility and do not foreclose future choices unnecessarily. Perhaps their greatest strength is their ability to identify both those actions that can be taken now, and those that should be delayed. The approach has been demonstrated for strategic flood risk management decisions in the Thames Estuary by McGahey and Sayers (2008). Figure 11.7 shows the decision tree developed for the Thames Estuary flood defence system, highlighting the choices to be made as sea levels rise. Depending on the degree of sea level rise that materialises in reality, the nature of the defence system required may be distinctly different. The decision tree outlined in Figure 11.7 supported decision-makers in implementing robust, cost-effective options, some actions now, while appropriately delaying expensive options or those actions that would foreclose future options.

Figure 11.6 Adaptive design keeps future options open without incurring unnecessary additional expenditure. Real options methods provide a means of valuing the efficiency of increased expenditure on initial investment to provide future flexibility in the context of an uncertain world

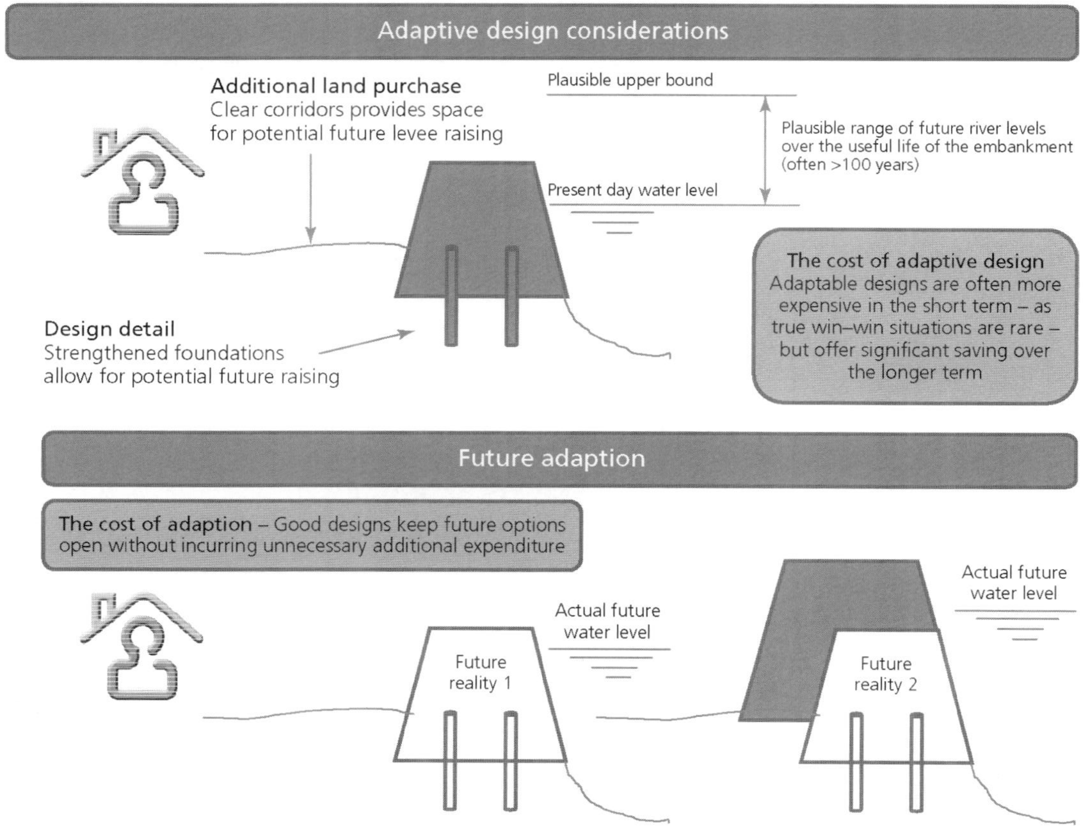

Figure 11.7 Decision trees were used as part of the Thames Estuary 2100 studies to construct adaptation options as a function of sea level rise. Associated robustness analysis can support the identification of the preferred strategy given future uncertainty
Adapted from Environment Agency, 2009

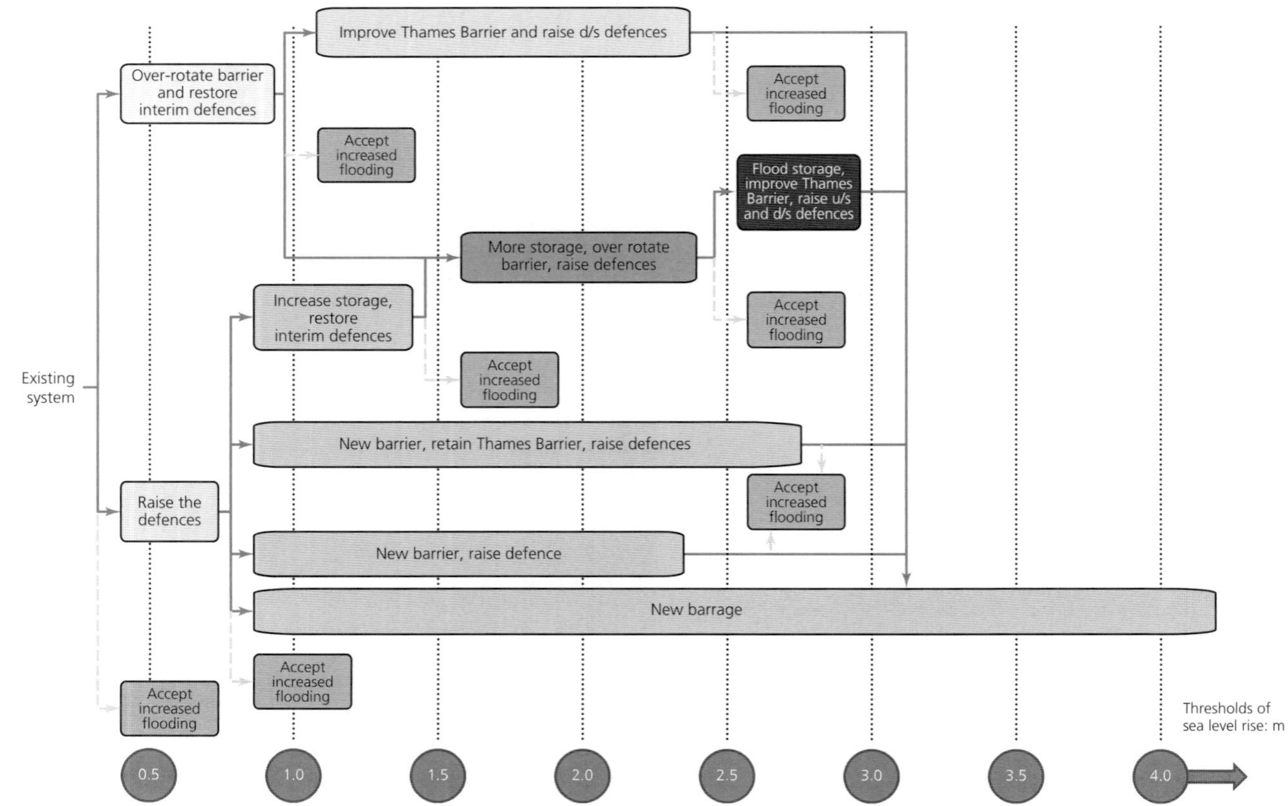

11.4.2 Automated optimisation methods

While the decision tree approach is intuitively appealing, the number of possible combinations and sequences of options rapidly increases, so it becomes impossibly time consuming to exhaustively evaluate every pathway through the decision tree. To overcome this problem, automated methods to optimise flood risk management strategies have recently started to appear in the context of asset management (Philips, 2006; Woodward *et al.*, 2010). The most promising methods make use of genetic algorithms (GAs), which have been widely used in other discrete optimisation problems, including problems with multiple objectives. The search for optimal solutions proceeds by a process that involves recombination of promising solutions with random variation to ensure that the search does not get stuck in local optima.

Real world constraints can be readily implemented in the GA, either by eliminating solutions that do not satisfy constraints, or by penalising options that do not satisfy constraints through the associated fitness function. Typical constraints might be annual budgetary ceilings, maintenance or inspection intervals, etc. Realistic cost functions (e.g. separating mobilisation costs from unit costs) can be used to reflect operational practice (e.g. combining work on adjacent sites into a single intervention). The progress of the GA in finding the best solutions (i.e. ones with higher net benefits) can be seen by plotting the performance of the best solution found at each successive generation (Figure 11.8). As shown in Figure 11.8 the optimum strategy (in

Figure 11.8 Net benefit of the investment strategy proposed by each successive generation (Phillips, 2006). Maximum and minimum values highlight the maximum and minimum values of the net benefit achieved by each generation – the difference between the maximum and minimum values reduces as the GA converges to an optimum strategy (given the single objective function)

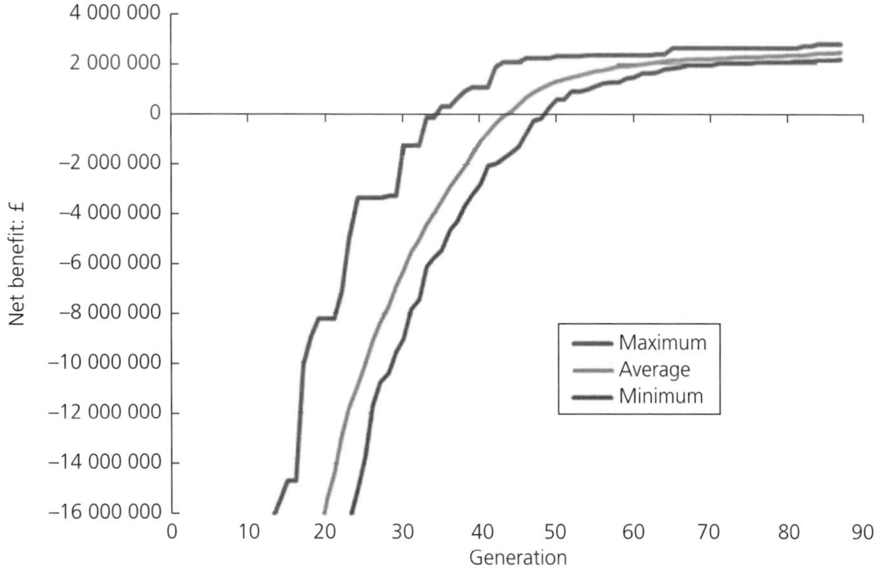

Figure 11.9 A Pareto set of optimal strategies based on an optimisation process – by comparing optimal strategies across a range of risk metrics the analysis can be used to highlight the trade-offs required in selecting one strategy over another and provides the evidence for decisions

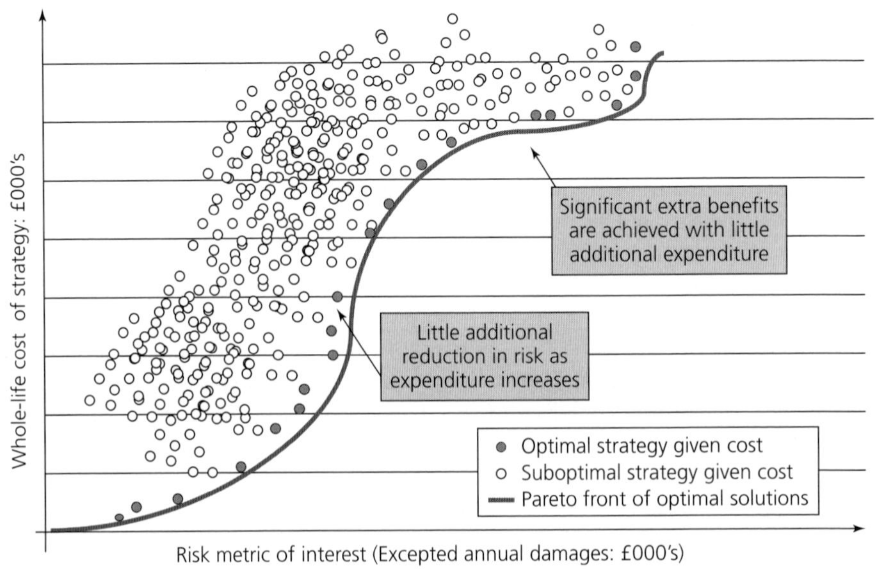

this case based on a single objective function of net benefit) emerges after around 60 generations.

Where multiple, and potentially conflicting, objectives are set (e.g. maximising net present value, minimising cost, minimising loss of life, maximising environmental gain), sets of optimal solutions will be developed, each optimal with respect to a single variable (e.g. the maximum benefit for a given level of expenditure). This set of solutions, or Pareto front (Figure 11.9), provides decision-makers with a graphical understanding of the trade-offs being made.

As with any automated analysis method, however, the outputs from an optimisation process need to be supplemented with engineering judgement. It is seldom possible to encode in the GA objective function all of the considerations that engineers will include in their design decisions. The identified solutions will need to be carefully scrutinised and, if necessary, modified, so that they satisfy all design criteria.

11.4.3 Real options analysis – formally valuing flexibility

The theory for valuation of financial options is well developed in financial economics. Real options analysis extends this theory to deal with real-life options, such as the decisions we are concerned with in this chapter, such as physical modifications to flood defence systems. Keeping one's options open will not, on the whole, be cost free. Real options analysis provides the theory required to estimate the financial value of having the option to do something in future. In other words, a real option is 'a choice that becomes available through an investment opportunity or action' (HM Treasury, 2009).

Real options analysis has, in recent years, been promoted as a means of evaluating climate change adaptation decisions (Ranger *et al.*, 2010), and is also being increasingly recognised as applicable to flood risk management (e.g. Appendix 1, HM Treasury, 2009). For example, consider a flood defence where we decide not to raise the defence right now, but may choose to incur the additional costs associated with the land purchase or foundation strengthening. The decision of whether to exercise the option to raise the defence is delayed to a future date, when more information regarding future climate change impacts, for example, will be known (lower part in Figure 11.6). Equally, a decision to continue to maintain an existing defence, while preventing development in the flood plain immediately behind the defence, enables a decision to refurbish/heighten the defence to be delayed. The disadvantage (increased) cost of maintaining this future flexibility is the increase in maintenance costs as the structure deteriorates. The advantaged (reduced cost) of this decision lies in the delay to major investments and the ability to implement a new design with greater certainty as much more becomes known. Following completion of new embankment projects there is frequently an intensification of land use behind the embankments, so that future acquisition of land not only becomes more expensive but creates a potential for social protest over actions that would result in relocations.

11.5. Conclusions

The need for resilient design has been demonstrated repeatedly in the past decade. It is evident that the climate is changing, that demographic shifts are resulting in land-use modifications, and that, as a result, there will be increasing pressure on both the flood plains and engineers to manage flood risk more efficiently than ever before, while reducing risk to acceptable levels. Adaptive management and resilient design will be at the heart of this effort, ensuring flexible strategies and readily modified designs that keep future options open. Engineering creativity will be at the heart of that process. Identifying resilient and flexible designs involves ingenuity and intimate understanding of the performance of design alternatives in particular contexts and locations. This is not something that can be taught in a textbook or design manual. Yet, having identified a set of promising solutions, engineers will then need to evaluate those solutions. In the presence of future uncertainties, it will be necessary to evaluate the performance of options with respect to those uncertainties and, where appropriate, to seek to justify the additional cost that enhancing robustness and resilience entails. The methods outlined in this chapter lend themselves to the process of testing robustness and making the economic case for resilience. They will not be necessary in every design situation – standardised methods can be used in routine situations. However, in situations where the costs (of investment or of failure) are high and where the uncertainties are great, the additional computation cost of analysis or risks, robustness and flexibility is bound to be well justified.

Underpinning the methods presented here is the concept of adaptive management, whereby flood risk management is a continuous process that takes place in the context of evolving risk, objectives and constraints. Our emphasis in this chapter has been on methods for evaluating flood risk management options. Alongside the analysis methods presented is a commitment to ongoing monitoring of the system and re-evaluation of risks, objectives and constraints. Without closing the loop through monitoring and management processes, the benefits of flood risk management are not going to be achieved.

REFERENCES

Ben-Haim Y (2006) *Info-gap Decision Theory: Decisions under Severe Uncertainty*, 2nd edn. Academic Press, London, UK.

Bertrand G, Michalski AL and Pench LR (1999) *Scenarios Europe 2010*. European Commission Forward Studies Unit. Working Paper 1999.

Bosher L, Carrillo P, Dainty A, Glass J and Price A (2007) *Realising a Resilient and Sustainable Built Environment: Towards a Strategic Agenda for the United Kingdom*. Blackwell, London, UK.

Dawson R and Hall J (2006) Adaptive importance sampling for risk analysis of complex infrastructures systems. *Proceedings of the Royal Society of London: Mathematical, Physical and Engineering Sciences* **462(2075)**: 3343–3362.

Department for Environment Food and Rural Affairs (Defra) (2005) Making Space for Water. Available at: http://archive.defra.gov.uk/environment/flooding/policy/strategy/index.htm (accessed 12 February 2012).

Environment Agency (2009) Thames Estuary 2100 Flood Risk Management Plant. Technical Report, September.

Evans E, Ashley R, Hall J *et al.* (2004a) *Foresight Future Flooding, Scientific Summary: Volume 1: Future Risks and their Drivers*. Office of Science and Technology, Department for Business Innovation & Skills, London, UK.

Evans E, Ashley R, Hall J, Penning-Rowsell E, Saul A, Sayers P *et al.* (2004b) *Foresight Future Flooding, Scientific Summary: Volume 2: Managing future risks*. Office of Science and Technology, Department for Business Innovation & Skills, London, UK.

Gouldby B, Sayers P, Mulet-Marti J, Hassan M and Benwell D (2008) A methodology for regional scale flood risk assessment. *Proceedings of the Institution of Civil Engineers: Water Engineering* **161**: 169–182.

Gouldby B, Sayers P, Panzeri M and Lanyon J (2010) Development and application of efficient methods for the forward propagation of epistemic uncertainty and sensitivity analysis within complex broad-scale flood risk system models. *Canadian Journal of Civil Engineering* **37(7)**: 955–967.

Hall JW and Harvey H (2009) Decision making under severe uncertainty for flood risk management: a case study of info-gap robustness analysis. In *Proceedings of the International Conference on Science and Information Technologies for Sustainable Management of Aquatic Ecosystems*, Concepcion, Chile. Electronic proceedings.

Hall JW and Solomatine D (2008) A framework for uncertainty analysis in flood risk management decisions. *Journal of River Basin Management* **6(2)**: 85–98.

Hall JW, Meadowcroft IC, Sayers PB and Bramley ME (2003) Integrated flood risk management in England and Wales. *Natural Hazards Review, ASCE* **4(3)**: 126–135.

Hall JW, Boyce SM, Wang Y, Dawson RJ, Tarantola S and Saltelli A (2009) Sensitivity analysis of hydraulic models. *Journal of Hydraulic Engineering, ASCE* **135(11)**: 959–969.

Harvey H, Hall J and Peppé R (2011) Computational decision analysis for flood risk management in an uncertain future. *Journal of Hydroinformatics*. In press, uncorrected proof.

HM Treasury (2009) *Supplementary Guidance to the Green Book: Accounting for the Effects of Climate Change*. HMSO, London, UK.

Hutter G and McFadden L (2009) *Strategies for Flood Risk Management*. EC Integrated Project FLOODsite Report T13-09-02. Available at: http://www.floodsite.net/html/partner_area/search_results3b.asp?docID = 694 (accessed 12 February 2012).

Intergovernmental Panel on Climate Change (2000) *IPCC Special Report. Emissions Scenarios*. Available at: http://www.ipcc.ch/pdf/special-reports/spm/sres-en.pdf (accessed 12 February 2012).

Klijn F, de Bruijn K, McGahey C, Mens M and Wolfert H (2009) *Towards Sustainable Flood Risk Management: On Methods for Design and Assessment of Strategic Alternatives Exemplified on the Schelde Estuary*. FLOODsite Report Number T14-08-02. Available at: http://www.floodsite.net/html/partner_area/search_results3b.asp?docID = 332 (accessed 12 February 2012).

Knight FH (1921) *Risk, Uncertainty, and Profit*. Hart, Schaffner and Marx.

Kouvelis P and Yu G (1997) *Robust Discrete Optimization and Its Applications*. Kluwer, Boston, MA, USA.

Lempert RJ, Popper SW and Bankes SC (2003) *Shaping the Next One Hundred Years: New Methods for Quantitative, Long-Term Policy Analysis*. Rand, Santa Monica, CA, USA.

Lindley DV (1990) *Making Decisions*, 2nd edn. Wiley, New York, USA.

McGahey C and Sayers P (2008) Long term planning – Robust strategic decision making in the face of gross uncertainty – tools and application to the Thames. In *Flood Risk Management: Research and Practice. Proceedings of FLOODrisk 2008*. Taylor & Francis, London, UK, pp. 1543–1553 (on CD).

National Institute of Building Sciences (NIBS) (2010) *Designing for a Resilent America: Summit on High Performance Resilient of Buildings and Related Infrastructure*. National Institute of Building Sciences, Washington, DC, USA.

Philips T (2006) Interim Project Report – Optimisation of Flood Risk Management Interventions. Flood Risk Management Research Consortium. Unpublished.

Ranger N, Millner A, Dietz S, Fankhauser S, Lopez A and Ruta G (2010) *Adaptation in the UK: a decision-making process*. Centre for Climate Change Economics and Policy, London and Leeds, UK. Available at: http://personal.lse.ac.uk/rangern/PB-adaptationUK-rangeretal.pdf.

Saltelli A, Chan K and Scott M (2000) *Sensitivity Analysis*. Wiley, New York, USA.

Sayers P, Hall J and Meadowcroft I (2002) Towards risk-based flood hazard management in the UK. *ICE Journal of Civil Engineering* **150(5)**: 36–42.

Sayers PB, Meadowcroft IC, Hall JW and Dawson R (2004) *Risk Assessment for Flood and Coastal Defence for Strategic Planning*. Environment Agency, Bristol, UK. R&D Technical Report W5B-030/TR, A Summary.

Sayers P, Galloway G, Penning-Rowsell E *et al.* (2011) *Flood Risk Management: A Strategic Approach*. UNESCO on behalf of WWF-UK/China and the General

Institute of Water Design and Planning, China. Consultation Draft (final publication due 2012).

Schweppe FC (1973) *Uncertain Dynamic Systems*. Prentice-Hall, Englewood Cliffs, NJ, USA.

Simon H (1956) Rational choice and the structure of the environment. *Psychological Review* **63**: 81–97.

UK Climate Impacts Programme (UKCIP) (2001) *Socio-economic Scenarios for Climate Change Impact Assessment: A Guide to Their Use in the UK Climate Impacts Programme*. UKCIP, Oxford, UK.

Wald A (1945) Statistical decision functions which minimize the maximum risk. *The Annals of Mathematics* **46(2)**: 265–280.

Willows RI and Connell RK (2003) *Climate Adaptation: Risk, Uncertainty and Decision-making*. UKCIP technical report. UKCIP, Oxford, UK.

Woodward M, Gouldby B, Kapelan Z, Khu ST and Townend I (2010) A real options approach to optimise flood risk management investments. *Proceedings of the Flood and Coastal Risk Management Conference*. Thomas Telford, London, UK.

USEFUL WEBSITES

Foresight toolkit: http://hsctoolkit.bis.gov.uk (accessed 12 February 2012).

UK Climate Impacts Programme (UKCIP), AdpatME (Adaptation monitoring & evaluation) toolkit: http://www.ukcip.org.uk/adaptme-toolkit (accessed 12 February 2012).

Flood Risk
ISBN: 978-0-7277-4156-1

http://dx.doi.org/10.1680/fr.41561.303

Chapter 12
Managing flood risk in the Thames Estuary – the development of a long-term robust and flexible strategy

Owen Tarrant* and Paul B. Sayers†
* Environment Agency, UK
† Sayers and Partners, Oxfordshire, UK

London is the economic powerhouse of the UK, yet much of its enabling infrastructure is within the flood plain of the Thames Estuary. Following the devastating floods of 1953, a system of flood defences was designed in the early 1970s and built in the 1980s to protect against the 1 in 1000-year extreme tide. The designs included an allowance for sea-level rise, which ensured the 1 in 1000-year standard of protection to the year 2030, after which standards are anticipated to drop. At the start of 2000, the Environment Agency felt that the time was right to plan for the future of London's defences in the context of possible future changes through the twenty-first century. This plan was developed under the name of the Thames Estuary 2100 Project.

To incorporate risk-based decision-making and adaptability at the heart of the strategy plan, it was necessary to develop innovative tools and techniques to assess the performance of different management options under future scenarios. Although the uncertainty within the climate change projections was severe, the condition and performance of the existing defence system enabled key investment decisions to be delayed by tens of years. This provided the freedom to develop an adaptable management strategy, enabling future investment to be triggered by the monitoring of key variables of change.

12.1. Managing tidal flood risk in the Thames Estuary
12.1.1 History of flood risk management in the Thames Estuary
12.1.1.1 Introduction
The Thames Estuary has always been at the centre of the British economy and culture, since the earliest settlement by Romans in around 50 AD. Since then, the estuary has grown to become one of the most significant financial and cultural capitals of the world, with a population today of around 7.5 million. It is estimated that there are approximately 0.5 million homes within the tidal flood plain, and this is set to increase, and that in any one working day there can be approximately 1.25 million residents tourists and commuters in the area. Figure 12.1 shows the extent of the 1 in 1000-year tidal flood plain that is currently defended.

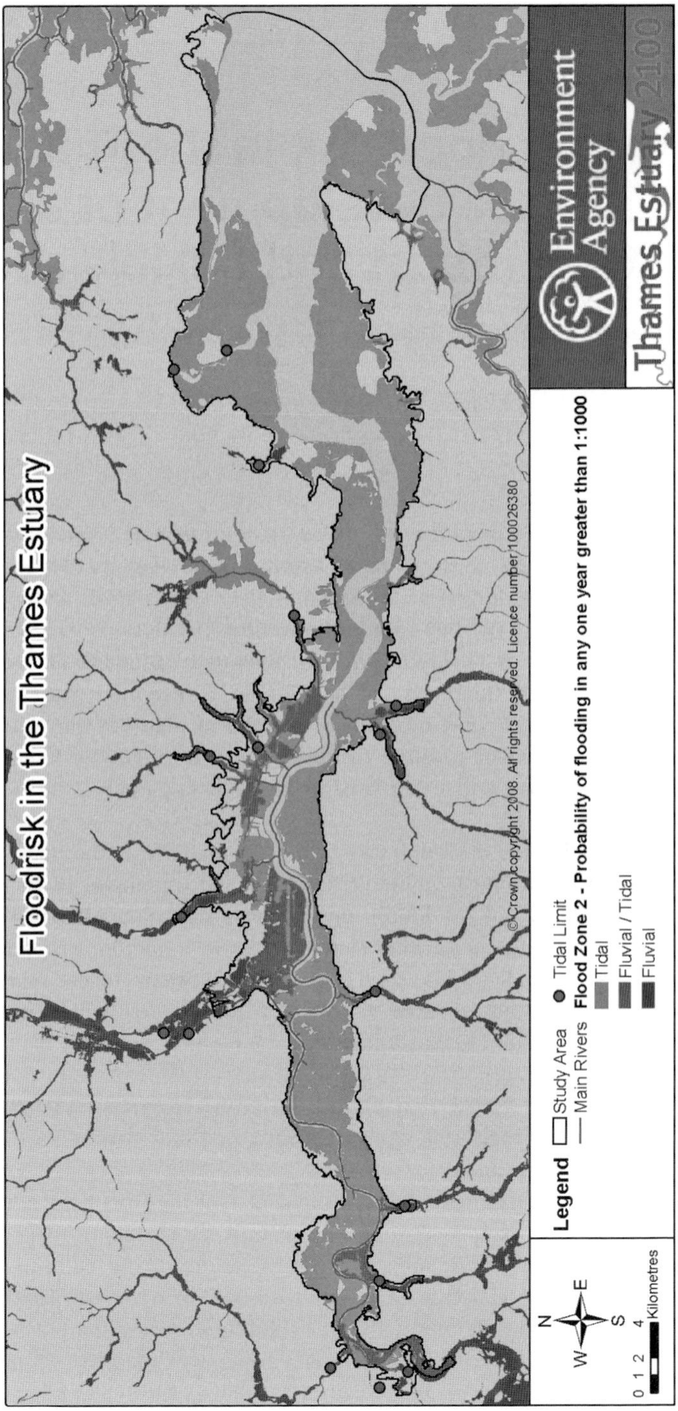

Figure 12.1 The extent of the 1 in 1000 year Thames Estuary flood plain that is currently defended
© Crown Copyright 2011. Contains Environment Agency information © Environment Agency and database right

London is the UK's largest centre of economic activity, contributing around £250 billion annually to the UK economy. A number of its business sectors are important players on the international stage. In particular, the financial and business service sectors operate in the global marketplace alongside those of other international centres such as New York and Frankfurt. Although the traditional 'square mile' of the City of London is outside the flood plain of the Thames, the more recent commercial centre of Docklands is, as its name suggests, wholly within it.

While the people and property of the Thames Estuary are currently well protected by a world-class system of flood defences, it is likely that this system will be increasingly relied upon and tested as the effects climate and socio-economic change act in tandem to increase both the probability and the consequences of any future flood event.

12.1.1.2 Historical approach to flood defence within the Thames Estuary

Flooding has always been an important topic for London. The earliest written record of flooding along the Thames Estuary is from an Anglo Saxon Chronicle of 1099, which recorded a flood on the 'first day of the new moon', and numerous floods have been recorded since then. Historically, the Thames tidal defences have been improved in a reactive manner in response to the most severe of these events. During the nineteenth century it was seen that the levels of exceptional surge tides were increasing and, after two record tides of 1874 and 1875, Parliament acted by passing the Metropolis Management (Thames River Prevention of Floods) Amendment Act 1879 (Gilbert and Horner, 1984). The Act set a statutory level for the flood defences in London. Following the 1928 event, the last major event to flood central London, the defences were raised again under the powers of a new Land Drainage Act, which was passed in 1930. Following the devastating North Sea surge of 1953, and 58 deaths on Canvey Island in the outer Thames Estuary after the defences were breached, the issue gained new prominence. This catastrophic flood (which affected much of the east coast of the UK as well as The Netherlands, Belgium and Germany) provoked an urgent interest in *protecting* London from tidal surges. In 1954, the government-appointed Waverley Committee reported on the disaster and studies into the natural phenomena that cause surge tides, and possible new flood defence options commenced. Eventually, the government Chief Scientist, Sir Herman Bondi, recommended, in 1966, that the best solution was a 'tidal surge barrier and raising the height of the river bank, backed up with a good system of flood warnings'. In 1972, legislation was provided through the Thames Barrier and Flood Prevention Act, to design and construct the system of River Thames tidal defences seen today. This approach is characteristic of a 'flood defence' led management paradigm, where events occur and defences are raised in response. This reactive approach is evident in the stratification of different materials used to construct the tidal defences as they exist today (Figure 12.2).

12.1.2 Present-day setting flood defence system for London

The Thames Barrier was inaugurated by the Queen in 1983, three decades on from the 1953 floods. The Barrier, its associated gates and defences were designed to protect London from the 1 in 1000-year combined tidal/fluvial flood event in the year 2030. This design standard included an allowance of 8 mm/year to account for the rate of

Figure 12.2 Historically, a flood defence approach has led to 'reactive' raising of the defences in response to major events
Contains Environment Agency information © Environment Agency and database right

change in mean sea level measured from the gauged record by the engineers and designers in the 1970s. In addition, this 8 mm/yr also allowed for and the known rate of local subsidence associated the rapid rate of groundwater abstraction that occurred during the industrial revolution and regional isostatic sinking following the last glaciation.

The final crest level chosen during the design included a further freeboard allowance, taking the actual standard to what is currently estimated to be approximately a 1 in 10 000-year return period (significantly higher than initial design standard).

Today the flood risk is managed in the estuary using a complete portfolio of approaches, the most obvious being the formal flood defence system consisting of barriers, tidal gates and flood walls and embankments as shown in Figure 12.3. However, less obvious risk-management measures, such as forecasting and warning, spatial planning and emergency exercises, all have a critical role to play. All these elements of the flood management system are described in more detail below.

- *Flood defence* – Flood defences remain the primary measure of flood management throughout London. This defence system consists of the Thames Barrier and five

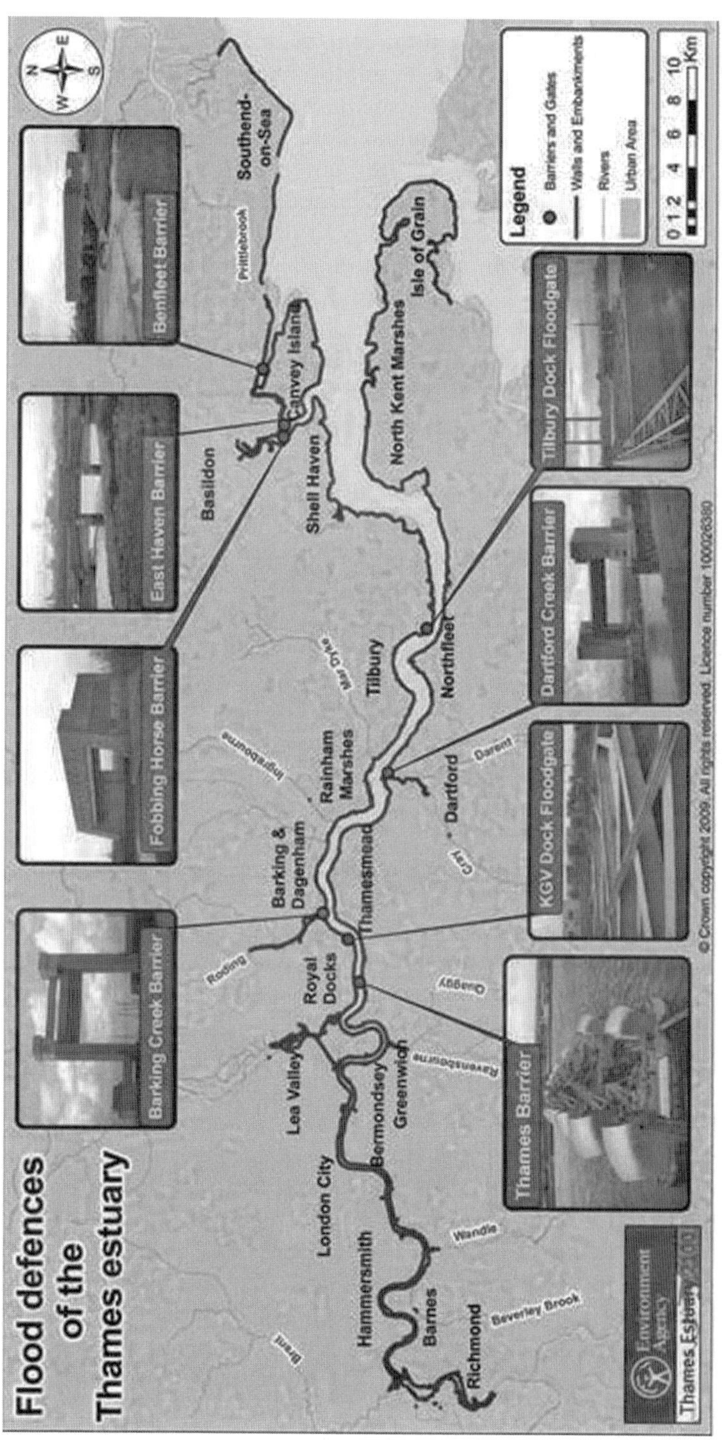

Figure 12.3 Thames Estuary Flood Defence System
© Crown Copyright 2011. Contains Environment Agency information © Environment Agency and database right

other barriers (Barking, Benfleet, Dartford, East Haven and Fobbing Horse) and two major flood gates at the entrance to the Royal and Tilbury Docks. The barriers and gates work alongside approximately 337 km of linear flood defences (walls and embankments), which are themselves punctuated by a further 300 industrial flood gates and 400 minor moveable structures.

- *Flood forecasting and warning* – Operation of the flood barriers along the tidal Thames is dependent on accurate forecast data. The potential of developing barrier closure conditions (storm surge, high tide and high fluvial flow) are monitored from the control room at the Thames Barrier, which is manned 24 h a day, 365 days of the year. The model (POLCS3) is used for operational surge forecasting in the UK, and is described in detail by Flather (2000). The model is forced by surface pressure and 10 m wind fields supplied hourly by the Met Office's meso-scale model (Horsburgh and Wilson, 2007). Four surge model runs are carried out each day, each comprising a 6-hour hindcast and a 36-hour forecast. The model is used to predict the storm-surge component, which accounts for interaction with the tides, by subtracting the model predicted tide from the tide-with-surge solution. The total water level is then estimated by adding the storm-surge component to the harmonically predicted tide based on tide-gauge observations. The Thames Barrier control room obtains forecasts for tide levels in the estuary mouth at Sheerness twice daily. These forecasts are then used to force a one-dimensional ISIS hydraulic model of the tideway from Mosely Lock (the upper tidal limit) to Sheerness (developed by Halcrow (1999)). Based on predefined trigger water levels at London Bridge, the Barrier is raised to prevent tidal surges propagating up the Thames, and the associated barriers downstream of the Barrier are closed to defend the tidal tributaries. For those levels below barrier closure, but which could cause flooding in the upper reaches of the estuary, typically through Richmond, flood warnings are issued to the public and the emergency services so that local resilience measures, such as individual property-level flood boards and gates, can be effectively deployed.
- *Spatial planning control and risk management* – Development pressure on London's tidal flood plain is constant and large, and therefore consideration of flood risk in the planning process is essential in order to:
  - avoid inappropriate development of the flood plain
  - ensure that any development seeks to manage flood risk;
  - avoid encroachment on the tidal foreshore
  - adopt sound engineering principles.

  Consideration of flood risk is made throughout the planning process, from the strategic assessment of risk at the spatial planning level, to the provision of detailed advice at the stage of individual planning applications.
- *Emergency and continuity planning* – Resilience forums operate to plan for and manage the potential consequences of flooding. Such planning activities include:
  - the production of business continuity plans
  - planning for the consequence of a flood event, emergency evacuation, shelter and clear-up operations
  - providing advice to local communities on what action they can take before, during and after a flood.

12.2. Habitat and environmental importance of the Thames Estuary

The Strategic Environmental Assessment (SEA) carried out by the Environment Agency (2009a) under the Thames Estuary 2100 (TE2100) Project provides a good description of the current environment characteristics of the estuary, and highlights the importance of the Thames as an estuarine corridor intrinsically connected to the (southern) North Sea and the east Atlantic fly-way. In this regard, it can be stated that the tidal Thames has a major influence on the life in the North Sea, as well as in the freshwater catchments upstream.

The importance of the Thames Estuary as a wildlife corridor must not be underestimated and, equally, the widely held perception of the Thames through London as a 'dirty river', which is perhaps due to high concentrations of suspended sediments and its industrialised past, are now simply unfounded. In fact, the river supports amongst the widest variety of animals of any estuary in Europe, specifically

- it is one of the five most important estuarine complexes for birds in Europe, supporting over 300 000 migratory waders and wildfowl
- 121 species of fish have been recorded since the 1960s, nine of which now spawn within the estuary.

Large parts of the estuary are included within statutorily designated wildlife sites, and support nationally and internationally important habitats and species (notably birds, but also including fish, invertebrates and marine mammals). Figure 12.4 shows the spread of habitats throughout the estuary.

12.3. Future drivers of change in the Thames Estuary

12.3.1 Climate change and sea-level rise

Climate change will be one of the most significant drivers acting to increase flood risk within the Thames Estuary. It is likely that the probability of future flood events will increase through changes in mean sea level, a possible increase in the magnitude and frequency of storm surge, and the increased fluvial flow associated with changes in rainfall intensity, seasonality and duration. Key findings from the studies undertaken in the TE2100 project (Environment Agency, 2009b) include

- the sea-level rise in the Thames over the next century due to thermal expansion of the oceans, melting glaciers and polar ice is likely to be between 20 cm and 90 cm.
- there remains a lot of uncertainty over the contribution of polar ice melt to increasing sea-level rise – at the extreme, it may cause the sea level to rise by a total of up to 2 m (including thermal expansion), although this is thought to be highly unlikely
- climate change is less likely to increase storm-surge height and frequency in the North Sea than was previously thought
- future peak freshwater flows for the Thames, at Kingston for instance, could increase by around 40% by 2080.

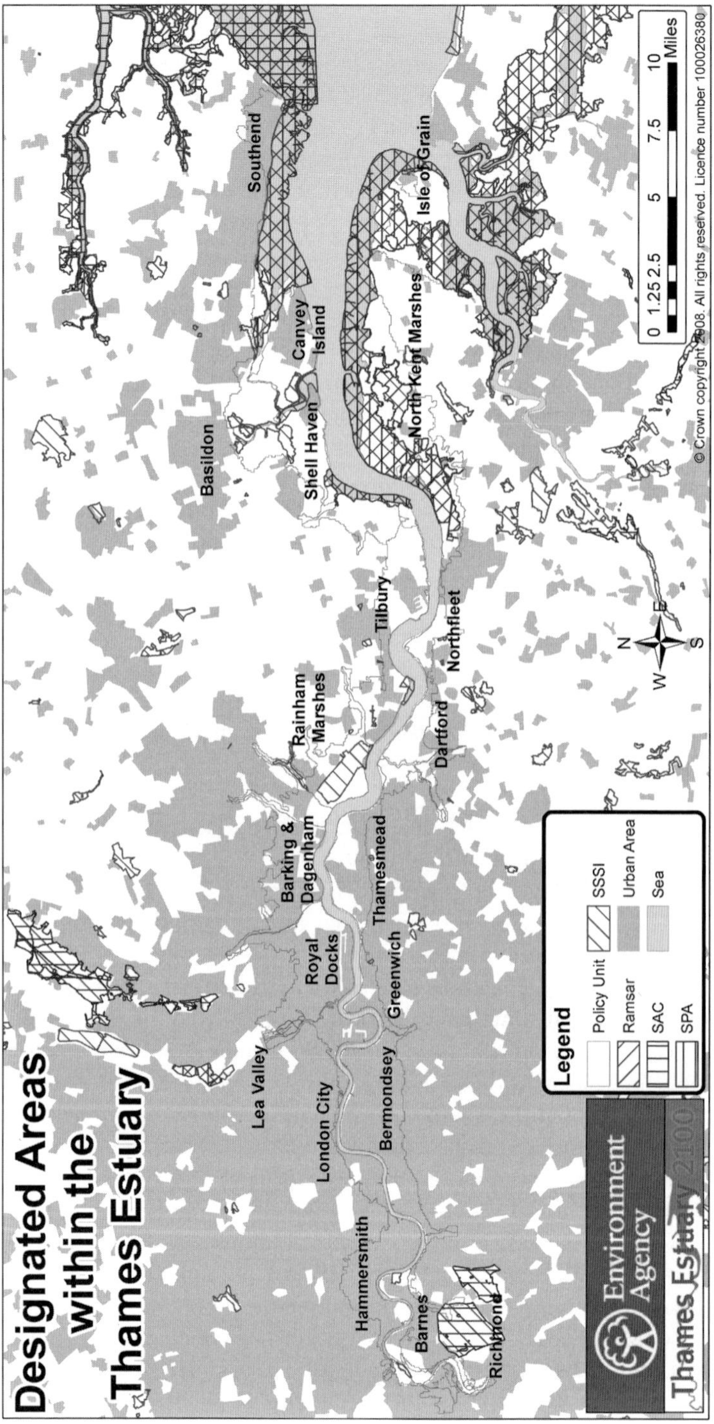

Figure 12.4 Classification of designated habitats along the Thames Estuary
© Crown Copyright 2011. Contains Environment Agency information © Environment Agency and database right

12.3.2 Socio-economic change

The Foresight Future Flooding Report (Evans *et al.*, 2004a,b) highlighted the importance of the socio-economic change in flood risk, both through changes in the value of and number of receptors at risk in the flood plain (e.g. residential properties and their contents) and through increased levels of urbanisation affecting rates of run-off and catchment response times.

Despite the recent recession, the long-term growth pressures on London and the South East remain, and the cost of doing business in these areas is high: office and staff costs are relatively high, there is transport congestion with overcrowding and delays, and housing costs are high (Communities and Local Government, 2010). The focus has been along the Thames corridor, with a desire to maximise the growth potential of 40 km of land stretching from the London Docklands to Southend in Essex and Sheerness in Kent. As part of this initiative, which is known as the Thames Gateway, up to 120 000 new homes will be built along the Thames, up to 75% of which could be located within the flood plain (Figure 12.5). It is essential that any future development in the Thames Gateway goes hand in hand with flood risk management, and takes account of future flood risk.

12.3.3 Ageing of the current flood defence infrastructure

Although the flood defence system currently provides an exceptionally high standard of protection (and it is now understood that it would continue to do so well past 2030), significant lengths of the wall and embankments will still reach their intended design life and require replacement or improvement works. Improving the condition of these defences is a large task that demands a strategic approach.

Given these drivers, there was a clear need to develop long-term plan for the future management of flood risk, which was adaptable to the potential changes facing the estuary.

12.4. Developing a strategy to manage tidal flood risk

12.4.1 The Thames Estuary 2100 Project (TE2100)

The TE2100 Project was established in 2002 with the aim of developing a long-term tidal flood risk management plan for London and the Thames Estuary. The project is led by the Environment Agency.

Studies undertaken during the development of the TE2100 plan (Environment Agency, 2009c,d) included: a detailed assessment and appraisal of the options available to manage flood risk; and their economic costs, benefits and environmental impacts. The plan set out the strategic direction for managing flood risk in discrete policy areas across the estuary, and contained recommendations on what actions will be needed in the short (next 15 years), medium (the following 35 years) and long term (to 2100). The plan was based on current guidance on climate change, but extended this to ensure that the plan was adaptable to changes in predictions of sea-level rise and climate change over the century.

Figure 12.5 Proposed Thames Gateway development zones in the tidal flood plain

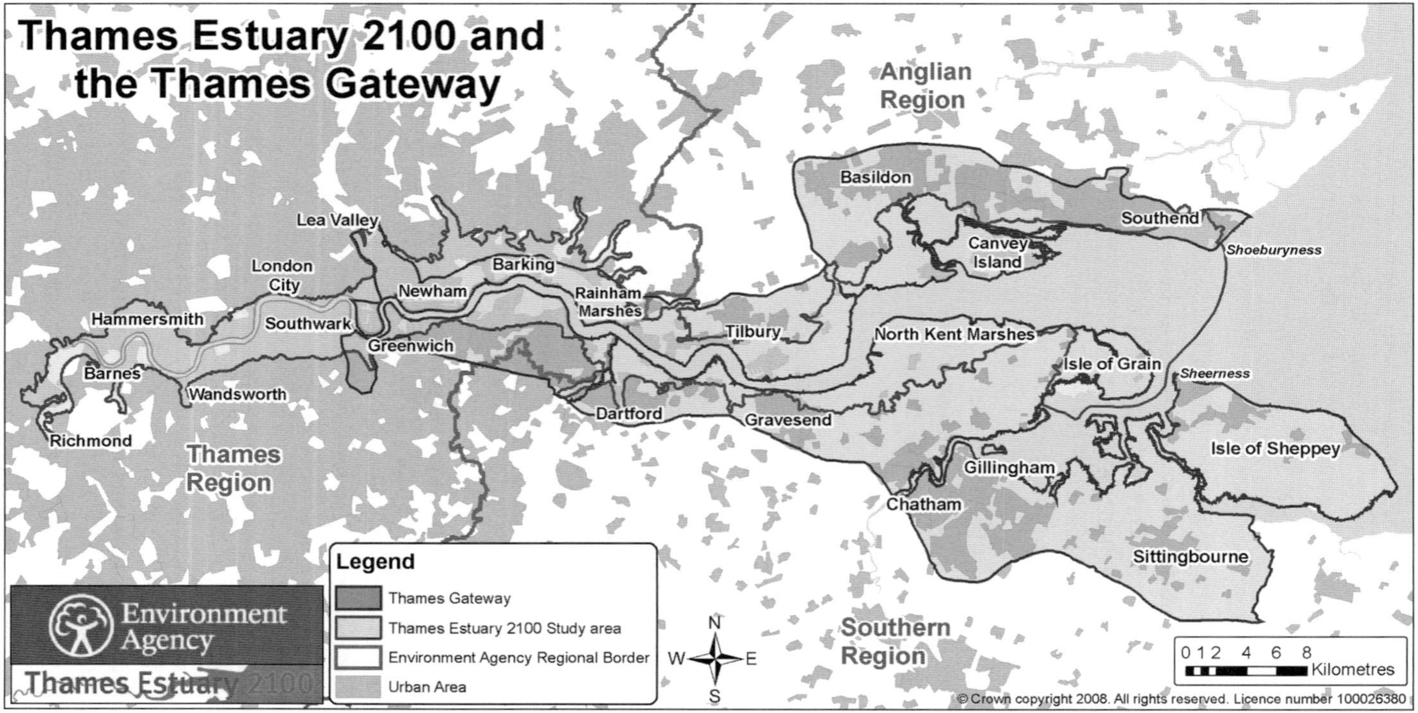

Figure 12.6 Use of the source–pathway–receptor concepts within a probabilistic analysis framework
Based on the RASP modelling approach (Sayers *et al.*, 2004)

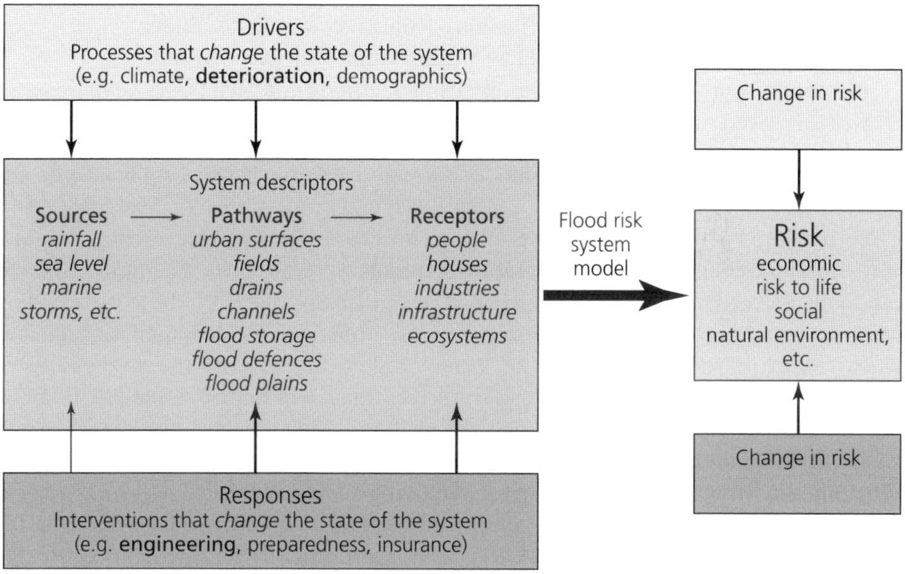

The scale of the study provided excellent momentum to the development of the plan and of innovative tools and techniques. Some of the key approaches used in, and findings and lessons from, the study are discussed in the following sections.

12.4.2 Tools and techniques for modelling and evaluating tidal risk

The risk analysis used to support the development of the TE2100 plan was considered state of the art at the time it was undertaken (reflecting the importance of the Thames Estuary within the UK, and the potential risks and high cost of any flood management work). The analysis adopted a probabilistic framework based on source–pathway–receptor concepts (Figure 12.6).

The probabilistic integration of these models was provided through a system risk model (so-called Risk Assessment for Flood and Coastal Defence for Strategic Planning (RASP) analysis (Gouldby *et al.*, 2008) which includes considerations of:

- *Sources* – Use of sufficient return-period events (60 in total, up to and including the 1 in 10 000-year return period event) to ensure that events which exceed the notional 1 : 1000-year design standard of the defences were considered. Incorporation of the spatial coherence with the tributary flows along the Thames was not considered, because at the time this was considered too difficult, and it was only addressed using simplified models.
- *Pathway* – The performance of the embankments, riparian gates and the most important barriers were all represented within the system risk model; including the

probability of both serviceability (e.g. overtopping without breach) and limit-state (e.g. breach) failures.

- *Receptors* – A wide range of simple to quantify impacts were considered, with primary risk metrics based on direct economic damages (based on residential and commercial property damages) (Penning-Rowsell *et al.*, 2005). Wider considerations, such as risk to the environment and people, and business disruption, were also included, but by using methods and techniques that are less well advanced and based largely on expert elicitation.

This RASP-based Thames Estuary model allowed a rich insight to be gained into the spatial distribution of risk to be developed. By changing the source and receptor terms in the model to represent the effects of climate change on extreme events flow and socio-economic change, respectively, future scenarios were explored. The model allowed risk to be tracked in the flood plain and then attributed back to the corresponding failed or overtopped flood defence lengths. This functionality was used to highlight where future improvements could be made. Figure 12.7 shows the probability of inundation in the Thames Estuary under present-day conditions, assuming that the Thames Barrier is fully operational and the wall and embankments are in the current condition. The majority of the flood plain has a lower than a 1 in 1000-year probability of inundation, which is as expected given the current level of defence. Certain hotspots can be seen in the tidal tributaries, and in the outer estuary where defence standards are lower.

The system risk model was then used to explore the influence of future change on the estimate of present-day risk. The source–pathway–receptor model provided a very useful framework to enable the impact of a wide range of future changes to be explored in a structured manner. This included the following.

- *Changes in the source terms* – Changes in climate, both in mean sea level, fluvial flows and potentially extreme surges are only important issues to explore. The system models provided a framework that made exploring the impact of such change relatively straightforward.
- *Changes in the pathways* – Modifications to the barriers and defences, either through raising, strengthening or repositioning, flood proofing of properties or more ambitious modification of the flood-plain topography, are considered as modifications to the pathway within the system model, and their individual influence on risk could be isolated.
- *Changes to the receptors* – Modifications to the number or location of the people or property in the flood plain or changes in their vulnerability (perhaps through enhanced warning or better preparedness), were reflected in the system model through changes in the quantified receptor damage functions.

12.4.3 Framework for decision-making with TE2100

The basic approach to decision-making in the TE2100 study was as follows.

- Step 1 – Develop the strategic alternatives (*generate options based on a portfolio of flood risk management interventions*).

Figure 12.7 Probability of inundation along the Thames Estuary under present-day conditions

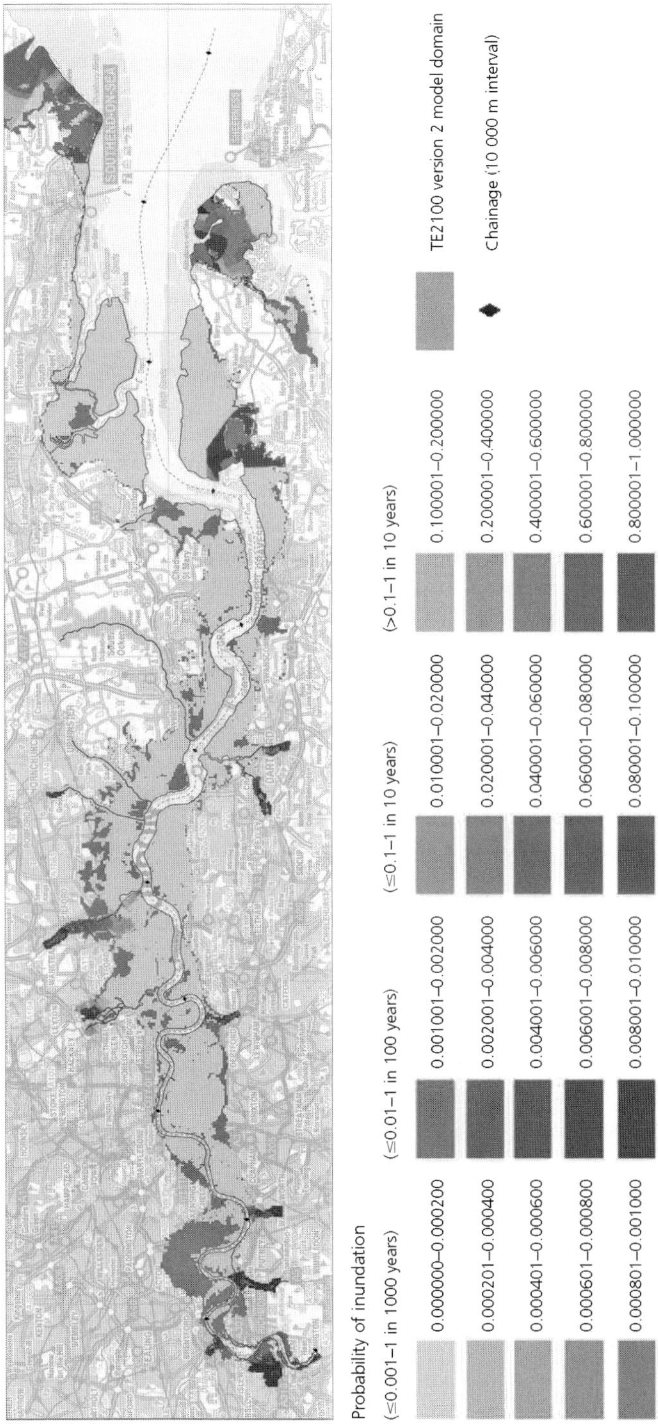

- Step 2 – Assess the performance of the alternatives against a range of risk metrics (*under present and future climates*).
- Step 3 – Explore the performance of the alternatives under a wide range of potential futures – climate and socio-economic (*assess performance against future uncertainty*).
- Step 4 – Identify a robust and adaptable option (*identify preferred sequence of interventions*).

Each of these basic stages is discussed in turn below.

Step 1: Developing strategic alternatives

First, a reference strategy was established:

- *Do nothing* – a walk-away option provided a baseline against which to judge how worthwhile the other strategies were.

Then, strategic options were considered for 23 policy units (representing notionally discrete flood areas along the Thames Estuary) and the Estuary as a whole. Each option included a combination of actions – raising river walls, modifying or building flood barriers or flood storage areas, applying resistance and resilience measures to buildings, and so on. The relative mix of options broadly reflected a range of management philosophies, as follows.

- *Resistant* – based on 'sweating' the current defences (i.e. ensuring that maximum value is obtained from past investment) and improving them where necessary – the focus throughout the estuary is on improving the present-day defence system and includes non-structural measures only where it is easy to do so.
- *Resilient* – based on a mix of maintaining the existing defence line and setting it back to provide flood storage and retention and biodiversity gains – the focus is on reconstructing the estuary defences to provide storage where possible, together with non-structural solutions.
- *Highly resistant* – based on improving defences and constructing major new physical infrastructure such as a new barrier or barrier with locks – the focus throughout the estuary is on defence and preventing flooding, both now and in the future.

Within each strategic option a number of different ways of achieving similar outcomes was explored (Box 12.1). To select the best way of providing the individual components of each strategic option, an assessment was made of the incremental benefits (if any) achieved with increasing cost.

Step 2: Assessing performance against a range of metrics

The performance of the strategic options were assessed against a wide range of criteria including (but not limited to)

- economic efficiency and effectiveness
- environmental performance
- people protection
- equity and fairness
- cultural heritage.

Box 12.1 Different ways of achieving similar outcomes with alternative options

Option 1. Improve the existing defences

1.1 Raise defences when needed
1.2 Allow for future adaptation of defences
1.3 Raise defences when they are replaced
1.4 Optimise defence repair and replacement

Option 2. Tidal flood storage

Four potential sites were identified that are in the right location to store tidal waters and reduce the level of storm surges

The sites identified are at Erith Marshes, Aveley and Wennington Marshes, Dartford and Crayford Marshes, and Shorne and Higham Marshes

Option 3. New barrier

3.1 Tilbury location
3.2 Long Reach location

The new barrier would be designed to resist the highest surge tides predicted under government climate change guidance

Option 4. Barrier with locks

4.1 Tilbury location
4.2 Long Reach location
4.3 Convert Thames Barrier to a barrier with locks when the operational limit of closures per year is reached

Step 3: Exploring performance under a wide range of potential futures

Each strategic alternative was assessed against four possible climate futures. These ranged from more moderate rates of change through to more extreme assumptions.

- 0.45 m rise in mean sea level by 2100 (so-called Medium–High).
- 1.0 m rise in mean sea level by 2100 (so-called Defra06-based on allowances outlined in Defra, 2006).
- 1.6 m rise in mean sea level by 2100 (so-called High+).
- 3.2 m rise in mean sea level by 2100 (so-called High+ +).

The difference in these assumptions make a significant difference to the estimated sea levels within the Thames Estuary over time, as shown in Figure 12.8. The magnitude of the uncertainty in sea level rise made a significant change in the estimate of future risk and the performance of different management options.

Step 4: Identifying a robust and adaptable strategy

The approach used in TE2100 followed the framework used for flood risk management as detailed in *Making Space for Water* (Department for Environment, Food and Rural Affairs (Defra), 2004). This requires that flood risk management plans are not only technically

Figure 12.8 Projected possible future mean sea-level rise and peak 1000-year surge tide level for the mouth of the Thames Esturay
TE2100 project, Environment Agency, 2009b

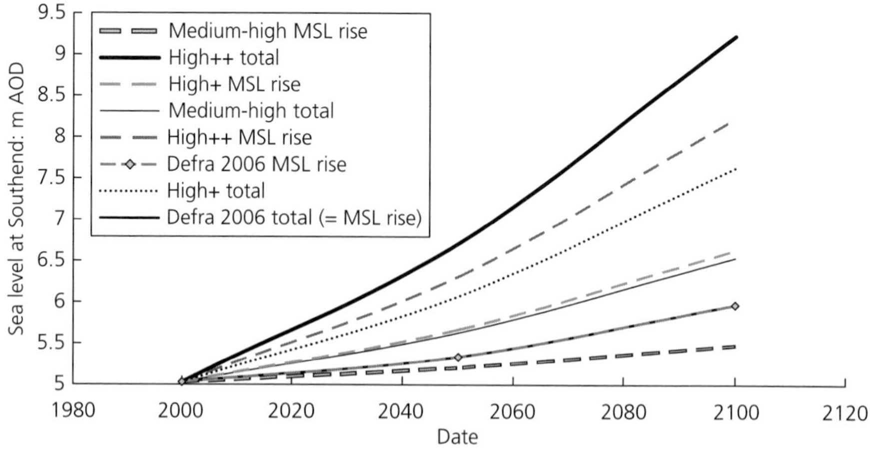

robust, but also integrate socio-economic and environmental factors. It puts greater emphasis on living with and adapting to flood risk and storage of flood water, rather than building ever bigger and higher defences to combat climate change, including

- a focus on people (and the economy as a proxy for people)
- the need to meet the requirements of statutory European Directives, including those on birds, the habitats (which aim to 'protect the integrity' of internationally important nature conservation sites) and the Water Framework Directive (which aims to 'maintain and improve the water environment')
- national economy efficiency (as defined by the Treasury Green Book on the investment of public money).

In addition to these core selection criteria, the TE2100 project sought to develop a strategy that was adaptable to any potential future change, including climate change, in order to avoid expensive 'white elephant' investments while managing risk to acceptable levels. No single strategic alternative provided the preferred management strategy under all climate futures. Effort was, therefore, directed towards providing an adaptable planning framework that could be reviewed and updated as the rate of change of a number of key indicators became increasingly certain. A strategy was developed based on an understanding of the performance of each strategic alternative under all four future scenarios, and assessed in terms of the following.

- *Flexibility* – reflecting the potential for the strategy to be adapted as the reality of future change becomes clear.
- *Robustness* – reflecting the performance of the strategy across a range of futures, ensuring a 'no regrets' approach.

The final strategy represented an amalgamation of all four strategic alternatives, with decisions on major infrastructure investment being delayed until the latest possible time (after allowing for design and detailed planning/consultation). Within the strategy, actions that continue to perform well no matter what future materialises were prioritised for early implementation – including continued investments in improving forecasting and warning and in spatial and emergency planning, but also in 'sweating' the existing defence system to ensure that maximum value from past investment was achieved. This included the development risk-based maintenance and improvement programmes.

12.4.4 How is adaptive capacity built into the plan?

Each option included interventions to cope with different levels of sea-level rise and the thresholds of change at which they would be required. A flexible strategy was then developed around the idea of a decision pipeline, were the most robust options (i.e. those that are able to perform well under all futures) were implemented first, leaving major irreversible (or highly scenario dependent) decisions as far as possible into the future to make best use of observed information on sea-level rise and other key indicators of change as these become available. A simple representation of the decision pipeline is shown in Figure 12.9.

The TE2100 plan also included a monitoring and ongoing re-evaluation process. The monitoring process provides the triggers (Box 12.2) for the decisions within the pipeline. For example, if monitoring reveals that climate change is happening more quickly (or slowly) than predicted, the strategy can be reappraised in light of the new information, and options can be brought forward (or put back). Other changes are also monitored and

Figure 12.9 Decision pipelines were used to show decision points within the strategy. The decision pipelines were also used to highlight the flexibility of different choices (McGahey and Sayers, 2008)

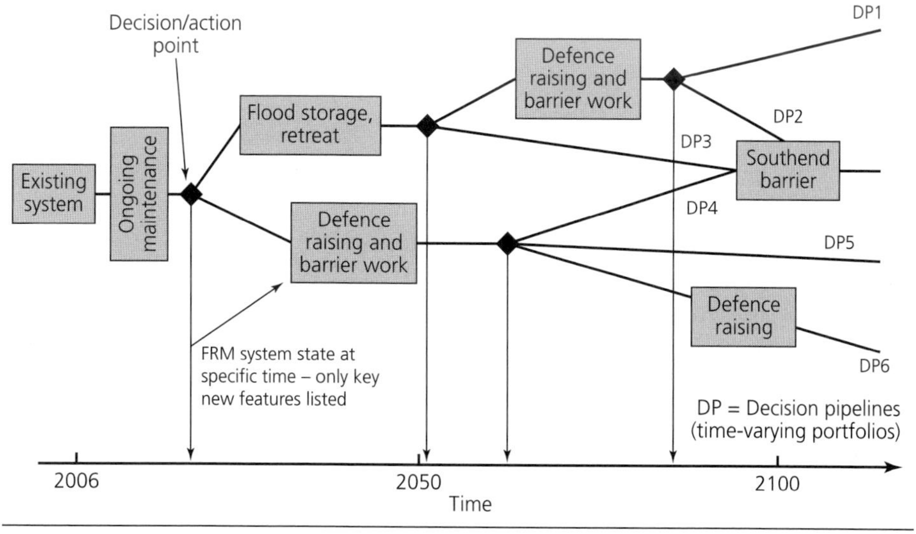

Box 12.2 Ten indicators of change considered in the TE2100 plan
Environment Agency, 2009d

1. **Mean sea level**
   The level to which protection may be required twice a day every day. The mean sea level is the level which determines the number of times a year that a barrier must be closed. It also has a major impact on the area of intertidal habitat in the estuary.
2. **Peak surge tide level**
   The extreme (but rare) tidal flood level that will have to be managed. It also determines the crest level of the defences, including those of the Thames and other barriers.
3. **Peak river (fluvial) flood flows**
   The combined tidal/fluvial flood risk at pressure points in west London and where tributaries meet the estuary.
4. **Condition of flood defence structures**
   This concerns how to optimise the repair and renewal of defences so that investment programmes yield the best value for money, while ensuring public safety and that the flood defence system will function as required. It involves the identification of the improvements needed to ensure the integrity of the system.
5. **Frequency of closure and reliability of the Thames and other barriers**
   This involves determining how much useful life remains for these important structures, and how efficient the structures are.
6. **Developed area and value/type of development**
   This covers the people and property at risk. Key social and economic information is required for flood risk management planning.
7. **Extent of erosion/deposition**
   Assessment of the health and stability of the intertidal habitat zone, and whether European regulations on habitats are being complied with.
8. **Intertidal habitat area including mudflat and saltmarsh**
   A measure of how well flood risk (i.e. safer flood plains) and opportunities for sustainability (e.g. the creation of green corridors) are being factored into the development. This includes predicting future needs of society and economic considerations.
9. **Land-use planning and development activities**
   A measure of how well flood risk (i.e. safer flood plains) and opportunities for sustainability (e.g. the creation of green corridors) are being factored into the development. This includes predicting future needs of society and economic considerations.
10. **Public/institutional attitudes to flood risk**
    The public (and hence political) appetite for risk, and institutional preparedness to manage risk and to plan for/respond to emergencies.

responded to, including the influence of potential major developments such as a new outer estuary airport or tidal energy scheme. Some decisions, of course, require a considerable lead time prior to implementation. The lag time between deciding to intervene and the delivery of change in the management was considered, allowing for up to 30 years for major decisions to come to fruition.

12.4.5 Exploring the effectiveness of non-structural options in managing flood risk

TE2100 explored a range of 'flood-plain management' options, including the following.

- *Flood forecasting and warning* – improving the lead time and accuracy of the flood-forecasting information used for operation of the Thames Estuary flood management system, and improving flood-warning information for emergency response and damage avoidance.
- *Pre-event measures* – raising public awareness and improving emergency/multi-agency response planning.
- *Responses affecting existing development* – for example, retrofitting resilience measures, provision of safe havens and high-level access for evacuation, and secondary defences around critical infrastructure.
- *Responses affecting new development* – for example, spatial planning, development layouts that take account of flood risk, resilient construction and moving vulnerable development.

To assess the benefits of such responses, an event-tree-type analysis was employed. This analysis built on the concept that many flood-plain management options can be viewed as a chain of dependent activities, with the effectiveness of each new action being dependent to some degree on the effectiveness of the preceding action. The cumulative impact of this chain was assessed in terms of the effective reduction in risk to life and reducing expected annual average damages.

While the benefits and the costs of flood-plain management were calculated separately and can be distinguished from the other options, it was felt that such improvements offered a no-regrets strategy, and so these were not considered as an alternative, but rather included as part of a portfolio of other structural options (Environment Agency, 2009c).

One of the greatest challenges facing the TE2100 plan is that the level of flood risk is currently very low due to the high standard of protection offered by the structural defence system, and awareness of the risk is low. This means that the potential of many flood-plain management responses is limited, particularly those responses that are reliant on the public taking effective action. Furthermore, the TE2100 studies indicated that it would be difficult to justify some flood-plain measures, because of the low level of residual risk. This experience is likely to be shared by other populated flood plains that are heavily defended.

However, the studies of defence performance showed that there is a significant chance of defence failure under load (i.e. high water levels in the estuary) for some defence types.

As the defence system has not been tested by a major surge tide event, there is uncertainty about how the defence system will perform. The plan therefore includes a precautionary approach to management of the residual risk in the flood plain, where some responses are provided for all areas within the estuary, and additional responses are provided where they can be justified. The improved responses included in the plan for all areas are

- flood forecasting and warning, for both tidal and fluvial sources of flooding
- high-level emergency planning by relevant authorities and agencies to ensure that there is a coordinated response across the estuary
- reducing potential impacts on new developments by adopting flood-resilient layouts, building designs and methods of construction
- setting back development from the defences, both to facilitate defence maintenance and improvement, and also to reduce the risk of damage if a defence fails
- promotion of flood insurance.

These responses are regarded as essential precautions for living and working on highly developed flood plains.

12.4.6 Communication and engagement

Communication and engagement with partners along the estuary was key to the success of TE2100, given that effective flood risk management requires a multi-partner approach. Three key vehicles were used.

- *Engagement with other strategic organisations* – throughout the life of the project, the TE2100 team engaged partner organisations along the estuary. The final plan draws heavily on information and suggestions made by these various groups. For example, through work with the Thames Estuary Partnership (a pre-existing forum where all issues related to living and the environment within the Thames were discussed) the flood risk management objectives of the TE2100 project were actively promoted in the broader context of the management of the Thames (e.g. navigation, nature conservation, tourism).
- *Public meetings and exhibitions* – the Thames Estuary is a large and diverse community. The TE2100 project sought to engage directly with the public frequently, in local borough halls or though local interest groups.
- *Provision of information on flood risk management* – a key contributing factor to the successful communication of the project was the development of numerous graphics and maps to convey effectively the present and future risks, possible management policies and options, and the effects of future uncertainties on the selection of those options. These graphics would often be required to distil down the complex risk information generated using the modelling methodologies described above. Figure 12.10 shows an example of the graphics used.

12.5. Summary of lessons learnt from the Thames Estuary

The Thames Estuary flood risk management plan is risk based, and attempts to direct effort where the best return can be achieved. This includes consideration of interventions

Figure 12.10 Communicating the preferred flood risk management policy choices for Canvey Island within the Thames Estuary
Contains Environment Agency information © Environment Agency and database right

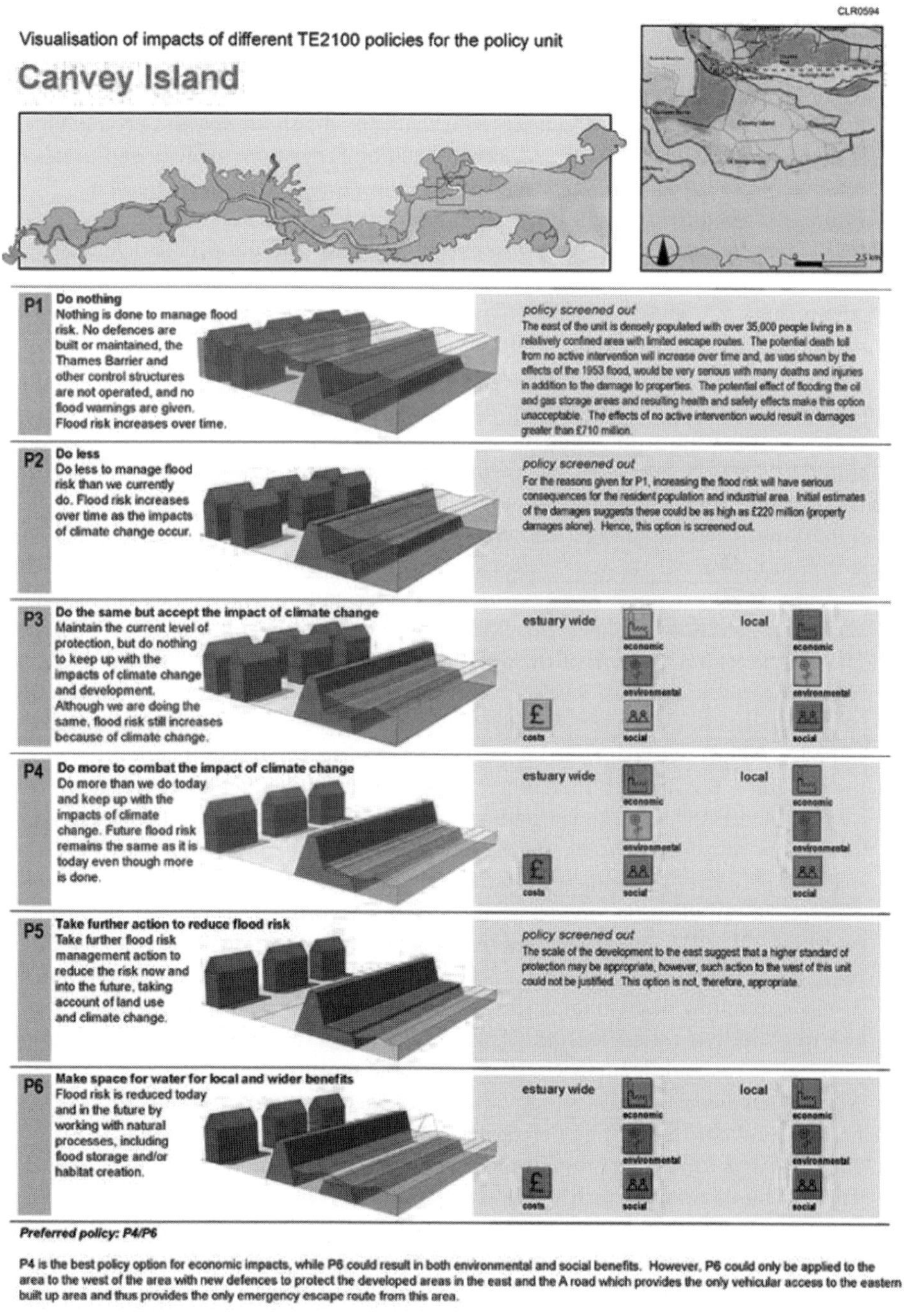

in the source, pathways and receptors of risk. The flood risk management plan for the Thames Estuary provides a number of lessons.

- *It is important to take a long-term view.* The Thames Estuary study sets out the strategic direction for managing flood risk across the estuary for the next 100 years, and contains recommendations on what actions will need to be taken in the short (next 25 years), medium (the following 40 years) and long term (to 2100). This long-term view enabled flood risk managers to challenge the *status quo* and promote more innovative approaches that are not constrained by present-day policies and practices.
- *Existing infrastructure can both constraint choices and provide opportunities.* Because of the high level of development on the Thames Estuary and the sunk investment in extensive defence systems, maintaining and improving defences turned out to be the only practical option.
- *Maintaining a broad-scale, strategic view is vital in developing a strategy.* Many competing demands are placed on the Thames Estuary – ranging from environmental interests to commercial development. Interests are both large in scale (the development of a port, with estuary-wide impacts) and local in scale (access to the waterfront). The scope of the TE2100 project was well defined, but this clarity was occasionally lost through excess focus on local-scale (non-strategic) issues. As the development of the strategy progressed, the importance of maintaining a strategic view become increasingly clear.
- *The need to deal with climate change and adopt a precautionary approach to investment.* Future uncertainties can have a profound influence on decisions, and the use of 'decision pipelines' provided a structured means of dealing with future climate uncertainty. Much of the data used to develop the plan was poor (and hence uncertain, e.g. crest levels), and future development in the flood plain could be significant. The change in sea level is potentially very large (+2 m) or relatively minor (<0.2 m) over the next 100 years. Developing a plan that is both meaningful – providing firm actions to avoid future risk – yet adaptable and avoids unnecessary investment and construction, has been shown to be possible.
- *The ability to attribute risk to individual components of the whole system was a major aid in the effective targeting of investment.* The use of a structured, estuary-wide, system risk model (that accounted appropriately for the performance of the defence system and the variation of consequences should an individual defence fail) provided a powerful tool for identifying which components of the system (e.g. individual defence) carried what risk. The ability to attribute risk to an individual component of the system was a major aid in helping identify high-risk areas and the defences that contribute most to that risk.
- *Replace and improve defences only where and when absolutely necessary.* A key finding was that the existing defence infrastructure continues to provide adequate and robust performance in most areas, and could be made to 'sweat'. The risk-based analysis provided a structured and transparent means of assessing the existing defence system. This allowed decisions about major upgrades of the infrastructure to be delayed, and presented an opportunity for alternative non-structural measures to be considered and programmed for implementation.

- *An effective and efficient strategy relies on a portfolio approach.* Flood risk management influences environmental, economic and people issues. The risk to each varies throughout the estuary, and the best meaning of 'managing the risk' also varies. The TE2100 study highlights the considerable opportunities that exist to enhance the environment, improve quality of life, and reduce risk. These goals have been shown not to be mutually exclusive. Building on the existing a mix of structural and non-structural flood risk management approaches, as well as estuary-wide (e.g. creation of wetlands in the outer estuary) and local (e.g. strengthening of a particular group of defences) solutions, was an important feature of the Thames study.

REFERENCES

Communities and Local Government (2010) Regeneration and economic growth. Available at: http://www.communities.gov.uk/regeneration/thamesgateway (accessed 12 February 2012).

Defra (Department for Environment, Food and Rural Affairs) (2004) *Making Space for Water*. Available at: http://www.floodprotectionassoc.co.uk/cms/documents_library/consultation.pdf (accessed 12 February 2012).

Defra (2006) FCDPAG3 Economic Appraisal. Supplementary Note to Operating Authorities – Climate Change Impacts.

Environment Agency (2008a) TE2100 – Floodplain Management Measures. Draft Technical Note – 'Existing' (unpublished).

Environment Agency (2008b) TE2100 – 'Enhanced', 'Best Case' Risk Reduction Factors (unpublished).

Environment Agency (2009a) *Managing Flood Risk Through London and the Thames Estuary: Strategic Environment Assessment – Summary Report*. Thames Estuary TE2100 Environment Agency, Bristol, UK.

Environment Agency (2009b) TE2100 – Climate Change Scenarios. TE2100 Report.

Environment Agency (2009c) Thames Estuary 2100 – Phase 3 Set 2 Options Appraisal. A Synthesis Report.

Environment Agency (2009d) Thames Estuary 2100: Managing flood risk through London and the Thames estuary. TE2100 Plan Consultation Document, April 2009. Available at: http://www.environment-agency.gov.uk/research/library/consultations/106100.aspx (accessed 4 May 2012).

Evans E, Ashley R, Hall J *et al.* (2004a) *Foresight Future Flooding, Scientific Summary: Volume 1: Future Risks and their Drivers*. Office of Science and Technology, Department for Business Innovation & Skills, London, UK.

Evans E, Ashley R, Hall J *et al.* (2004b) *Foresight Future Flooding, Scientific Summary: Volume 2: Managing future risks*. Office of Science and Technology, Department for Business Innovation & Skills, London, UK.

Flather RA (2000) Existing operational oceanography. *Coastal Engineering* **41**: 13–40.

Gilbert S and Horner R (1984) *The Thames Barrier*. Thomas Telford, London, UK.

Gouldby B, Sayers P, Mullet-Marti J, Hussan MAAM and Benewell D (2008) A methodology for regional-scale flood risk assessment. *Water Management* **161(3)**: 169–182.

Halcrow (1999) Tidal Thames, flood risk to London upstream of the Barrier. Prepared for Association of British Insurance (unpublished).

Horsburgh KJ and Wilson C (2007) Tide–surge interaction and its role in the distribution of surge residuals in the North Sea. *Journal of Geophysical Research* **112(C8)**: Art. No. C08003. 10.1029/2006JC004033.

McGahey C and Sayers P (2008) Long term planning – Robust strategic decision making in the face of gross uncertainty – tools and application to the Thames. In *Flood Risk Management: Research and Practice. Proceedings of FLOODrisk 2008*. Taylor & Francis, London, UK, pp. 1543–1553.

Penning-Rowsell E, Johnson C, Tunstall S *et al.* (2005) *The Benefits of Flood and Coastal Risk Management. A Handbook of Assessment Techniques*. Flood Hazard Research Centre, Middlesex University, London, UK.

Sayers P, Meadowcroft I, Hall J *et al.* (2004) *RASP – Risk Assessment for Flood and Coastal Defence Strategic Planning*. Environment Agency, Bristol, UK. HR Wallingford Report SR603.

USEFUL WEBSITES

Thames Estuary 2100 project (TE2100): http://www.environment-agency.gov.uk/homeandleisure/floods/104695.aspx (accessed 12 February 2012).

Thames Barrier history: http://www.environment-agency.gov.uk/homeandleisure/floods/117047.aspx (accessed 12 February 2012).

Thames Barrier operations: http://www.environment-agency.gov.uk/homeandleisure/floods/38359.aspx (accessed 12 February 2012).

Flood Risk
ISBN: 978-0-7277-4156-1

http://dx.doi.org/10.1680/fr.41561.327

Chapter 13
The 2001 flood of Houston and Harris County, Texas, USA

Jack Davis,* Steve Fitzgerald† and Austin Davis*
* US Army Engineer R&D Center, Vicksburg, USA
† Harris County Flood Control District, Harris County, USA

The Tropical Storm Allison rainfall and flood event of east-central Texas affected southern and eastern regions of the USA, becoming the costliest tropical storm in US history. Unprecedented amounts of rain and record flood levels were associated with Allison, as portions of the Harris County region received the 0.2% annual-chance rainfall event. The storm cost $5 billion in damages and the lives of 22 individuals, and severely inundated downtown Houston, Texas. The Federal Flood Control Act of 1936 allows state and local governments to partner with the federal government to study, design, construct and maintain a variety of flood mitigation strategies, both structural and non-structural in nature. The Harris County Flood Control District, US Army Corps of Engineers, landowners and developers improved most of the channels and flood control systems, many of which were in place prior to the flood event. The consequences of the storm and the modern flood management practices used in the event have been reviewed to examine the success of the Allison flood response, in order to better allocate current and future resources to improve flood management policies, planning and design.

13.1. The flood

Tropical Storm Allison (Allison) stalled over, and eventually departed, east-central Texas, and affected south and eastern regions of the USA, becoming the costliest tropical storm in US history. Allison was responsible for the deaths of 22 people and US$5 billion in direct damages (Risk Management Solutions Inc., 2001). The President of the USA declared official disaster areas in 31 Texan counties, 25 Louisiana parishes, nine counties in Florida, five counties in Mississippi and two counties in Pennsylvania (Harris County Flood Control District (HCFCD), 2002). The storm's impacts were far reaching, but Allison's greatest impact culminated in a major flood event in Houston and Harris County, Texas (HCFCD, 2002).

Allison formed in the Gulf of Mexico, 80 miles offshore, south of Harris County and Houston, as shown in Figure 13.1. Allison's slow movement and shifting track direction resulted in a 5-day stall over east-central Texas. The storm drifted onshore and north on 5 June 2001, before drifting south and slightly west on 7 June 2001. The south-western

Figure 13.1 Track of Tropical Storm Allison across east-central Texas
Reproduced courtesy of David R. Maidment

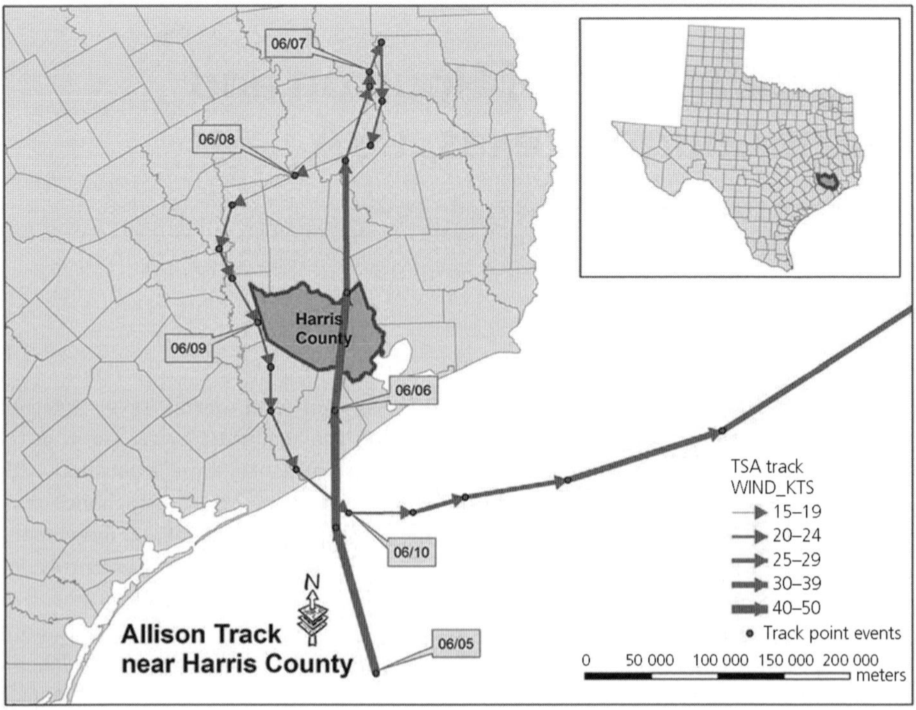

drift direction of Allison led to increased Gulf moisture being precipitated over the region, creating a significant rain event that saturated the landscape. Allison tracked south and slightly east on 9 June 2001, and back into the Gulf of Mexico on 10 June 2001. The sum total of rainfall in and around Harris County and Houston on 5–9 June 2001 is shown in Figure 13.2.

During Allison's first pass northward, rains saturated soils, filled channels and storm water detention basins, and inundated over 900 homes and businesses on two separate days. Just before turning south, Allison was downgraded to a tropical depression, but the most intense rains fell during the storm's second pass over the region. At one point, in its southward track, the storm dumped more than 28 in. of rain in 12 h, which, due to surface saturation, became mostly runoff. Storm sewers, drainage ditches, creeks, bayous and storm water detention basins that had filled and emptied during the first pass of the storm, filled again during the second pass, and in some cases overflowed (HCFCD, 2002).

As much as 80% of the area's average annual rainfall fell during Allison, and one half of Harris County received rainfall that equalled or exceeded the 1% annual-chance

Figure 13.2 Total rainfall across Houston and Harris County, Texas, 5–9 June 2001
Figure courtesy of Harris County Flood Control District

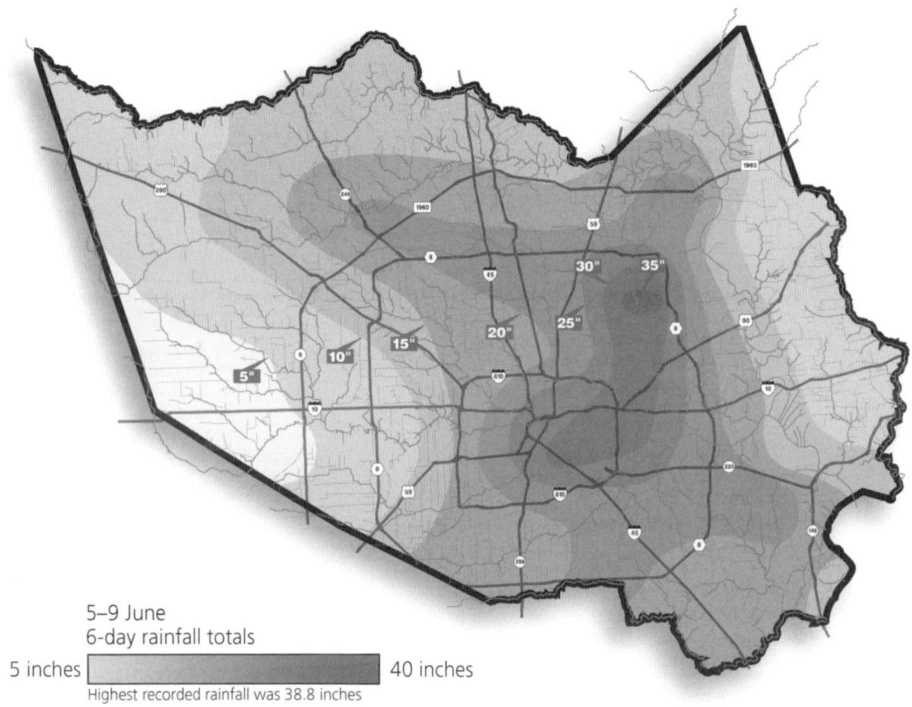

rainfall; significant portions of the area received the 0.2% annual-chance rainfall. Two-thirds of all flood losses occurred outside the 1% annual-chance flood plain, due to minimal topographic variation of the relic coastal plain (historically, the region was covered by bottomland forests, marshes, swamps and prairie). As floodwaters grew deeper, they flowed overland, following the subtle contours in the landscape rather than the general paths of channels (HCFCD, 2002).

The storm was devastating to the population and commerce of Harris County and Houston, with rainfall rates and magnitudes rarely experienced by USA metropolitan centres. Harris County is the third largest county in the USA, and the city of Houston makes up about half of the county population. The county was, and is, an economic centre for shipping, oil and gas production, and medical research (Risk Management Solutions Inc., 2001). Some 2 million people were affected by the storm during the night; 73 000 residences were flooded; 30 000 people were left stranded in temporary shelters; and 95 000 vehicles were damaged (HCFCD, 2002). Surprisingly, only 22 people lost their lives, a number significantly less than would be predicted for a proportional flood on such flat terrain. Despite unusually deep flooding (over 5 in. deep) in many areas, no people drowned in their homes. Instead, drowning occurred when people attempted to walk or drive through deep or fast-moving water.

Figure 13.3 Tropical Storm Allison flooding of Interstate Highway 10 in Houston, Texas
Figure courtesy of Harris County Flood Control District

While the consequences were widespread, a vast majority of the storm's damages occurred at the Texas Medical Center (TMC) complex and downtown areas of Houston, where floodwaters inundated hospital facilities, offices, stores, campuses, municipal buildings and facilities, roads, underground tunnel systems, basement areas and parking garages. Approximately 1000 patients were evacuated from the TMC's many hospitals when power was lost and floodwater rendered backup generators inoperable. Two trauma centres in the medical centre complex were affected – one was affected directly by the floodwater, and the other was rendered unreachable because access roads were under floodwater for almost 10 h (Risk Management Solutions Inc., 2001).

Despite the many losses Houston sustained, it is fortunate that the worst of Allison's rain fell in less-populated areas where runoff had a relatively short distance to travel to Galveston Bay. Had the rains fallen west of Houston, such that runoff had a longer travel distance to the bay, through the most populated areas of the region, the damages would have been significantly higher. Figure 13.3 shows an image of flooding in Houston along Interstate Highway 10.

13.2. Physiography

Harris County has 22 defined watersheds (Figure 13.4). When it rains, the unsurfaced, largely clay landscape, may absorb up to an inch of rain before the ground is fully

Figure 13.4 Watersheds in Harris County, Texas
Figure courtesy of Harris County Flood Control District

saturated. Upon full saturation, water flows off the landscape and streets into storm sewers and ditches, typically designed to handle rainfall rates up to 1–2 in./hour. The sewers and ditches route water into the larger channels and bayous, before it is conveyed to the Gulf. The roads themselves store water when the sewers and ditches along roadways are overwhelmed. Once a roadway's storage capacity for water is exceeded, water will overflow and follow the contours of the landscape.

The HCFCD system consists of approximately 2500 miles of channels and 160 excavated detention basins, compared with the natural system prior to development, which had approximately 800 miles of natural channels. Landowners and developers constructed and improved most of the channels. The HCFCD and US Army Corps of Engineers (USACE) improved channel conveyance on most main channels and some major tributaries. The two largest detention basins were constructed by the USACE in western Harris County, in the 1940s, using embankments to impound water.

Figure 13.5 illustrates the gently sloping landscape (Figure 13.5(a)), the network of natural channels (Figure 13.5(b)), the network of constructed flood-control channels (Figure 13.5(c)), and the total flood-control network that includes storm drains and streets (Figure 13.5(d)) which are intended to store and convey flood waters during an event like Allison.

Figure 13.5 (a) Elevations across Harris County, (b) natural channel network, (c) natural and constructed open channel network, and (d) total drainage network
Figure courtesy of Harris County Flood Control District

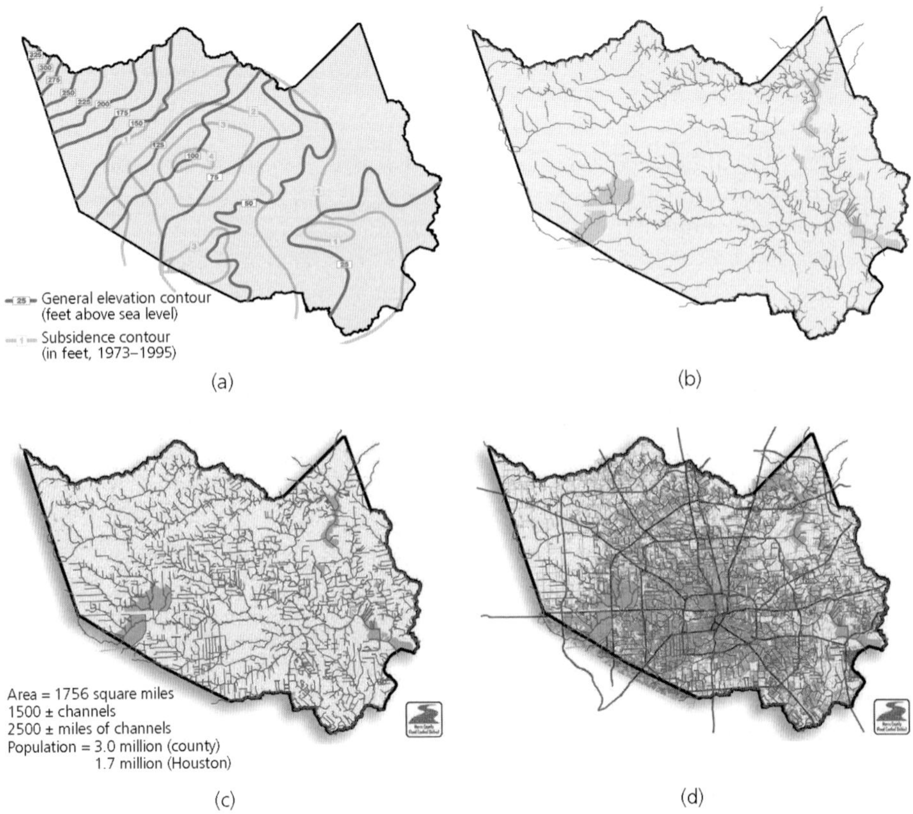

During Allison, the central and south-central watersheds received the most severe rainfall, including Greens Bayou, Hunting Bayou, White Oak Bayou, Brays Bayou, Sims Bayou and Clear Creek (Chen, 2006). The following descriptions of Allison's impacts in those watersheds are taken from HCFCD (2002), which was published by the HCFCD and the US Federal Emergency Management Agency (FEMA) to summarise the storm and its impacts, as well as to assist with recovery. The following extract is reproduced courtesy of the HCFCD.

- *Greens Bayou.* The Greens Bayou watershed, and its major tributary Halls Bayou, experienced the most severe rainfall in the county during Allison. Up to 28 in. of rain were recorded in a 12-hour period at two locations, which was greater than the 0.01% annual-chance rainfall, and close to the probable maximum precipitation. As a result, the bayou experienced some of the most devastating flooding ever recorded in Harris County. Over 28 000 residences flooded within the watershed. Floodwaters were so high in several places that water actually flowed overland from bayou to bayou – a very rare occurrence. This overland flow caused flooding of many residences outside the 1% annual-chance flood plain.

Greens Bayou has had multiple channel conveyance improvements and has three large regional detention basins supported by numerous smaller basins constructed to offset excess storm water runoff from new land development. These facilities worked well during the storm but were no match for the volume of rain that fell.

- *Hunting Bayou.* In the Hunting Bayou watershed, during a 12-hour period, 14–22.5 in. of rain fell. The recorded figure of 22.5 in. is two times the 1% annual-chance rainfall. Flooding was extensive, with severe, and mostly record, flood levels recorded. Over 8000 residences along the length of the Bayou flooded. In addition, deep overland flow inundated most of the upper watershed.
- *White Oak Bayou.* Areas within the White Oak Bayou watershed received rainfall amounts of 8–15 in. over a 12-hour period, approaching or exceeding the 1% annual-chance rainfall. In the bayou, flooding varied from moderate to record levels, and over 11 000 residences flooded. White Oak Bayou could not contain the intense rainfall of Allison despite its enlarged downstream channel's ability to carry relatively high flows and its six large regional detention basins. Flooding would have been far worse without these structural improvements.
- *Brays Bayou.* The Brays Bayou watershed experienced some of the most extensive damages witnessed during Allison. Rainfall amounts in the watershed during 8 and 9 June varied greatly, from 2 in. in the upstream watershed to 13 in. near downtown Houston. Overall, more than 6000 residences flooded near and along the Bayou. The area near downtown was the hardest hit, including the Texas Medical Center (TMC).

  The TMC experienced record flooding from Allison after the area received nearly 8 in. of rain in a 3-hour period and 12 in. in a 12-hour period. Although Brays Bayou floodwaters did not inundate the TMC, the backwater effects did slow drainage from the storm sewer system, causing the water to back up and flow overland. The TMC was inaccessible for many hours, and several hospitals were shut down for days. The bayou had been straightened and enlarged along its length in the 1950s and 1960s, which made it a very efficient channel for conveying flows to the Houston Ship Channel and ultimately to Galveston Bay. Two large regional detention basins were in a state of partial construction during Tropical Storm Allison.
- *Sims Bayou.* The Sims Bayou watershed received three rounds of rainfall during Allison: 4–8 in. in a 12-hour period, 3–5 in. in a 6-hour period, and 6–13 in. or more in a 12-hour period. While Sims Bayou is a fast-draining watershed, some 6000 residences flooded (900 of them twice), with the highest flood levels occurring with the last round of rain. While the bayou stayed within its banks, the flooding occurred because the local sewers and drainage ditches were unable to handle the intense rainfall.

  The Sims Bayou watershed has two large regional detention basins. These facilities were able to store some of the runoff, helping to reduce damages. Also, about half the channel length had been constructed on the Sims Bayou Federal

Project (the second largest flood-damage reduction project in the USA), sponsored by the HCFCD and managed by the USACE. It prevented a few thousand homes from flooding.

- *Clear Creek*. The Clear Creek Watershed is one of the largest watersheds in the Houston Area. This watershed received two large rainfall events during Allison. One rainfall event was 5–11 in. in a 6-hour period, and during this 300 residences flooded. The other event was 5–11 in. in a 12-hour period, and this second rainfall event severely aggravated the flooding situation on Clear Creek, a slow-draining stream previously inundated by the earlier rain event.
  Flood levels along the creek were generally the second highest on record, and overall approximately 2000 residences flooded, including homes flooded earlier in the week. Damages were reduced, however, by the effectiveness of a secondary channel constructed to convey floodwaters out of Clear Lake.

13.3. Flood defence

The HCFCD uses a variety of structural and non-structural techniques to manage floodwaters. Many were in place prior to Allison, and some were incorporated after the storm. Some were completed in partnership with FEMA and USACE. The duration of flooding during Allison was minimised due to the well-designed flood management infrastructure in Harris County. The HCFCD had been improving its flood defences in the years leading up to Allison, and the flood defences functioned efficiently and suffered minimal damage during the event. Had flood defences not been in place, the Houston and Harris County region may have remained inundated for a much longer period of time (HCFCD, 2002) and thousands of more homes and businesses flooded.

The most frequent structural solutions used in Harris County are channel conveyance improvements and floodwater storage facilities. Levees are infrequently used in flood defence in Harris County, because they remove valuable flood plain storage, create interior drainage issues on the flat landscape, and create the potential for catastrophic, or sudden, failure. The channel conveyance improvements include channel widening, deepening and straightening, reducing channel resistance (by removing woody vegetation and providing concrete lining where necessary), improving bridge crossings (by raising them above flood levels and minimising the number of bridge-pier obstructions that increase water levels and trap debris), and constructing bypass channels (that shorten travel distances for flows during high water events) (HCFCD, 2002).

While channel improvements make channels more efficient at moving water downstream, flood-plain storage is another effective structural solution in the flat regional landscape of Harris County. On a small scale, regulations require developers to build localised detention ponds that effectively reduce the rate of run-off from developed property. On a larger scale, the HCFCD uses regional detention basins, located along major channels to store floodwaters. Generally, detention basins are excavated to increase storage capacity and are tens of acres in size. Weirs built into the banks of the channels divert storm water into the basins. After the flood wave passes, and water levels in the channel subside, water in the basins slowly drains over the weir and through conduits back into the channel.

The non-structural techniques used to minimise loss and reduce the damage impact to the population and commerce are buy-outs (government purchase of property) and insurance. The HCFCD has an active buy-out programme that utilises FEMA's grant programmes. The programme is used to buy properties from willing sellers that are located deep in the flood plain, where structural alternatives for flood protection are unreasonable. Most of these properties were developed prior to the establishment of flood-plain mapping and improved flood-plain management practices. Quite simply, once these structures are removed from the flood plain, they are no longer a source of loss. For new homes and businesses within the 1% annual-chance flood plain, insurance is required. From 1989 to 2010, over 2000 homes were purchased using HCFCD and FEMA funds. However, the HCFCD is active in promoting the purchase of insurance by all homeowners in Harris County, including those properties outside the 1% annual-chance flood plain; as was seen during Allison, two-thirds of the damages occurred outside the 1% annual-chance flood plain.

13.4. Flood policy

In the USA, flood defence, land-use planning, flood-plain management and emergency management are the responsibility of individual states and local governments. However, when the US Congress passed the Flood Control Act of 1936, it established a federal role in flood defence. The Flood Control Act (as well as subsequent authorisations) allowed state and local governments to partner with the federal government to study, design, construct and maintain flood defences, and develop effective land-use, flood-plain and emergency management plans. When a clear federal interest is confirmed, and studies show the total benefits exceed total costs, the project may be authorised for federal investment in cost-share with local governments. Typically, once constructed, projects are maintained and operated by the local governments. Such was, and still is, the case in Houston and Harris County. Several major projects in the area that are part of the HCFCD's flood defence system were, and are, partially funded by the federal government. The HCFCD was created in 1937 primarily to be a local sponsor for USACE projects.

Today, in the USA, national flood insurance is available in most urban counties and cities. While interest in a national flood insurance programme was voiced in the 1950s, it was the National Flood Insurance Act of 1968 that authorised the creation of a National Flood Insurance Program (NFIP). If a community adopts and enforces a flood-plain management ordinance to reduce future flood risk, the federal government will make flood insurance available within the community as a financial protection against flood losses. Subsequent to the 1968 Act, the US Congress passed the Flood Disaster Protection Act of 1973, which prohibits federal agencies from providing financial assistance for the acquisition or construction of buildings, and certain disaster assistance, in the flood plains of any community that does not participate in the NFIP. The Act further stated that federal agencies and federally insured or regulated lenders had to require flood insurance on all grants and loans for acquisition or construction of buildings in areas within the 1% annual-chance flood plain. Therefore, the Flood Disaster Protection Act essentially made flood insurance mandatory in 1% annual-chance flood plains (King, 2005).

While emergency management response and recovery operations are the responsibility of states and local governments, the federal government recognises that some disasters may overwhelm local resources. Therefore, state governors may request federal assistance in response to an event. After approval by the President of the USA of a governor's request for disaster assistance, the federal government may mobilise resources through the FEMA for search and rescue, electrical power supply, and food, water, shelter and other basic human needs. Long-term assistance for recovery may also be requested to help disaster victims, businesses and public entities in various forms, including monetary loans, temporary housing, crisis counselling and legal assistance. The communities of Houston and Harris County are part of the NFIP programme and, following Allison, nearly 30 000 flood insurance claims were submitted to FEMA. During the 6 months following Allison, nearly 120 000 Texas residents applied for various forms of federal disaster relief aid, totalling more the US$1 billion (Risk Management Solutions Inc., 2001).

13.5. Conclusions

The effect of Allison was of such a magnitude its name was retired to preserve its place in history – an action usually reserved for powerful hurricanes. No other tropical storm's name has ever been retired. Allison brought unprecedented amounts of rain that resulted in record flood levels and damages. While the region's flood defences were designed well and functioned as planned, they were quite simply overwhelmed by the volume of rain. In conjunction with aggressive flood-plain management practices by the 32 local flood plain administrators, the HCFCD has studied the storm, evaluated the storm's consequences, updated 20-year-old hydrology and hydraulic information flood-plain maps, and benefited from expedited local and federal funding for implementing continued improvements to flood defences. For the 2 million residents who were directly or indirectly affected by Tropical Storm Allison flooding, they learned a lesson that no public education campaign could come close to achieving. However, as the regional population grows and changes, the need for clear, concise and continued public education is important to ensure understanding of the flood risks and mitigation options available to each individual and family. Through consistent public education and awareness, it is possible to ensure minimised future risks.

REFERENCES

Chen B (2006) Tropical Storm Allison and its impact on Harris County. University of Texas at Austin, TX, USA. Available at: http://www.crwr.utexas.edu/gis/gishydro07/Introduction/TermProjects/Chen.pdf (accessed 12 February 2012).

Harris County Flood Control District (HCFCD) (2002) *Off the Charts. Tropical Storm Allison Public Report*. Tropical Storm Allison Recovery Project (Harris County Flood Control District and Federal Emergency Management Agency). Available at: http://www.hcfcd.org/downloads/reports/TS-Allison_PubReportENGLISH.pdf (accessed 12 February 2012).

King RO (2005) *Federal Flood Insurance: The Repetitive Loss Problem*. CRS Report for Congress, Washington, DC, USA. Available at: http://digital.library.unt.edu/ark:/67531/metacrs7693 (accessed 12 February 2012).

Risk Management Solutions Inc. (2001) *Tropical Storm Allison, June 2001*. RMS Event Report. Available at: http://www.rms.com/Publications/TS_Allison.pdf (accessed 12 February 2012).

Flood Risk
ISBN: 978-0-7277-4156-1

http://dx.doi.org/10.1680/fr.41561.339

Chapter 14
Design of an environmentally aligned flood alleviation scheme: the Burn of Mosset, Moray, Scotland

Colin R. Thorne,* Matthew Hardwick,† Paul Winfield† and Carina Oliver†
* University of Nottingham, Nottingham, UK
† Royal Haskoning, Peterborough, UK

Forres, Scotland, has a long history of flooding from the Burn of Mosset, and it was forecast that, without investment in flood risk management, flood damages were likely to exceed £43 million over the next 50 years. The decision was taken to implement a flood alleviation scheme (FAS) featuring an earth-fill dam upstream of Forres, capable of storing up to 3.8 million m^3 of water and limiting peak discharges into the urban area to 8.5 m^3/s. It was recognised that sediment- and debris-related risks to the performance of the structure had to be managed and Burn Management Works were designed using cutting-edge river restoration techniques for channel rehabilitation and habitat creation to enhance the capacity of the flood basin to store sediment and debris. The FAS was opened on 28 August 2009 and the dam impounded flood water for the first time on 4 September 2009, avoiding approximately £9 million of flood damages to the town of Forres. The Burn Management Works have evolved mostly as expected, although some features have not functioned as intended, and lessons have been learnt concerning the need to allow for unexpected developments when using 'prompted recovery' in flood risk management and river rehabilitation.

14.1. Introduction
Forres is a market town (population 9000) located 16 km west of the city of Elgin in Moray, Scotland (Figure 14.1). The Burn of Mosset is a small but geomorphologically active, gravel-bed stream that drains an area of 49 km^2. It flows northwards through the centre of Forres, before confluencing with the River Findhorn in Findhorn Bay. Forres has a long history of flooding from the Burn of Mosset, with six events causing serious property damage and disruption of daily life within the last 50 years.

14.2. Background
The most recent severe flooding occurred in 1997, but significant events also occurred in 2000 and 2002. The 1997, 2000 and 2002 events had estimated return periods of 50, 5 and somewhat less than 5 years, respectively. Base flow in the Burn of Mosset at Forres is less than 1 m^3/s, and the 1 in 100-year flow is estimated to be 68 m^3/s, making a 15%

Figure 14.1 Forres location map

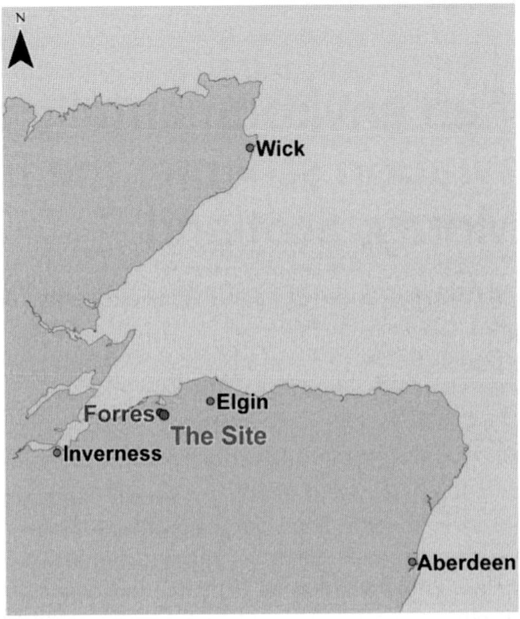

allowance for climate change up to 2080. Peak flow during the 1997 event was estimated at 48 m^3/s. During that event, approximately 430 homes and 27 commercial properties were inundated (Figure 14.2) and regional communications were seriously disrupted, with both the A96 trunk road and the Inverness to Aberdeen mainline railway being closed due to flooding.

Economic losses caused by the 1997 flood alone were of the order of £3.7 million, and it was forecast that without investment in a flood alleviation scheme (FAS), flood damages in Forres would be likely to exceed £43 million over the next 50 years. The economic case for a FAS for Forres was, therefore, clear.

However, geomorphological reconnaissance surveys (performed using methods described by Thorne *et al.* (2010)) performed as part of feasibility studies for a FAS revealed that the Burn of Mosset is highly active geomorphologically, and that it transports substantial quantities of sediment ranging in size from cobbles to sand and silt-sized material. Most of the cobble and gravel-sized bedload is deposited in the channel upstream of Forres, in the form of both shifting gravel bars and semi-stationary riffles, which have historically been managed for agricultural land drainage by dredging, with the dredge spoil used to build flood embankments on an *ad hoc* basis (Figure 14.3).

Substantial quantities of suspended sand/silt load, derived from field erosion associated with agricultural practices in the upper catchment, are also transported as wash load, but

Figure 14.2 Flooding in Forres during the 1997 event
Photographs supplied by Moray Council

these are carried into urban area of Forres, where they are mostly deposited in Sanquhar Loch, which is an artificially impounded water body located on the southern edge of the town. The field survey further revealed that the Burn carries considerable quantities of organic debris, including leaves, twigs and large wood. The findings of the geomorphological assessment indicated that, to reduce the unacceptably high flood risk in Forres sustainably, it was essential to design a scheme that accounted for the dynamics of sediment and organic debris, as well as those of the floodwaters.

14.3. Option development and appraisal

A variety of potential solutions were considered for addressing the flooding problem. These included a number of contrasting flood risk management strategies. Options ranged from hard engineering solutions (involving construction of a diversion channel around Forres, channel enlargement, or the raising of existing flood defences within

Figure 14.3 Gravel removal and *ad hoc* embanking of the channel of the Burn of Mosset upstream of Forres

the town) to flood storage, removal of properties in vulnerable areas, and changes in catchment management. The criteria by which the options were appraised were developed by the stakeholders at a value management workshop.

- *Environment* – The FAS will aim to create sustainable development that recognises the importance of the environment, minimises construction impacts and enhances the environment where possible. The scheme will aim to preserve the local character of the town and its surroundings.
- *Sustainability* – The FAS will aim to ensure the best achievable balance between environmental, social and economic well-being through use of the Moray Flood Alleviation Group's sustainability criteria and indicators. In the absence of any locally accepted methodologies available at that time, a document entitled *Rethinking Sustainability: Design Guidance and Procedures* was produced in November 2002 to inform management of scheme development.
- *Geomorphology* – The FAS will aim to minimise disturbance to natural geomorphological processes.
- *Stakeholder consultation* – The public, statutory consultees and interest groups shall be consulted at key stages of the development of the FAS. The views and opinions of stakeholders shall be taken on board and reflected in the development process and its outcome. Community values shall be central to the scheme.
- *Landowners* – The FAS will aim to minimise the impact on landowners and property occupiers.

- *Planning development* – The FAS will, where possible, accord with the Moray Development Plan (2000), planning policies and objectives. The FAS will avoid constraining future development opportunities.
- *Operation and maintenance* – The selected option will be designed with the aim of minimising future operation and maintenance requirements. The design shall accommodate safe access and egress for all operations and maintenance activities.
- *Opportunities* – The FAS design shall fully consider opportunities for improving recreation and amenity values.

Three options judged to be technically feasible and capable of delivering a minimum flood defence standard of service of 1 in 100-years were appraised according to these criteria.

14.3.1 Option 1: Channel diversion

The principal works would include the following.

- Construction of a diversion channel from the Burn of Mosset upstream of Forres to the River Findhorn, including:
  - 1.9 km long diversion channel
  - culvert or bridge under the A940 road
  - cascade to grade the diversion channel into the Findhorn valley.
- Works through Forres, including:
  - rehabilitation of floodwalls and embankments
  - removal of existing weirs and channel bed reprofiling
  - flood-plain lowering.

14.3.2 Option 2: Floodwalls and embankments through Forres

The principal works would include

- raised floodwalls and embankments throughout the urban area of Forres
- modifications to seven bridges (including the A96 trunk road bridge and the Inverness to Aberdeen mainline railway bridge)
- removal of existing weirs and channel bed reprofiling
- flood-plain lowering at selected locations
- raising the dam at Sanquhar Loch upstream of the town centre.

14.3.3 Option 3: Flood storage dam at Chapelton

The principal works would include

- construction of an earth-fill dam situated on the Burn of Mosset at Chapelton and capable of storing up to 3.8 million m^3 of water, including a flow-control structure to limit peak discharges into the urban area to a maximum of 8.5 m^3/s
- construction of a sediment and debris trap upstream of the dam
- earth embankments to protect properties close to the perimeter of the storage area from flooding.

14.4. Preferred option

The preferred approach for the FAS was identified by evaluating the three options using the sustainability criteria outlined above, while also considering the benefit–cost ratio for each option. As noted earlier, sustainability was evaluated with reference to a guidance document *Rethinking Sustainability*, which provided a framework for assessing sustainability based on economic, social and environmental factors.

This appraisal resulted in option 3 (flood storage dam at Chapelton) emerging as the preferred option. The main structural element in this option is a dam at Chapelton (Figure 14.4), which restricts peak flows passing through it during flood events to a maximum of 8.5 m^3/s, with the excess flow being stored temporarily in a flood reservoir upstream of the dam, to provide a standard of service to Forres of at least 1% annual exceedance probability (i.e. something in excess of the 1 in 100-year flood event).

One of the major concerns in designing the dam was to avoid the possibility of blockage of the flow control structure by sediments or floating material, such as large woody debris (LWD). The design selected for the flow control structure was novel, being a baffled crump weir. This is a passive flow control system (i.e. it has no moving parts and, therefore, meets the criteria for low operation and maintenance requirements) that had not been used in a dam of this size before.

Like any hydraulic control structure, the hydraulic performance of the baffled crump weir may be compromised if it were to be partially obstructed, or even completely blocked due to the excessive accumulation of sediment or debris. Consequently, the control of sediment and floating debris being conveyed downstream by the Burn

Figure 14.4 Chapelton Dam
Photographs supplied by Moray Council

during high-flow events had to be considered, and the associated risk to the performance of the structure managed very carefully.

Conventionally, sediment and debris management involves construction of a sediment trap immediately upstream of the control structure. If designed correctly, a gravel trap can be highly effective in catching and retaining sediment and debris, provided that it is emptied by excavation whenever a significant amount of material has accumulated within it. In the Forres FAS, this option was, however, not favoured, because the need for repeated excavation of the trap meant that it failed to meet the criterion that the design should minimise future operation and maintenance requirements. Also, such a trap would pose a potential barrier to fish passage and disrupt geomorphological processes. Consequently, it was decided to seek a softer solution that would be consistent with the requirement to minimise the maintenance and the environmental and geomorphological impacts of the scheme.

Hence, in place of a conventional sediment trap, Burn Management Works were designed for the reaches of the Burn of Mosset within the flood storage basin immediately upstream of the new dam. The principle underpinning these works was to take maximum advantage of the natural processes and landforms adjacent to the Burn. The dam is located at the downstream limit of a large, morphological basin, which should naturally act as a storage area for sediment and debris. However, the natural functioning of the basin in storing sediment and debris had been degraded by past realignment and channelisation of the watercourse to improve agricultural land drainage. It was, therefore, decided to perform Burn Management Works within the flood storage basin, reinstating and enhancing the capacity of both the channel and its flood plain to store sediment and debris. This approach relied on cutting-edge river restoration techniques for channel rehabilitation and habitat creation, to produce a sustainable solution to managing the sediment and debris-related risks to the baffled crump weir in the dam.

14.4.1 Burn Management Works

Most of the flood storage basin comprises a broad, low relief area of peat bog. Prior to construction of the FAS, the Burn of Mosset flowed through this area along two straight reaches connected by a right-angle bend, which is also the confluence with a tributary, the Rafford Burn (Figure 14.5).

This pre-FAS alignment was clearly unnatural, the Burn of Mosset having been realigned and resectioned for land drainage. The channel was also constrained on both banks by *ad hoc* flood embankments, and along much of the upper, straight reach within the flood storage basin the bed of the channel was perched above the elevation of the surrounding flood plain. This was the outcome of historical dredging and dumping of dredge spoil on the adjacent banks by landowners (see Figure 14.3). The frequency and extent of this practice was, however, reported to have declined more recently.

It was recognised that, during a major flood event, the *ad hoc* embankments constraining the perched channel would overtop and/or breach, with the potential for the channel to

Figure 14.5 Aerial view of dam, flood basin, controlled breaches (1 and 2) and Burn Management Works
Photograph courtesy of Scotavia Images

avulse onto a new alignment. If breaching of the embankments occurred on the left descending bank, the new alignment would cut off the right-angle bend (see Figure 14.5), taking flow towards the dam structure. If, however, a breach occurred on the right bank, the channel would have no obvious connection with the channelised reach approaching the control structure, and reconnecting it would require significant engineering works (see Figure 14.5). It was, therefore, decided to manage the alignment of the channel by constructing controlled breaches in the left descending bank (Figure 14.6), which ensured connectivity between the point where flow enters the flood basin and the control structure within the dam, where it leaves it.

In light of the pre-scheme alignment of the Burn within the area of the flood basin, and in the context of the wider aspirations for the FAS, the main aims of the Burn Management Works were to reduce the risk of blockage of the baffled crump weir in the dam by sediment and/or LWD to an acceptable level, and to reduce to a tolerable level the risk of the Burn avulsing onto an alignment that would disrupt connectivity between the channels entering and exiting the basin. Nested within these broad aims were objectives to

1. create an anastomosed planform (i.e. a watercourse with multiple subchannels, each of which operates semi-independently as a dynamically adjusting alluvial

Figure 14.6 Engineered breach as constructed in November 2008 to divide flow between the new channel (left) and the existing Burn (right). Note the block stone and riprap scour protection

stream) that would act as a natural sediment accretion zone with a large capacity to store material ranging in size from cobbles to sand

2. create a wet woodland around the anastomosed channel that would retain LWD while enhancing the habitat value of the area
3. configure the anastomosed channel and wet woodland so as to require minimal future management and little, or preferably no, maintenance throughout the project life of the Forres FAS
4. position the anastomosed channel so that channel shifting or avulsion would be unlikely to disrupt connectivity in the fluvial system.

Consideration of these aims and objectives according to the criteria for options appraisal developed by the stakeholders indicated that it was the most sustainable option for managing sediment, LWD and morphologically related risks to the FAS.

14.5. Implementation

Rather than attempting to construct an anastomosed channel system in the centre of the basin to restore the Burn to a more natural planform, it was decided to use a form of 'prompted recovery' in which natural processes would be encouraged to produce the desired morphological outcomes. This was initiated through controlled breaching of the existing artificial embankments at two locations along the left, descending, bank to allow flow to spill out across the flood plain and return to the original channel, while leaving it to nature to establish the pathways joining the two breaches (see Figures 14.5 and 14.6). The breaches were sized to divert 10–20% of the volume of run-off during low flow conditions, rising to a much larger proportion during floods. The aim was to avoid too rapid abandonment of the pre-FAS channelised course, which might strand fish and degrade the existing in-stream habitat faster than new habitats were being created in and around the new course. In this regard, a pilot channel was constructed between the two breaches, to ensure that passage for migratory fish was maintained (see Figure 14.5). A programme of tree planting was undertaken to encourage development of a wet woodland and associated aquatic, riparian and flood plain habitats along the new course. Only tree species native to eastern Scotland and of local provenance were selected for planting. These included

- common alder – *Alnus glutinosa* (25%)
- silver birch – *Betula pendula* (25%)
- sessile oak – *Quercus petraea* (25%)
- rowan – *Sorbus aucuparia* (15%)
- goat willow – *Salix caprea* (10%).

The specification was for bare-rooted transplants (whips) with lengths of 450–600 mm, as these tend to establish better and grow more quickly than older specimens. The whips were planted in clumps of 3–5 of the same species, spaced at 2 m centres. The planting density was selected to allow for some failures. A decision was taken against installing mesh guards to protect the whips from grazing rabbit or deer due to the risk that the guards might be washed off during a flood event and pose a hazard to wildlife further downstream in the Burn of Mosset.

Design and implementation of the Burn Management Works were aligned with the objectives of the EU Water Framework Directive, and were intended to improve the ecological status of the watercourse. This was possible because they replaced the artificially straightened, single-threaded, embanked course of the Burn with a multi-channel, anastomosed system that is fully connected to its flood plain. The works were also planned to work with, rather than against, natural processes. In this respect, it was recognised that the majority of coarse sediment delivered to the flood storage area during a flood event would be deposited where the energy gradient decreases as flow enters the flood basin. The breaches and area planted for wet woodland were, therefore, positioned to take advantage of and enhance this natural sedimentation process. Consequently, the works were approved by Scottish Natural Heritage and permitted under the relevant Controlled Activities Regulations by the Scottish Environment Protection Agency.

Creation of a wet flood-plain woodland is also consistent with a range of environmental and habitat targets for the region, including the North East Scotland Biodiversity Action Plan (North East Biodiversity Steering Group Partnership, 2000). In addition, the Burn Management Works not only facilitated restoration of a previously heavily modified reach of the Burn of Mosset, but also created a valuable area of wetland habitat in an area of the flood plain that was previously disconnected from the channel by artificial embankments.

It was anticipated that the pilot channel and its surrounding wetland and wet woodland would develop slowly under the action of natural processes to form a larger wet woodland feature that would, in time, link to adjacent areas of woodland and the area of wet woodland found less than 1 km downstream, at Sanquhar Loch.

14.6. Post-project performance and appraisal

When the left embankments of the Burn were artificially breached in November 2008, flow was divided between the pre-FAS, artificial course of the Burn of Mosset and the pilot channel cut through the Burn Management Works. Initially, the upstream breach did in fact divert 10–20% of the base flow, as planned. It was anticipated that the Burn Management Works would prompt recovery of a more natural form to the watercourse relatively slowly, with the new course taking perhaps 3–5 years to develop a multi-channel, anastomosed morphology that would convey the majority of flow formerly in the artificial channel. This was intended to allow time for the fluvial corridor to adjust progressively, thus avoiding a dramatic change in the geomorphology and ecology, which might lead to even a temporary loss of biodiversity.

The dam and associated FAS were officially opened on 28 August 2009, albeit that construction had not quite been completed at that time. The dam impounded flood water for the first time just a few days later, on 4 September 2009, and the FAS functioned well during what was estimated to be a 1 in 20-year flood event (Figure 14.7). The 'instantaneous' peak discharge entering the storage area was estimated to be around 36 m^3/s. Based on existing flood magnitude versus damage curves, implementation of the FAS avoided approximately £9 million of flood damages to the town of Forres.

The 4 September event supplied a considerable quantity of coarse sediment to the Burn Management Works, which was deposited as flow expanded across the flood plain. It also allowed floating debris to leave the artificial channel, and several pieces of LWD grounded in association with bars developing in the flood plain. Accretion of sediment and LWD triggered significant changes to the pilot channel, through the initiation of anabranching behaviour downstream of the first breach (Figure 14.8).

The Burn Management Works performed as anticipated during the flood, although failure of rock and riprap protection and enlargement of the upstream breach during the event resulted in abandonment of the old course in favour of the new one more quickly than intended (Figure 14.9).

Post-project monitoring has shown that the realigned channel is rapidly becoming an established feature within the fluvial environment, with flow, wetlands and areas of open water being evident throughout most seasons (see Figure 14.5). Natural functioning of the low-energy system is trapping coarse sediment and preventing it from

Figure 14.7 First impoundment by the Chapelton Dam on 4 September 2009 – less than a week after implementation of the FAS

Figure 14.8 Anabranching downstream of the upstream breach following the 4 September 2009 flood event. Note the development of anabranching upstream of the straight, pilot channel in the left foreground

Figure 14.9 (a) Controlled breach 1 in the left embankment following the 4 September 2009 flood. Note the disturbance to rock scour protection, widening of the breach, accretion of coarse sediment and accumulation of LWD. (b) The abandoned channel, viewed looking downstream of breach 1

(a)

(b)

Figure 14.10 Interaction between grounded LWD, fluvial processes and sediment dynamics in the Burn Management Works. Note the remnant of the pilot channel in the upper left of the photograph, the scour pool in the foreground, the anabranch on the left and the extensive coarse sediment deposition

approaching the control structure within the dam. LWD permanently grounded in the Burn Management Works is also interacting with fluvial processes and sediment dynamics to drive local scour, deposition, bar building, channel evolution and habitat diversification (Figure 14.10).

The wet woodland habitat is in the early stages of establishment, but it has already attracted a diverse range of flora and fauna (Tables 14.1 and 14.2). Following a site visit, a representative of Scottish Natural Heritage commented:

> I was really impressed to see this area of work and just how well the water is spreading out across the area. The area provides a really good example of . . . measures that can be taken to try and reconnect our rivers with their flood plains. (Scottish Natural Heritage, personal communication, 13 November, 2009)

In 2010, the Saltire Society of Scotland (http://www.saltiresociety.org.uk/3) in association with the Institution of Civil Engineers awarded the Forres FAS its 'environmentally sustainable construction' commendation in recognition of the innovative way that the Burn Management Works were designed and constructed with 'high regard for the environment and surroundings'.

Table 14.1 Fauna observed in the Burn Management Works area

	Latin name	Common name
Birds	*Anas platyrhynchos*	Mallard
	Anas crecca	Teal
	Ardea cinerea	Grey heron
	Gallinago gallinago	Snipe
	Carduelis carduelis	Goldfinch
	Troglodytes troglodytes	Wren
	Parus major	Great tit
	Corvus corone corone	Carrion crow
	Buteo buteo	Buzzard
	Motacilla cinerea	Grey wagtail
Insects	*Sympetrium striolatrum*	Common darter
	Sympetrium danae	Black darter
	Sympetrium fonscolombii	Red darter
	Cynthis cardui	Painted lady butterfly
	Lestes sponsa	Emerald damselfly
	Cordulegaster boltonii	Golden ringed dragonfly
	Inachis io	Peacock butterfly
	Bombus sp.	Bumble-bee
	Pentatomidae	Shield bug
	Carabidae	Ground beetle
	Syrphidae	Hoverfly
	Tipuloidea	Cranefly
	Myrmicinae	Ant
	Lycosidae	Wolf spider
	Deilephila elpenor	Elephant hawk moth
Fish and amphibians	*Salmo truta* sp.	Trout
	Rana temporaria	Common frog
Mammals	*Capreolus capreolus*	Roe deer
	Lutra lutra	Otter

14.7. Knowledge transfer and lessons learned

Notwithstanding the award winning success of the FAS and the Burn Management Works to date, there are lessons that should be learnt concerning the design and construction of measures intended to meet multiple goals for flood control and environmental enhancement. It is also important that knowledge gained through experience with the Burn of Mosset is transferred to other practitioners so that they can take advantage of it when future schemes are designed and implemented.

1. It was anticipated that the new channel would capture the majority of flow in the Burn relatively slowly, over a period of perhaps 3–5 years. In fact, a significant,

Table 14.2 Flora observed in Burn Management Works area

	Latin name	Common name
Grasses	*Agrostis canina*	Velvet bent-grass
	Agrostis capillaris	Common bent-grass
	Alopecurus geniculatus	Marsh foxtail
	Arrhenatherum elatius	False oat-grass
	Dactylis glomerata	Cock's foot
	Deschampsia cespitosa	Tufted hair-grass
	Glyceria fluitans	Floating sweet-grass
	Holcus lanata	Yorkshire fog
	Lolium perenne	Rye-grass
	Phleum pratense	Timothy
	Poa trivialis	Rough meadow-grass
Sedges	*Carex rostrata*	Bottle sedge
Rushes	*Juncus articulatus*	Sharp-glowered rush
	Juncus bufonius	Toad rush
	Juncus effusus	Soft rush
	Luzula sylvatica	Great wood rush
Ferns	*Dryopteris felix-mas*	Male fern
	Pteridium aquilinum	Bracken
Flowering Plants		
Ranunculaceae	*Ranunculus repens*	Creeping buttercup
Ulmaceae	*Ulmus glabra*	Wych elm
Urticaceae	*Urtica dioica*	Common nettle
Fagaceae	*Quercus petraea*	Sessile oak
Betulaceae	*Alnus glutinosa*	Alder
	Betula sp.	Birch
Caryophyllaceae	*Cerastium fontanum*	Common mouse ear
	Sagina procumbens	Procumbent pearlwort
	Silene latifolia	White campion
	Spergularia arvensis	Corn spurrey
	Stellaria alsine	Bog stitchwort
Polygonaceae	*Persicaria amphibia*	Amphibious bistort
	Rumex obtusifolius	Broad-leaved dock
Clusiaceae	*Hypericum perforatum*	Perforate St John's wort
Salicaceae	*Salix cinerea*	Grey willow
	Salix sp.	Willow
Brassicaceae	*Cardamine hirsute*	Hairy bitter cress
	Rorippa nasturtium-aquaticum	Watercress

Table 14.2 Continued

	Latin name	Common name
Rosaceae	*Potentilla anserina*	Silverweed
	Rubus fruticosus	Bramble
	Sorbus aucuparia	Rowan
Fabaceae	*Cytisus scoparius*	Broom
	Trifolium campestre	Hop trefoil
	Trifolium hybidrum	Alsike clover
	Trifolium pratense	Red clover
	Trifolium repens	White clover
	Ulex europaeus	Gorse
	Vicia cracca	Tufted vetch
Onagraceae	*Epilobium ciliatum*	American willowherb
	Epilobium palustris	Marsh willowherb
Aceraceae	*Acer pseudoplatanus*	Sycamore
Geraniaceae	*Geranium robertianum*	Herb robert
Apiaceae	*Anthriscus sylvestris*	Cow parsley
	Heraclium mantegazzianum	Giant hogweed
	Heraclium sphondylium	Hogweed
Lamiaceae	*Galleopsis tetrahit*	Common hemp nettle
	Mentha sp.	Mint
	Prunella vulgaris	Selfhealt
	Strachys palustris	Marsh woundwort
	Strachys sylvatica	Hedge woundwort
Boraginaceae	*Myosotis scorpioides*	Water forget-me-not
	Pentaglottis sempervirens	Green alkanet
Callitrichaceae	*Callitriche* sp.	Water starwort
Plantaginaceae	*Plantago lanceolata*	Ribwort plantain
Oleaceae	*Fraxinus excelsior*	Ash
Scrophulariaceae	*Digitalis purpurea*	Foxglove
	Mimulus guttatus	Monkey flower
	Mimulus luteus	Blood-drop emlets
	Parentucellia viscosa	Yellow bartsia
	Scrophularia nodosa	Common figwort
	Veronica beccabunga	Brooklime
Rubiaceae	*Galium aparine*	Cleavers
Asteraceae	*Achillea millefolium*	Yarrow
	Arctium minus	Lesser burdock
	Cirsium arvense	Creeping thistle

Table 14.2 Continued

	Latin name	Common name
Asteraceae	*Chrysanthemum segetum*	Corn marigold
	Hypochaeris radicata	Cat's ear
	Lapsana communis	Nipplewort
	Leontodon autumnalis	Autumn hawkbit
	Matricaria discoidea	Pineapple weed
	Senecio jacobaea	Common ragwort
	Senecio vulgaris	Groundsel
	Tanacetum vulgaris	Tansy
	Tripleurospermum inodorum	Scentless mayweed
	Tussilago farfara	Coltsfoot
Typhaceae	*Typha latifolia*	Reedmace

but not particularly extreme, flood event that occurred in September 2009 (less than 2 years after implementation of controlled breaching) diverted the entire flow of the Burn of Mosset into the new course. This occurred partly due to deposition of cobbles and gravel on the bed of the old channel just downstream of the first breach, but principally because failure of rock and riprap scour protection, followed by erosion that enlarged the dimensions of breach 1, allowed it to capture the entire discharge of the Burn. These morphological changes accelerated what was intended to be a more progressive, evolutionary trend. While abrupt switching of the flow did not trigger any unforeseen morphological developments or deviation from the longer term plan for evolution of the Burn Management Works, rapid transfer of the flow into the new channel must have led to some degree of unintended environmental degradation in the reach of the artificial, pre-FAS channel, which is now dry except during high-flow events.

During the September 2009 flood, block stone and riprap scour protection installed at the upstream breach were unable to prevent bank erosion, with unintended consequences. This is similar to experience on the Jubilee River when it experienced a flood soon after construction, in January 2003 (Thamesweb, 2003). It is, therefore, recommended that additional attention should be paid to ensuring the stability of diversions or outlets in order to reduce the risk of their being damaged or bypassed, especially soon after construction and before the bank vegetation that provides natural erosion resistance has been able to recover from the disruption that is inevitable during project implementation.

It is further concluded that, when using 'prompted recovery' to drive channel restoration and environmental enhancement, it is impossible to predict the timing of flow events of sufficient magnitude and duration to trigger significant geomorphological changes to either the pre-existing channel, the new channel or both. Hence, the risks associated with unintended and abrupt fluvial and morphological adjustments must be carefully considered and accounted for when 'prompted recovery' is selected as a component in the design approach.

2. In late 2009, significant erosion was observed on the right descending bank opposite breach 1, the upstream of the two controlled breaches. Appraisal of the erosion suggested that, left unaddressed, this could lead to breaching of the right embankment, and the possibility of channel avulsion to the right of the old channel and away from the Burn Management Works. This would have been highly undesirable, as it would have resulted in the avulsed flow having no obvious pathway to the baffled crump weir in the dam. Also, it would have arrested, or at least significantly impeded, evolution of the anastomosing channel, wetland and wet woodland system in the Burn Management Works area.
   The cause of the bank erosion was identified as being a large, fallen tree that had grounded close to the right bank in the vicinity of breach 1. It was decided to move this tree away from the bank (although not to remove it from the watercourse) in order to reduce its effect in constricting the flow, locally increasing near-bank velocities and elevating fluid shear stresses on the right bank. Continued monitoring shows this management intervention to have been effective. Some bank erosion is also occurring opposite breach 2 (the downstream breach) and, while this does not currently pose a significant risk of right-bank avulsion, monitoring is, and must be, ongoing.
   The point is that, without post-project monitoring and adaptive management, bank erosion opposite breach 1 would probably have continued, perhaps to the point of jeopardising the effective operation and intended development of the Burn Management Works. It follows that appropriate monitoring and adaptive management are integral and essential components of any river works that involve, or allow for, sediment transport and a degree of channel evolution. This is the case because the magnitude, timing and sequencing of future morphological adjustments are not just unknown, they are unknowable. This is because channel changes are the outcomes of flow and sediment transporting events that have not yet happened and the ways that they interact with random factors (e.g. grounded trees) – which cannot be predicted deterministically.
3. The detailed drawings for excavation of a pilot channel as part of the Burn Management Works specified a sinuous course, aligned to connect the upstream and downstream breaches via low areas identified using light detection and ranging (LiDAR) imagery. However, this element of the intended design was not implemented accurately by the contractor and site staff who actually constructed the channel.
   The outcome was a channel alignment comprising relatively long, straight reaches connected by short-radius bends, as it was easier for the excavator driver to dig a channel with this planform rather than the continuous curves of the intended design. It was always intended that the pilot channel would evolve away from its constructed form and should respond morphologically to the occurrence, sequence, magnitude and duration of channel-forming flow events. Consequently, it was concluded that the fact that the channel was not built according to plan would not be crucial to the long-term success of the Burn Management Works. In hindsight, it would have been wise for the design team to have given more specific setting-out instructions to the contractor and also to have briefed site staff on the significance of design details in the planform of the pilot channel.

Figure 14.11 Low boulder weirs constructed in the pilot channel and intended to encourage formation of a pool and riffle sequence. These have been successful in trapping LWD, but have not been effective in promoting pool and riffle formation

These steps would have resulted in a better functioning and more aesthetically pleasing initial channel form.

This experience should be borne in mind when restoration requires construction of a channel with the morphological complexity required to provide the diverse habitats necessary to support a sustainable ecosystem. This is likely to include many environmentally aligned channels constructed within the spirit or the letter of the EU Water Framework Directive.

4. Finally, it is now realised that more attention should have been given to the design of microtopography and bedforms within the pilot channel. The design implemented included low, boulder weirs constructed at intervals along the pilot channel that were intended to encourage formation of a pool and riffle sequence (Figure 14.11).

   Post-project monitoring shows these structures to have been ineffective and, with hindsight, it would have been preferable to vary the cross-sectional geometry and conveyance capacity along the pilot channel to create in-channel non-uniformity and facilitate exchange of momentum and sediment between the channel and its flood plain. This would have better coupled in-stream and flood-plain processes, encouraging localised scour and deposition and stimulating interactions between water, sediment and vegetation over a wider area.

The general point here is that, in an era when stakeholders expect to see positive results in months rather than years, anything that can be done to accelerate morphological and ecological development in a constructed or restored channel must be welcome. However, to optimise this type of 'process forcing', highly detailed, topographic data, extending along the project reach and across the active flood plain, are essential in identifying locations where the terrain is amenable to flow leaving or returning to the channel and mapping the pathways suitable for overbank flow and/or storage. Successful delivery of morphological complexity and the diverse habitats it supports early in a restoration also depend not only on a deep understanding of hydromorphology and ecohydraulics on the part of the design team, but also a recognition by the contractor that design details are important, and effective application by the machine operator of specialist, practical skills.

The transferrable message is that the best designs are bespoke with reference to both the catchment context and the local terrain, and that, unless designers have the luxury of reporting to stakeholders who are both understanding and patient, the devil actually does lie in the detail.

Acknowledgements

The research reported in this chapter benefitted from support from the Flood Risk Management Research Consortium (http://www.floodrisk.org.uk), which is funded by the Engineering and Physical Sciences Research Council under grant number EP/FP202511/1 with contributions from the Department of Environment, Food and Rural Affairs/Environment Agency Joint Research Programme, United Kingdom Water Industry Research, the Office of Public Works Dublin and the Northern Ireland Rivers Agency. The authors also wish to thank Gary Brindle of Scotavia Images (http://www.scotaviaimages.co.uk) for permission to include the aerial photograph reproduced in Figure 14.5. Finally, the active support of Moray Council and cooperation of Morrison Construction are gratefully acknowledged.

REFERENCES

Moray Council (2002) *Rethinking Sustainability: Design Guidance and Procedures*. Council Office, High Street, Elgin, Moray IV30 1BX, Scotland.

Moray Development Plan (2000) Available at: http://www.moray.gov.uk/devplan2000/index.html (accessed 12 February 2012).

North East Biodiversity Steering Group Partnership (2000) *North East Scotland Local Biodiversity Action Plan*. Moray Council, Moray, UK. Available at: http://www.nesbiodiversity.org.uk/publications/plan.pdf (accessed 12 February 2012).

Saltire Society of Scotland (a non-political body concerned with all aspects of Scottish life and culture) Available at: http://www.saltiresociety.org.uk/3 (accessed 12 February 2012).

Thamesweb (2003) *January 2003 Floods and Jubilee River Bank Collapse*. Available at: http://www.thamesweb.co.uk/floodrelief/2003_JRbankcollapse.html (accessed 12 February 2012).

Thorne CR, Soar P, Skinner K, Sear D and Newson M (2010) Driving processes II: Investigating, characterising and managing river sediment dynamics. In *Guidebook of Applied Fluvial Geomorphology* (Sear D, Newson M and Thorne C (eds)). Thomas Telford, London, UK, pp. 120–195.

FURTHER READING

Hoey TB, Smart DWJ, Pender G and Metcalfe N (1998) *Engineering Methods for Scottish Gravel-bed Rivers* (Leys K (ed.)). Scottish Natural Heritage, Review No. 47. Available at: http://www.snh.org.uk/pdfs/publications/review/047.pdf (accessed 12 February 2012).

Sear DA, Newson MD and Thorne CR (2010) *Guidebook of Applied Fluvial Geomorphology*. Thomas Telford, London, UK.

Skinner KS and Thorne CR (2005) *Review of Impact Assessment Tools and Post Project Monitoring Guidance*. Review performed for the Scottish Environmental Protection Agency, Haycock Associates Limited with Nottingham University Consultants. Available at: www.sepa.org.uk/water/regulations/guidance/idoc.ashx?docid=4d7c1fef-4fad-4ced-a80a-72cc241b008f&version (accessed 12 February 2012).

Thorne CR, Hey RD and Newson MD (1997) *Applied Fluvial Geomorphology for River Engineering and Management*. Wiley, Chichester, UK.

Flood Risk
ISBN: 978-0-7277-4156-1

http://dx.doi.org/10.1680/fr.41561.361

Chapter 15
Glossary of terms

The following definitions are based on Sayers *et al.* (2012) and are adapted from a variety of international sources including:

- ADRC: Asian Disaster Reduction Center (in United Nations University)
- Defra: Department for Food Environment and Rural Affairs (England and Wales)
- Environment Agency (England and Wales)
- FEMA: US Federal Emergency Management Agency
- FLOODsite: an EC Integrated Project
- ICOLD: International Commission of Large Dams
- UNESCO–IHE: United Nations Educational, Scientific and Cultural Organization Institute for Water Education
- UNU–EHS: United Nations University Institute for Environment and Human Security
- USGS: US Geological Survey
- WHO: World Health Organisation

Acceptable risk The level of risk a society or community considers acceptable given existing social, economic, political, cultural, technical and environmental conditions. An understanding of acceptable (and hence unacceptable) risk helps guide the level of investment that may be appropriate to reduce the risk (where possible).

Adaptability The ability to modify a particular measure (structural or non-structural) or instrument (policy or regulation) as the reality of the future becomes known or future projections change.

Adaptation The ongoing adjustment in natural, engineered or human systems in response to actual or changing expectations in climate or other drivers of risk. Adaptation may be either autonomous (and achieved through natural change) or planned (and achieved through purposefully adaptation planning; replacing the reactive adaptation often seen in response to an extreme flood that has invariably been characteristic of traditional flood control approaches).

Afflux The increase in water-surface elevation in a watercourse due to the presence of a constriction in flow (e.g. arising from a structure such as a bridge or culvert), relative to that which would exist without the constriction in place.

Alternative When making a choice, the decision maker selects from available alternatives (and holds options for future selection).

Anabranched (anastomosed) channel A multi-threaded channel that resembles braiding but in which the individual subchannels (anabranches) are separated by islands that are not inundated at bankfull stage, allowing them to vegetate and stabilise as semi-permanent landscape features. In low-energy streams, this channel planform is sometimes referred to as *anastomosed.* Due to their morphological complexity and diversity, anabranched channels provide valuable and diverse habitats for a wide range of aquatic, riparian and flood-plain species.

Asset management Systematic and coordinated activities through which an organisation manages its assets and asset systems.

Baffled crump weir A passive structure to control the flow of water through a dam, enabling migratory fish to pass, over a full range of retained water levels and without the need for power. Can be used to minimise the height of a dam, by storing the smallest volume of water necessary to meet the design criteria. In terms of water elevation versus discharge characteristics, the performance of a baffled crump weir is often significantly superior to a simple fixed orifice and combines high levels of reliability and low maintenance (as it has no moving parts).

Biodiversity A measure of the health of ecosystems – it can be readily destroyed or enhanced by management choices. Biodiversity, as most commonly used, is the totality of genes, species and ecosystems within a region – in context this may refer to a single river reach through to a river basin, and even a network of basins. Biodiversity provides a unified description of the traditional three levels of biological variety – namely, species diversity, ecosystem diversity and genetic diversity – all of which are important considerations in flood risk management.

Capacity The combination of all the strengths, attributes and resources available within a community, society or organisation that can be used to achieve agreed goals.

Catchment (river) The area of land surface that drains to a given point in the river system.

CCTV Closed-circuit television.

Combined sewer A sewerage system that conveys both foul sewage and surface water run-off.

Combined sewer overflow (CSO) A device that allows untreated wastewater to be discharged from the belowground combined sewer system during storm conditions that exceed the capacity of the sewer system.

Consequence An impact such as economic, social or environmental damage/improvement that may result from a flood. Consequence can be expressed in many valid forms, either quantitatively or qualitatively by category (e.g. high, medium, low) or description. The magnitude of the consequence will be influenced by the inherent vulnerability of the receptor and the value society places upon the harm caused, and may be expressed in monetarised form, the native form of the impact (e.g. hectares of habitat lost) or in more abstract units.

Control (flow) A means of modifying (typically limiting) the peak flow to the downstream system.

Conveyance (flow) The process by which water (or effluent within a sewer system) is transferred from one location to another.

Coping capacity The degree to which people, organisations and systems, using available skills and resources, are able to face and manage adverse conditions, emergencies or disasters.

Culvert A fully enclosed section of watercourse, commonly within an engineered section (including invert).

Debris Solid natural and anthropogenic material, carried through a watercourse by the flow, that has the potential to increase flood risk (either through the blockage (at bridges and culverts) or through collision with people and buildings). Debris can range significantly in size – from large woody material and shopping trolleys through to individual leaves and bags. In natural channels, natural vegetation is a positive and important contributor to biodiversity – in these settings such material is not considered debris. Typically, inorganic sediments carried by rivers are also excluded from the term 'debris', except when they are part of a hyperconcentrated phenomenon termed a 'debris flow'.

Disaster A serious disruption of the functioning of a community or society, causing widespread consequences (including human, material, economic or environmental losses) that exceed the ability of the affected community to cope using its own resources.

Discharge The volume of water passing a given point of an open channel or closed structure per unit time. Synonymous with the terms 'flow' or 'flow-rate', and typically described in cubic metres per second (m^3/s).

Effectiveness The risk is reduced as expected/desired. In general, effectiveness is increased by adopting a portfolio approach, where the advantages of one option compensate for the disadvantages of another, to minimise risk and maximise opportunities.

Efficiency Requires avoiding wasting resources of time, effort, money, environmental capital, etc. In general, efficiency management seeks to develop measures that are synergistic such that the sum effect is greater than the individual parts. In more specific terms, resources are said to be used inefficiently when it would be possible, by using them differently, to make at least one person or community better off without making any other person or community worse off. Conversely, resources are used efficiently when it is impossible, by using them differently, to make any one person or community better off without making at least one person or community worse off.

Environmental impact assessment (EIA) A systematic assessment of possible positive and negative impacts that a proposed project may have on the environment, considering all natural, social and economic aspects. The purpose of the assessment is to ensure that decision-makers consider the ensuing environmental impacts prior to major decisions being taken and commitments made.

Erosion Removal of inorganic particles (sand, gravel, etc.) from their place of rest through the action of wind, flow, wave or ice.

Exceedance (sewer) flow Flow that has surcharged from a sewer system and is then conveyed overland.

Exposure The people, property, habitats, networks and other receptors (see *Receptor*) that may be flooded and thereby subjected to potential harm/losses.

Flexibility The ability of a given management strategy to be changed as the reality of the future unfolds and/or projections of the future change.

Flood The temporary covering by water of land not normally covered by water.

Flood control Measures taken to modify the behaviour of the flood wave and reduce the probability of flooding in some areas and to increase the probability of flooding others. Typically these are structural measures, either on a large scale (e.g. barriers and levees) or on a small local scale (e.g. run-off attenuation).

Flood plain The generally flat areas adjacent to a watercourse or the sea where water flows in time of flood or would flow but for the presence of structures and other flood controls. The limits of a flood plain are notionally infinite, and are therefore defined by the maximum flood extent (based on the worst case, or with or without flood control) resulting from a given return-period storm.

Flood-plain maps (flood) Maps that typically indicate the geographical areas which could be covered by a flood (during a given return-period storm or extreme event) in the absence of control structures. The maps may be complemented with: type of flood, the water depths or water level and, where appropriate, flow velocity, and often with simplified hazard categories.

Flood risk management (FRM) The process of data and information gathering, risk analysis and evaluation, appraisal of options, and making, implementing and reviewing decisions to reduce, control, accept or redistribute flood risks. It is a continuous process of analysis, adjustment and adaptation of policies and actions taken to reduce flood risk (including modifying the probability of flooding and its severity as well as the vulnerability and resilience of the receptors threatened). FRM is based on the recognition that risks cannot be removed entirely but only partially, and often at the expense of other societal goals.

Flood zone An area of the flood plain that has a given chance of flooding in any one year, in the absence of flood control structures (e.g. raised defences). Often used to support precautionary planning processes (see *Planning policy statement*).

Fragility (curve) The relationship between the conditional probability of failure (e.g. the chance of a breach levee) and a given loading condition (e.g. the water level in the river). The fragility curve provides a graphical representation of this relationship over a range of loading conditions.

Fluvial geomorphology The academic study of channels and landforms associated with rivers and the processes that form, alter and maintain them.

Foul sewer A sewer system that conveys effluent from toilets, sinks, etc.

Froude number A dimensionless number defined as the ratio of a body's inertia to gravitational forces. In context, the Froude number is used to determine the resistance of sediments/objects to flows. The higher the Froude number, the greater the resistance.

Hazard (flood) The potential for inundation that threatens life, health, property and/or natural flood-plain resources and functions. The flood hazard is composed of three elements: severity (magnitude, duration, and extent of flooding), probability of occurrence, and speed of onset. A hazard does not necessarily lead to harm.

Hazard zoning (flood) Delineation of areas with different possibilities and limitations for investments and development, based on flood hazard.

Head Total energy per unit weight of fluid expressed as metres above a datum.

Head loss Difference in head between two points along a watercourse due to frictional or other features that result in energy loss.

Headwall Retaining wall at a culvert inlet or outlet that provides support to the embankment.

Highway drain A surface sewer system that conveys surface water run-off from the highway.

Individual risk The risk faced by a particular individual (as distinct from societal – or group – risk, discussed below).

Integrated flood risk management (IFRM) An approach to dealing with flood risk that explicitly recognises the interconnection of flood management actions within broader water resources management and land-use planning; the value of coordinating across geographic and agency boundaries; the need to evaluate opportunities and potential impacts from a system perspective; and the importance of environmental stewardship and sustainability.

Integrated urban drainage (IUD) An understanding of how different elements of the drainage system interact, how they perform as a whole system, and how they can be managed to effectively and efficiently manage risk.

Invert Lowest internal point of any cross-section in a culvert.

Landscape Human perception of the land conditioned by knowledge and identity with place.

Landscape assessment An umbrella term for the description, classification and analysis of landscape.

Landscape enhancement Improvement to the landscape through restoration, reconstruction or creation.

Landscape quality (or condition) Based on judgements about the physical state of the landscape, and about its intactness, from visual, functional and ecological perspectives. It also reflects the state of repair of individual features and elements that make up the character in any one place.

Light detection and ranging (LiDAR) A technology that employs a scanning laser range finder to measure surface relief. Used extensively for topographical surveys. The resolution and accuracy are controlled by the surface cover, instruments used, distance between device and the surface, and the post-processing employed.

Mitigation Measures and instruments, including any process, activity or design, to avoid, reduce, remedy or compensate for adverse impacts of development proposals or other decisions.

Multiple deprivation measure Index of social deprivation used by the Northern Ireland Statistics and Research Agency.

Non-structural measures Any measure not involving physical construction that uses knowledge, practice or agreement to reduce risks and impacts, in particular through policies and laws, public awareness raising, training and education.

Option When holding an *option*, the opportunity to choose between *alternative* (see previous) actions in the future remains. The option-holder can delay making the final decision, rather than having to make it now.

Outcome measures Measures used to express, in quantified terms, the desirable outcomes that are considered important. This might include the reduction in annual expected lives lost, economic risk reduced, biodiversity gained, etc.

Overflow Flow over a structure, such as a flood embankment or sea wall, by a progressive increase in water level.

Overtopping Periodic flow over a structure, such as a flood embankment or sea wall, through wave action.

Pathway (of the risk) The connection between a particular initiating event (see *Source (of risk)*–) and the receptor that may be harmed or experience loss (e.g. a property; see *Receptor*). For example, the pathway may consist of the upland land surfaces, the river channel, the levees and the flood plain between an upstream inflow boundary (the source) and a particular house (the receptor).

Planning Policy Statement 25 (PPS 25) Planning policy for development in flood risk areas used within England.

Policy and regulatory instruments Policies and regulations provide the principles and rules that guide the framework within which flood risk management strategies are developed, decisions made and, in some instants, delivered on the ground.

Pool and riffle Persistent bed configuration consisting of pools (deep hollows) interspersed with riffles (topographic highs) at a spacing of five to seven times the channel width. Pool and riffles are especially marked in gravel-bed rivers, where they provide the morphological diversity that supports a wide range of valuable habitats.

Portfolio approach A management approach to reducing risk that relies upon the implementation of a wide range of options, in space and in time. Within a portfolio approach the aim is to develop a strategy that considers many options, where the advantages of one option compensates for the disadvantages of another, and synergies provided by combinations of options are exploited (e.g. wetland creation and support, flood warning).

Post-project appraisal (PPA) Evaluation of the performance and success of a river project based on systematic monitoring. PPA allows scientists and engineers to ascertain whether a project was built as designed, to evaluate the degree to which it has achieved its objectives, and to identify whether adaptive management and maintenance should be performed to address undesirable developments and/or unintended consequences of the project.

Probability (general) Probability is a measure of the degree of the belief that a flood will occur within a given time frame (annual, lifetime, etc.) or during a given event. This measure has a value between zero (impossibility) and 1.0 (certainty). There are two main interpretations:

- *Statistical frequency* – The outcome of a repetitive experiment of some kind, such as flipping coins. It includes also the idea of population variability. Such a number is called an 'objective' probability because it exists in the real world and is, in principle, measurable by experiment.
- *Subjective probability* – A quantified measure of a belief in, judgement of or confidence in the likelihood of an outcome, obtained by considering all available information honestly, fairly and with a minimum of bias. Subjective probability is affected by the state of understanding of a process, judgement regarding an evaluation, or the quality and quantity of information. It may change over time as the state of knowledge changes. The majority of 'probabilities' of interest to the

flood manager/analyst are subjective and cannot therefore be formally validated by observation. For example, the probability of a particular river water level being exceeded relies on a combination of observation of past events and extrapolation/interpolation using expert selected statistical models.

Probability of flooding The chance of a particular part of the flood plain experiencing flooding after taking into account the performance of any associated flood control infrastructure (including both failure and non-failure possibilities). The chance of flooding must be linked explicitly to an associated reference timescale (e.g. annual or lifetime probability) and a specific characteristic (or combination of characteristics) of the flood (e.g. depth or velocity). The probability of flooding is not simply related to the chance of the driving storm.

Receptor The entity that may be harmed by a flood. For example, in the event of heavy rainfall (the source) flood water may propagate across the flood plain (the pathway) and inundate housing (the receptor), which may suffer material damage (the harm or consequence).

Residual risk The risk that remains after accounting for the performance of all flood risk management actions (i.e. measures to reduce the chance of flooding and those taken to reduce vulnerability or improve resilience). To avoid confusion, the date at which the residual risk has been assessed should be communicated. Typically, the stated residual risk of relevance to the public is associated with the present day. For planners, however, understanding how the residual risk varies over time due to climate or other changes is crucial.

Resilience The ability of an individual, community, city or nation to resist, absorb or recover from a shock (e.g. an extreme flood) and/or successfully adapt to adversity or a change in conditions (e.g. climate change, economy downturn) in a timely and efficient manner.

Resilient design Resilient design fosters innovative approaches to the design, construction and operation of buildings and infrastructures that are resilient to natural and man-made disasters. Adopting an integrated approach incorporates resilience as one of the primary goals during building design. In addition to protecting the lives of building occupants, buildings that are designed for resilience can absorb and rapidly recover from a disruptive event. Continuity of operations is a major focus.

Return period A statistical measure denoting the average recurrence interval that an event of a given magnitude (e.g. an extreme in river water level) is exceeded (when considered over an extended period of time). While it is true that a 10-year event will, on average, be exceeded once within any 10 year period, the chance of encountering such an event in the next 10 years is approximately 65% – the so-called encounter probability.

Risk The combination of the chance of a particular event (e.g. a flood), with the impact that the event would cause if it occurred. Risk therefore has two components – the chance (or probability) of an event occurring, and the impact (or consequence) associated with that event. The consequence of an event may be either desirable or undesirable. Generally, however, flood risk management is concerned with protecting society, and hence a risk is typically concerned with the likelihood of an undesirable consequence and our ability to manage it. (Note: Opportunities for positive gains should also be sought, but recorded as 'opportunities gains', not risks).

Risk analysis (flood) The application of tools and techniques to determine risk objectively by analysing and combining probabilities and consequences. It involves the use of available (and, by definition, uncertain) information to estimate the risk to individuals or populations, property or the environment, from hazards. Risk analyses generally involve the following steps: (i) scope definition; (ii) hazard identification (including source and pathway terms); (iii) receptor identification; and (iv) risk estimation. Risk analysis involves the disaggregation or decomposition of the flooding system and sources, pathways and receptors of risk into their fundamental parts at a resolution appropriate to understand the nature of the risk and determine its essential features at the scale of interest.

Risk-based or risk-informed decision-making An approach to decision-making that supplements information on risk (both probability and consequence) with subjective trade-offs and issues of equity and opportunity gains. (In some countries, a more restricted definition of 'risk-based' is used and relates to a decision made only through consideration of benefits and costs.)

Risk evaluation (flood) The process of examining and judging the significance of risk estimated through the process of risk analysis. The risk evaluation stage is the point at which values (societal, regulatory, legal and owners) and value judgements enter the decision process, either explicitly or implicitly. Within risk evaluation consideration is given to the significance of the estimated risks and the associated social, environmental, economic and other consequences, together with an understanding of the investment needed to reduce the risk in order to develop an appropriate flood risk management strategy.

Risk identification (flood) A qualitative process of determining what could go wrong, why and how.

Risk management (flood) See *Flood risk management*.

Risk maps Maps that combine information on probability and consequences in order to spatially differentiate risk. The mapped risk is often expressed in terms of expected annual risk (integrating all possible storm events and possible system responses that may occur in a year) or event risks (i.e. the expected damages associated within a specified storm event). Risk maps typically display: (i) the number of potential deaths or serious injuries; (ii) economic damages (notional or financial); and (iii) secondary impacts (e.g. arising from accidental pollution caused by flooding, or loss of power to non-flooded properties).

Risk mitigation (flood) A selective application of options (both structural and non-structural) to reduce either the likelihood of a flood or its adverse consequences, or both.

Robustness The degree to which an option or strategy continues to perform well across a range of possible future scenarios.

Scour Local erosion of the river or sea bed, shoreline or river banks by the action of moving water. Scour may act to undermine structures, leading to collapse.

Screen bar A purposefully designed screen to prevent debris entering a culvert. The spacing of the bars is often a critical design consideration, and is defined as either the clear spacing or centre to centre.

Screen platform A purpose-designed horizontal platform to allow operatives to stand safely when cleaning a screen. Such platforms require either solid or open-tread flooring, and cannot contribute to the screen area in terms of conveying flow.

Screen rake A purpose-designed device used by maintenance teams to remove trash and debris from a screen. Rakes are not capable of removing large or heavy items.

Security screen Screen for which the primary purpose is to prevent unauthorised or accidental access to an enclosed section of watercourse.

Societal concerns Concerns engendered by those hazards that, if realised, have the potential to impact on society as a whole. The evaluation of a risk will reflect the degree of societal concern.

Societal risk Widespread or large-scale consequences arising from the realisation of an extreme hazard, the implication being that the consequence would be on such a scale as to provoke a socio-political response, and/or that the risk provokes public discussion and is effectively regulated by society as a whole through its political processes and regulatory mechanisms. Such large risks are typically unevenly distributed, as are their attendant benefits. For example, the construction of a dam may increase the risk to those close by but provide a benefit to those remote from the dam, or an action/decision may harm a future generation more than the present one (e.g. to tie a future generation in poor, and expensive, planning decisions). The distribution and balancing of such major costs and benefits is a classic function of government, subject to public discussion. The results of such a debate shape the evaluation of risk and the nature of the management policies and approach adopted.

Source (of risk) The event(s) that is/are considered to initiate a potential flood (e.g. heavy rainfall, strong winds, surge, or even human error/attack such as accidental opening of a gate or aircraft collision into a dam).

Stakeholder Any one person or group of people with a legitimate interest in the decisions being made.

Strategic flood risk assessment (SFRA) A specific form of assessing the likelihood of flooding and its consequences used in England and Wales to support appropriate development and mitigation planning.

Strategy (flood risk management) A coherent plan(s) that sets out goals, specific targets, decision points, and the mix and performance of both structural and non-structural measures to be employed. Flood risk measures within the strategy are then grouped into coherent packages as the basis for further development and implementation.

Structural measures Any physical construction for reducing the chance or severity of flood waters reaching a receptor (see *Receptor*). Structural measures range from large-scale infrastructure responses, such as barriers and levees, through to local responses to improve the resistance and resilience of individual homes or critical installations.

Surface flow pathway Urban surfaces along which surcharged flows can be conveyed.

Surface water management plans (SWMP) A particular planning document required in England that is used in the investigation of flooding issues from all sources of flooding. It includes the identification and appraisal of options to appropriately manage flood risk now and in the future.

Surface water sewer A sewerage system that conveys only surface water run-off.

Sustainability First defined as 'development which meets the needs of the present without compromising the ability of future generations to meet their own needs', sustainability is a simple yet powerful concept. In particular, it emphasises the interlinkages between economic development, environmental health and social well-being – not as three separate objectives but one. Agenda 21 reinforces the notion of integration, and stresses the need to move away from sector-centred ways of working to new approaches that involve cross-sectoral coordination and integration. Broad public participation in decision-making as a fundamental prerequisite for achieving sustainable development is also emphasised. Flood risk management is inextricably linked with issues of sustainability. Not only does flood risk management impact the physical environment, through the development of control structures and spatial planning measures, but also provides opportunities for, and constraints on, human and natural activities in the long term.

Sustainable urban drainage systems (SUDS) Sometimes known as sustainable drainage systems (SuDS), these seek to reduce the potential impact of new and existing developments with respect to surface water drainage discharges. The aim is to replicate natural (pre-development conditions) through cost-effective solutions with low environmental impact to drain away dirty and surface water run-off through collection, storage and cleaning, before allowing it to be released slowly back into the watercourse network.

Sustainable urban drainage system approval body (SAB) An approval body that will review, approve, adopt and maintain sustainable drainage systems serving more than one property.

System (flood risk) In the broadest terms, the social and physical domain within which risks arise and are managed. An understanding of the way a system behaves and, in particular, the mechanisms by which it may fail, is an essential aspect of understanding risk. This is true for an operational system, such as flood warning, as well as for a more physical system, such as a series of flood defences protecting a flood plain, and, importantly, the system as a whole.

Telemetry Use of connected networks to convey data from a remote site to an operational data centre (often by radio, mobile networks, etc.).

Tolerable risk The degree of residual risk that society is prepared to tolerate in order to secure certain net benefits (e.g. environmental improvement, power generation, urban development, limited expenditure on flood management). Tolerable risk varies from situation to situation, and is not negligible or something we might ignore. The associated residual risks must be keep under review, and reduced further where appropriate.

Trash The term 'trash' is often used to distinguish anthropogenic material from natural organic matter. (Beware, however, as the terms 'debris' and 'trash' are also often used interchangeably.)

Trash screen Any screen at the entrance to a culverted watercourse, the primary purpose of which is to prevent trash and debris from entering an enclosed section of the watercourse.

Unacceptable risk A level of risk that cannot be justified except in extraordinary circumstances. Typically, where the continuation of the risk has been authorised by government or a regulator in the wider interests of society, and either further reduction of risk is simply not possible (e.g. all levees and dams, regardless of their design and maintenance regimes, have some, albeit small, chance of failure) or the resources required to reduce the risk are disproportional to the additional benefits secured.

Uncertainty Any situation (timing of a storm, data, model or decision) that is not accompanied by complete certainty, whether or not described by a probability distribution. Uncertainty can be attributed to: (i) the inherent variability in natural properties and events (aleatory uncertainties), and (ii) incomplete knowledge of parameters and the relationships between model (both quantitative and qualitative models) input and output values (epistemic uncertainties).

Visual amenity The perceived and stated value (in monetary or non-monetary terms) of a particular area or view in terms of what is seen.

Vulnerability A combination of the inherent susceptibility of a particular group, people, property and/or natural feature to experience damage during a flood event and society's preferred means of valuing the harm experienced. For example, the vulnerability of a property is expressed as a flood depth versus economic damage relationship, the vulnerability of an individual may be expressed through a relationship between flood depth/velocity and the chance of dying or being seriously injured. Vulnerability may, therefore, be modified through actions that reduce a receptor's susceptibility to experience harm (e.g. improved speed of recovery after a flood).

Watercourse Any river, stream, brook, beck or drain that acts to convey rainfall run-off and/or groundwater flow.

Wetland A permanently moist and seasonally flooded area the formation of which has been dominated by water, and the physical characteristics of which are largely controlled water, and which supports a rich and diverse ecosystem that is specifically adapted to the prevailing hydrological regime.

Wet woodland An area of trees growing on land that is seasonally or permanently waterlogged. Characteristic species include alder, birch and willows, but ash, oak, pine and beech may also be found. Succession from an open herbaceous wetland to a wet woodland results in a wide range of habitats and ecological compositions, depending mainly on climate and nutrient availability.

Wingwalls A pair of retaining walls to support the channel bank at a culvert inlet or outlet.

Flood Risk
ISBN: 978-0-7277-4156-1

http://dx.doi.org/10.1680/fr.41561.375

Index